D1253252

Advance Praise for
McGraw-Hill's National Electrical Code® 2014
Grounding and Earthing Handbook

"Although it doesn't resemble a wire, the Earth's soil is a conductor, one that you can't ignore. Wherever electricity is delivered, some of it has a good chance of going to ground, especially when it has nowhere else to go. Understanding the relationship between electrical fields and earth is essential to any engineer's success in circuit design and energy delivery. Using concise text and crystal-clear diagrams, David Stockin's book spells it out for you: how to connect wiring to make the circuit you're using work, while earth becomes conductor, capacitor, and antenna. It's a subtle business involving moist soil, dry rocky dirt, metal conductors, complex connectors, and long runs of conduit. If you are responsible for delivering power, or if you just want to know what's going on beneath your feet, read these pages; ground your circuit as he suggests, and you can close the master switch with confidence."

—Bill Nye, "The Science Guy"

"Grounding and bonding are often subjects that are not treated in accordance with their importance. While the Code addresses minimum requirements necessary for safety, the installation is often not optimal from an economic or operational standpoint. Before exceeding the Code's minimum requirements, often a good practice and cost-justified, it is important to understand what those minimum requirements are. David Stockin's book develops a clear understanding of Code requirements through extensive use of illustrations and clear text."

—David Brender, P.E. [Mr. Brender is one of the 15 full-voting members of Code-Making Panel No. 5, which covers Articles 200, 250, 280, and 285 of the National Electrical Code (NEC)]

"The subject of grounding is considered by many in the electrical industry to be 'voodoo science,' translating to a lot of misunderstanding. In his new book, David Stockin does a magnificent job of explaining away most of the shadows and dark territory in the science. In his book, David adds his own experiences to the fundamentals of NEC 250, thus allowing the technical-minded person to visualize practical applications of routine rules in a new and novel way."

—Travis Lindsey

McGraw-Hill's

NATIONAL ELECTRICAL CODE® 2014 GROUNDING AND EARTHING HANDBOOK

David R. Stockin
Manager of Engineering
E&S Grounding Solutions
Hermosa Beach, California

Illustrations by Gilbert Juarez

Contributing Authors:
Michael A. Esparza
Jeffrey D. Drummond
Dennis Langerud
Christopher Clemmens

New York Chicago San Francisco
Athens London Madrid
Mexico City Milan New Delhi
Singapore Sydney Toronto

McGraw-Hill Education books are available at special quantity discounts to use as premiums and sales promotions or for use in corporate training programs. To contact a representative, please visit the Contact Us page at www.mhprofessional.com.

McGraw-Hill's National Electrical Code® 2014 Grounding and Earthing Handbook

1 2 3 4 5 6 7 8 9 0 DOC/DOC 1 2 0 9 8 7 6 5 4

ISBN 978-0-07-180065-5
MHID 0-07-180065-4

The pages within this book were printed on acid-free paper.

Sponsoring Editor
Judy Bass

Editorial Supervisor
Stephen M. Smith

Production Supervisor
Pamela A. Pelton

Acquisitions Coordinator
Amy Stonebraker

Project Manager
Sheena Uprety,
Cenveo® Publisher Services

Copy Editor
Mary S. Curioli

Proofreaders
Mary E. Kanable and Irina Burns

Indexer
Robert Saigh

Art Director, Cover
Jeff Weeks and Gil Juarez

Composition
Cenveo Publisher Services

About the Author

David R. Stockin is Manager of Engineering at E&S Grounding Solutions. A full-time grounding engineer for more than 12 years, he has been the lead electrical grounding/earthing engineer for dozens of data centers used by major search engine companies and government agencies, located from the Americas to Asia and Europe. Mr. Stockin has also been the lead grounding and electrical safety engineer for over a thousand projects involving human safety in high-voltage environments, in both the United States and Canada. He has authored numerous publications on electrical grounding and earthing.

Contents

Acknowledgments

Writing this book has not only been a difficult task, it has also been a great personal learning experience. While researching this book, I learned a great deal about engineering issues, but what I learned most was about the people who support me, care about me, provide for me, and generally enable me to be my best. This book, like my life, is literally unimaginable without you.

This book is dedicated to Michael A. Esparza (Principal/Director of Sales, E&S Grounding Solutions), the man who has helped me, more than anyone other than my parents, become who I am today. You wrote, edited, reviewed, and contributed to every single page in this book. You have put more effort into helping me make this book become a reality than anyone else, and I hope you are as proud of it as I am. Thank you, Michael, for everything you do for me. My poor words here will never be able express the depth of my gratitude for what you have contributed to my life.

To Gilbert Juarez (Creative Director, RevDesign), who has always believed in me, stood by me, and given up his time and hard work to help me achieve my dreams. My friend, it seems that no matter what comes up, whatever crazy project falls in our laps, you are always there, standing beside me, working away, and doing everything you can to make it a success. Thank you for always being there. I hope to do the same for you.

I have been very fortunate to have the best book-writing team anyone could ever ask for, and I would be very remiss in not thanking them.

To Jeffrey D. Drummond, M.Eng., P.E. (Electrical Engineer, E&S Grounding Solutions), thank you for the countless late night phone calls, discussions, reviews, and research projects regarding an innumerable number of complex subjects. Jeff, thank you for all your thoughtful guidance and suggestions, but most importantly, and perhaps you do not know that you do this, for inspiring me to do it better, to get it right, and to make it clear. I am proud to have you as a partner, but even happier to have you as a friend.

To Dennis Langerud, P.E. (Director, Electrical Engineering, PDC Corporation), thank you for all of the excellent guidance over the last 10 years. Your considerate and thoughtful work on this book was crucial, and it is a much better work because of you. I look forward to working with you for many more years to come. And, of course, thank you for being there when I needed you most, even when it was not the best timing for you. Your dedication through such adversity makes your contributions all the more worthwhile.

To Christopher Clemmens, M.B.A., M.Acc. (Chief Financial Officer, E&S Grounding Solutions), who has been my friend and mentor since before I even started working in grounding. Thank you for your time and efforts in helping the business grow and for all you have done for me and my family through the years; I am glad that you were able to help contribute to this effort by writing a chapter in this book.

I would like to thank my daughter Kristin Stockin, who sat and expertly edited nearly every single page of this book during her college breaks. You amaze me every single day, and I am so proud of you. I love you.

A big thank you goes to my editors Judy Bass and Amy Stonebreaker at McGraw-Hill and the entire staff at Cenveo for all the encouragement and guidance over this long project. Your kind words of support were more valuable than you know.

If I can take a few more moments, I would also like to express my deepest gratitude to the other people who were there to support me.

To my daughter Courtney Stockin, who also helped to edit parts of this book, but mostly was there to support me by picking up the slack and taking care of the many things I had little time for. I am so proud of the person you have become. I love you.

My thanks to Elizabeth Esparza and Dana Silva, who have tirelessly read and reread the manuscript, helping to get the wording and punctuation just right.

Thank you, Bill Nye (yes, "The Science Guy"), for all the great conversations, encouragement, and thoughtful science discussions. But mostly for inspiring me to take a stand and speak my mind about issues that are important. It is because of you that there are parts of this book that are critical of the current electrical code. The hope is that thoughtful criticism will help to "Change the world!".

I would like to thank Joe Anderson and the management team at the YMCA of the Rockies in Estes Park, Colorado, for providing me with that special writing getaway. Thank you, Joe, for all the great conversations; you influenced this book more than you know.

My thanks to the staff of the Infusion Center at Children's Hospital of Orange County for their great care of my son and for allowing me to type much of this book during our time with them.

To my good friends Andy Yoon, Joe Nguyen, MG and Krista Kelly, Kia Harirchian, Ruben Espinoza, Tambre Leighn, and Juliana Yoon, thank you for all your support and for being such great friends. My life is better because of you.

I ask for the forgiveness of all those who have supported me, consulted with me, and assisted me over the years and whose names I have failed to mention. Thank you.

Finally, to my son Zachary … you are my heart and my soul.

David R. Stockin

I would like to thank the [illegible] and the [illegible] team [illegible] [illegible] [illegible] [illegible]. [illegible] [illegible] [illegible] [illegible] [illegible] [illegible] [illegible].

My thanks to [illegible] of the [illegible] Children's Hospital [illegible] for their [illegible] [illegible] for allowing me to [illegible] of this book [illegible] with them.

[illegible] good friends [illegible] [illegible] [illegible] [illegible] [illegible] [illegible] [illegible] [illegible] [illegible] [illegible] [illegible] of [illegible].

[illegible] who have supported [illegible] with [illegible] over the years and [illegible] [illegible] [illegible].

Finally [illegible] my heart and my soul.

[illegible]

Introduction

The science of electrical earthing/grounding goes back as far as Benjamin Franklin's day. From the great disputes between Thomas Edison and the City of New York, to modern questions regarding the technological requirements of sensitive electronic equipment, electrical grounding is still a vastly misunderstood science. Even with updates to the grounding section of the National Electrical Code® (NEC®; NFPA 70®, Art. 250), which are done on a regular basis, grounding questions are still some of the most commonly fielded questions in the entire electrical industry. Electricians and engineers want to have a one-stop resource that is easy to understand, uses common language, and gives real-world examples related to electrical grounding.

Electrical grounding involves two components: above-grade grounding and below-grade earthing. As below-grade earthing is seldom, if ever, discussed in any book, having both the above-grade grounding and the below-grade earthing in a single resource should prove to be much more useful. To have a book with professional, informative, and easy-to-read graphics will surely be an invaluable tool for a wide array of electricians, electrical engineers, engineering students, and professionals throughout the world.

There are multiple handbooks available that are designed to help guide you through the complexities of the NEC; all of these guides minimize the grounding and earthing information. They shrink the

74 pages of grounding information found in the *NEC Handbook* down to a dozen or so pages. This leaves many engineers and electricians in need of more information regarding grounding and earthing. The intent of this book is to expand on the Code and provide the relevant information regarding the Code, along with common-language explanations. If you are looking for more information regarding earthing and grounding, and not less, then this is the book for you.

You will find that the book is broken down into four main sections:

1. Summary of grounding and earthing: This is a brief overview of the grounding and earthing requirements found in the NEC. The goal is to summarize the Code's requirements for grounding in a few pages, using common language.
2. An in-depth discussion of each and every one of the Code's requirements, article by article, with both the requirement and a common-language discussion. The necessary definitions and other referenced materials are provided in each area to eliminate any need to flip back and forth from page to page to understand what is required of you.
3. The earthing section gives you more information regarding the below-grade requirements of earthing that are so often misunderstood and overlooked.
4. A question-and-answer section covering the most relevant grounding and earthing questions many electrical professionals may have, and a very interesting section that discusses the economic benefits of proper earthing and grounding.

It is a sincere hope that you find this book to be a relevant source of information and that it becomes your go-to source and guide for grounding and earthing information in regard to NEC compliance.

Chapter One

SUMMARY OF GROUNDING AND EARTHING REQUIREMENTS IN THE 2014 NATIONAL ELECTRICAL CODE

INTRODUCTION

The National Electrical Code® (NEC®) has grounding and earthing requirements that extend from the substation to the power poles that bring electrical energy to your premises, to your electrical service, to each branch circuit and receptacle, and down to the final ground rods (see Fig. 1.1). It would not be a stretch to say that Art. 250 of the NEC is one of the most important sections in the entire code book.

The primary purpose of grounding and earthing is related to safety in two areas: (1) providing a common reference point for the electrical and (2) providing electrical safety mechanisms for both people and equipment. For the purposes of this short summary, we will concentrate on the safety aspects and not discuss the common reference point, other than to say there are valid scientific and engineering reasons why a connection to earth/soil is needed at each and every electrical service, and we will just leave it at that.

This summary will also *not* be discussing delta-type systems, ungrounded systems, high-voltage systems (1000 V and higher), impedance-grounded neutral systems, isolated grounding circuits, and other less common electrical systems. This summary will only discuss a grounded wye-type electrical system. Details on the other systems can be found in the appropriate chapters of this book.

LEGAL

The NEC has been adopted into law in all 50 states and the District of Columbia in the United States and is similar to the electrical

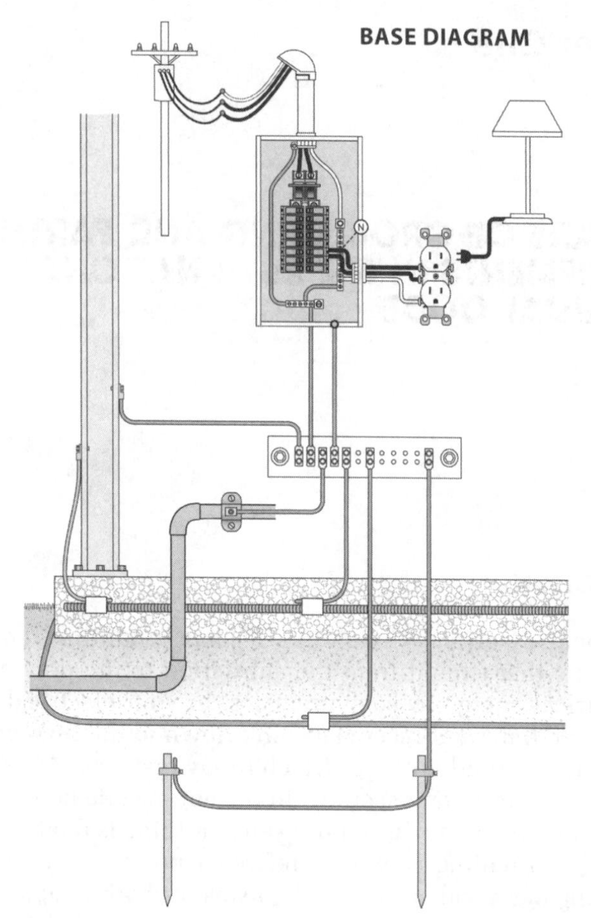

Figure 1.1 Grounding and bonding for a typical building.

requirements found in Canada, the European Union, and nearly all other parts of the world. In the United States, the Code is generally adopted state by state, with 30 of the states typically adopting the Code the same year it comes out. However, approximately 12 of the states are holdouts and are often one Code edition behind (i.e., currently on the 2008 edition), with only one state (Connecticut) and the District of Columbia still using the 2005 edition. Arizona, Illinois, Kansas, Maryland, Mississippi, Missouri, and Nevada adopt the Code at the local level (municipalities). The Sate of California actually re-writes the NEC and adopts it's own version of the Code called "The California Electrical Code." There are some excellent websites available with regularly updated information regarding Code adoption by state.

OVERVIEW

Article 250 of the NEC covers grounding and earthing requirements for your electrical system, and it is broken down into 10 parts. This book has a chapter dedicated to each of these 10 parts and follows each and every article of the Code.

However, despite how the Code breaks down grounding systems, one can generally think of grounding as being broken down into two main areas, above-grade grounding requirements and below-grade (buried) requirements (see Fig. 1.2).

The above-grade grounding requirements of the Code can be further broken down into four additional areas: system bonding, equipment-grounding conductor, bonding, and the grounding electrode conductor.

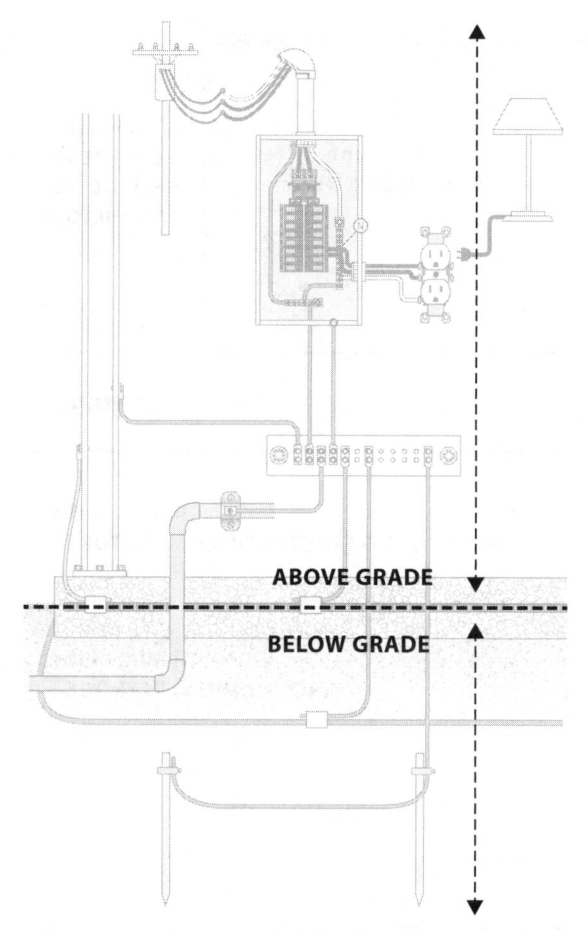

Figure 1.2 Above-grade grounding and below-grade earthing.

The below-grade grounding requirements are contained in a single area of the Code covering the grounding electrodes.

Here is a list of each breakdown we will be discussing in this summary, with the corresponding part of Art. 250 of the Code:

- System bonding: Parts II and III
- Equipment-grounding conductor: Parts IV, V, and VI
- Bonding: Parts III and V
- Grounding electrode conductor: Parts III and V
- Grounding electrode: Parts III

Please see Figs. 1.3 and 1.4 for a visual schematic showing the general breakdown of how Art. 250 of the National Electrical Code is broken down. The terminology within the Code can be very confusing as many

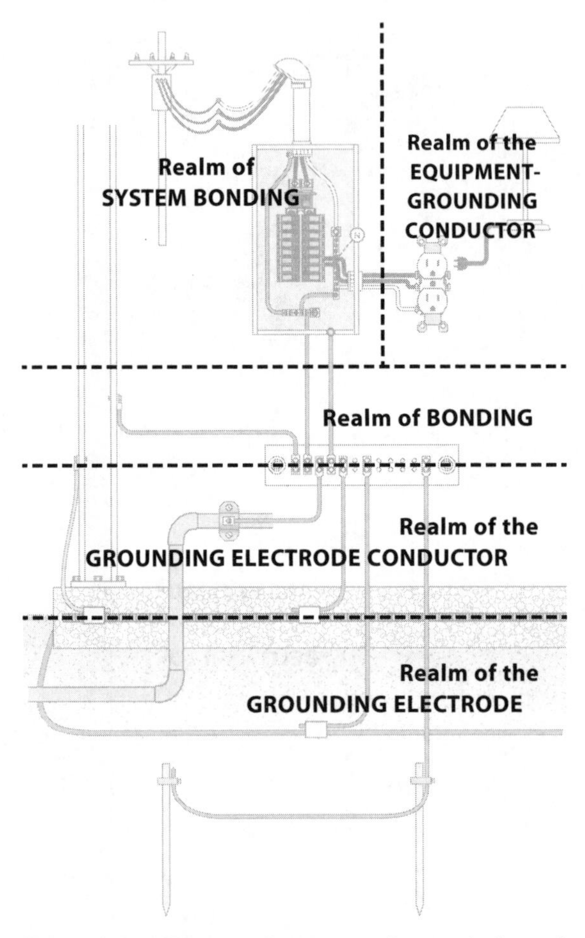

Figure 1.3 NEC grounding terminology and physical location.

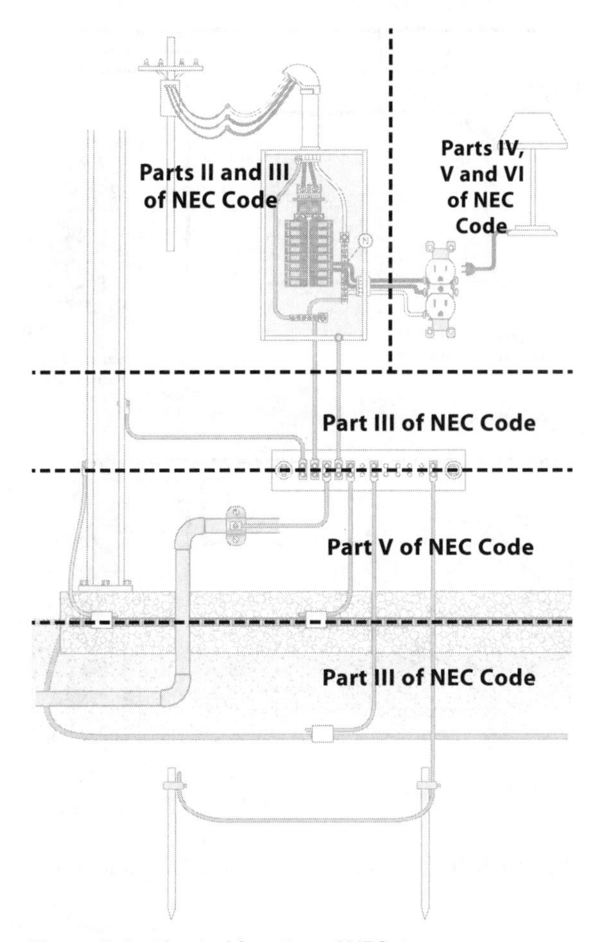

Figure 1.4 Physical location of NEC Art. 250 parts.

of the terms are quite similar. These illustrations will help you to visually understand how the various sections of grounding/earthing equipment relate to the individual parts of the Code and the terminology that is used to describe those components.

SYSTEM BONDING (NEUTRAL-TO-GROUND BONDING)

System bonding is one of the longest and most complex sections found anywhere in the entire NEC. The primary issue in system bonding involves the neutral-to-ground bond. This is where the critical connection

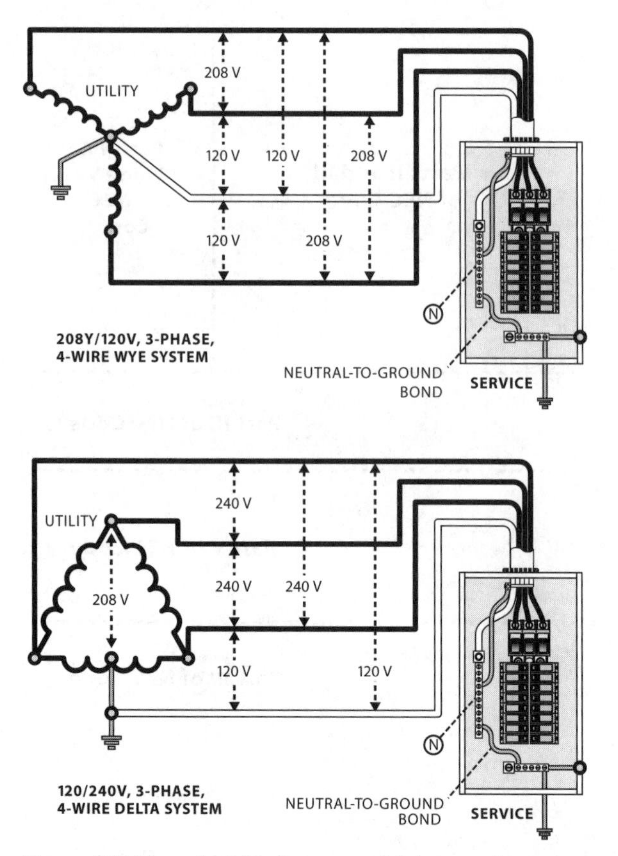

Figure 1.5 Neutral wiring for wye and delta transformers.

occurs between the neutral conductor and earth; hence, the term *grounded conductor* is used to indicate the neutral wire (Fig. 1.5).

The *neutral conductor* is the wire that is connected to the XO terminal in a wye-type transformer, and it is the point where the transformer winding(s) has 0 V. The Code mandates that there be a connection between the ground bar in the electrical cabinet and the neutral bar, thus establishing the neutral-to-ground bond. *At no other point downstream from the first service disconnect is an additional neutral-to-ground connection permitted.*

First service disconnect The very first electrical panel that will disconnect (turn off) the power coming in from the utility company. Sometimes it is used to describe the first electrical panel that can turn off power on the secondary or low-voltage side of a transformer.

This single neutral-to-ground connection is designed to enable ground-fault currents a path back to the winding of the transformer through the third-wire green ground (grounding electrode conductor system) should an electrical fault occur on a circuit-phase wire. In other words, should the black hot wire of a 120-V receptacle come in contact with the metal chassis of the equipment, the third-wire green ground conductor would be able to take the fault current back to the ground bar, where the fault current would travel through the neutral-to-ground bond, up the neutral wire to the transformer, through the transformer winding, down the phase conductor, and to the circuit breaker in the electrical cabinet, disconnecting power to the circuit.

Neutral-to-ground connection (bond) Where a grounded current-carrying conductor, the neutral, is intentionally bonded to earth/ground. In general, this should only occur at the transformer and at the first service disconnect(s). The Code uses several names for neutral-to-ground connections: system bonding jumper, main bonding jumper, and supply-side bonding jumper.

If the neutral-to-ground connection did not exist, the same fault currents would travel back via the third-wire green ground to the ground bar, where, because there is no neutral-to-ground bond, the fault currents would be forced to go down through the grounding electrode, where the fault currents would have to travel through the earth/soil, up the transformer's ground rod (the resistance across the earth between the two ground rods is unknown) to the XO terminal, where hopefully there is still enough current to travel through the winding and down the phase wire to trip the remote circuit breaker/fuse, thus disconnecting the power. The problem is that the resistance of the earth is almost always great enough to prevent the circuit breaker/fuse from tripping.

This is why you must have a copper-to-copper conductor path back to the winding of the transformer, which is often only possible through the neutral conductor. When utility companies only bring in phase conductors and neutral to your building, you must have a neutral-to-ground bond in your first service disconnect.

When you are installing a step-down transformer, such as a 480-V to 208-V system, you will have phase conductors, a neutral conductor, and at least two ground paths: both a green ground wire and the metal conduit connecting the 208-V panel to the 480/208-V transformer. In this case, you may *not* have a neutral-to-ground connection in the panel, as the neutral-to-ground connection already exists in the transformer.

As stated earlier, the area of system bonding is one of the most complex and misunderstood sections of the Code. One of the best summaries regarding the complex neutral-to-ground rules was written by Donald W. Zipse, P.E., and has become known as *Zipse's Law*, which is stated roughly as follows:

Zipse's Law In order to have and maintain a safe electrical installation: All continuous flowing current shall be contained within an insulated conductor or if a bare conductor, the conductor shall be installed on insulators, insulated from earth, except at one place within the system and only one place can the neutral be connected to earth.

EQUIPMENT-GROUNDING CONDUCTOR

The equipment-grounding conductor is the third-wire green ground that is run to your typical 120-V outlet. Its purpose is to provide a dedicated ground-fault current path back to the electrical panel, as stated previously in the section above on system bonding. The realm of the equipment-grounding conductor is the electrical circuits downstream from the circuit-breaker box (first service disconnect).

The Code has an entire section dedicated to this conductor. In essence, you must have a wire-type conductor routed with the phase and neutral wires to each circuit and receptacle. Now, the Code does allow you to use other metal objects as equipment-grounding conductors, some examples being the structural metal members of the building and the metal conduit (raceway) in which the circuits are routed. However, this practice is not recommended and certain industrial standards actually forbid it. The short answer for why you should not use the metal conduit as an equipment-grounding conductor is that the metal conduit (raceway) is electrically noisy and is unreliable as a ground-fault current path (consider the total impedance of the conduit as all the ground-fault current passes through all the joints and connections across the total distance from the circuit breaker to the outlet).

The basic principle is that for each of your single-phase 120-V circuits, you have one circuit breaker, one phase wire, one neutral wire, and one ground wire, plus a properly installed and electrically continuous metal conduit (raceway) system running from the disconnect enclosure all the way to the branch circuit receptacles. The Code does allow a number of exceptions for this basic principle. But these exceptions to the rules will generally only make your electrical system less safe and less effective.

BONDING

The bonding section of Art. 250 of the Code is concerned with ensuring that the normally non-current-carrying metal parts of equipment and enclosures (the exposed metal chassis of equipment and electrical components) are bonded together and connected to earth ground.

The reason the Code is concerned about ensuring that the metal objects you can touch are connected together and tied to earth ground is twofold: (1) to provide a ground-fault current path for any accidental electrical energy that may be on the surface of metal objects during a fault; and (2) to remove harmonics, transients, and other objectionable currents that naturally form on metal raceways and enclosures during the normal operation of the circuit.

For these reasons, the Code has an entire section dedicated just to ensuring you install your electrical system with all of the metal enclosures bonded together, in the proper manner, to guarantee that there is an electrically continuous path for all metal objects that can be touched by a person, all the way back to the main circuit-breaker panel.

In other words, the Code mandates that all of the metal enclosures and raceways involved in the electrical system must be made electrically continuous throughout the system. If there are plastic components, or other components that may not provide a good conductive path, you must provide additional physical jumpers (wire) around the objects to ensure conductivity.

The NEC does not simply mandate that only your electrical system must be bonded back to your main electrical panel, it also mandates that other systems be bonded back to the main electrical panel as well: water pipes, building steel, the rebar in your concrete foundation, fire sprinklers, lightning-protection systems, Cable Television (CATV), broadband systems, alarm systems, Telco (telephone company) systems, optical fiber systems, and gas pipes are all included in the NEC, and they must be bonded to your building's common ground.

The best way to bond all these systems together is to install a copper ground bar directly below your electrical panel that can be used not only to bond the various systems together but also to connect the grounding electrode conductors together (see next section).

The goal is to ensure that any hazardous electrical currents that may come in contact with the touchable metal objects will have a fault current path back to the circuit breaker through the metal raceway and not through a person.

GROUNDING ELECTRODE CONDUCTOR

The grounding electrode conductor section in Art. 250 of the Code discusses the wiring mechanisms used to connect your main electrical panel to the earth. The term *grounding electrode conductor* is specifically the conductor network that ties your electrical service to the grounding electrodes buried in the earth (see next section).

Once the system bonding, bonding, and the equipment-grounding conductor are properly installed in your electrical system, how do you go about connecting the electrical system to earth/soil? The grounding electrode conductor is the wiring system below your electrical panel that ties the panel to the grounding electrodes.

The basic premise of the grounding electrode conductor section is that the path from the main electrical panel to the grounding electrodes is of sufficient size to handle the likely fault currents your electrical system may encounter. If a critical electrical fault were to occur, your grounding electrode conductor system must be of sufficient size to safely conduct the high current levels to the grounding electrodes.

The Code allows you to use three things as a grounding electrode conductor: wire(s), copper water pipe (only within 5 ft of the entrance to the building), or the structural steel frame of your building (Fig. 1.6).

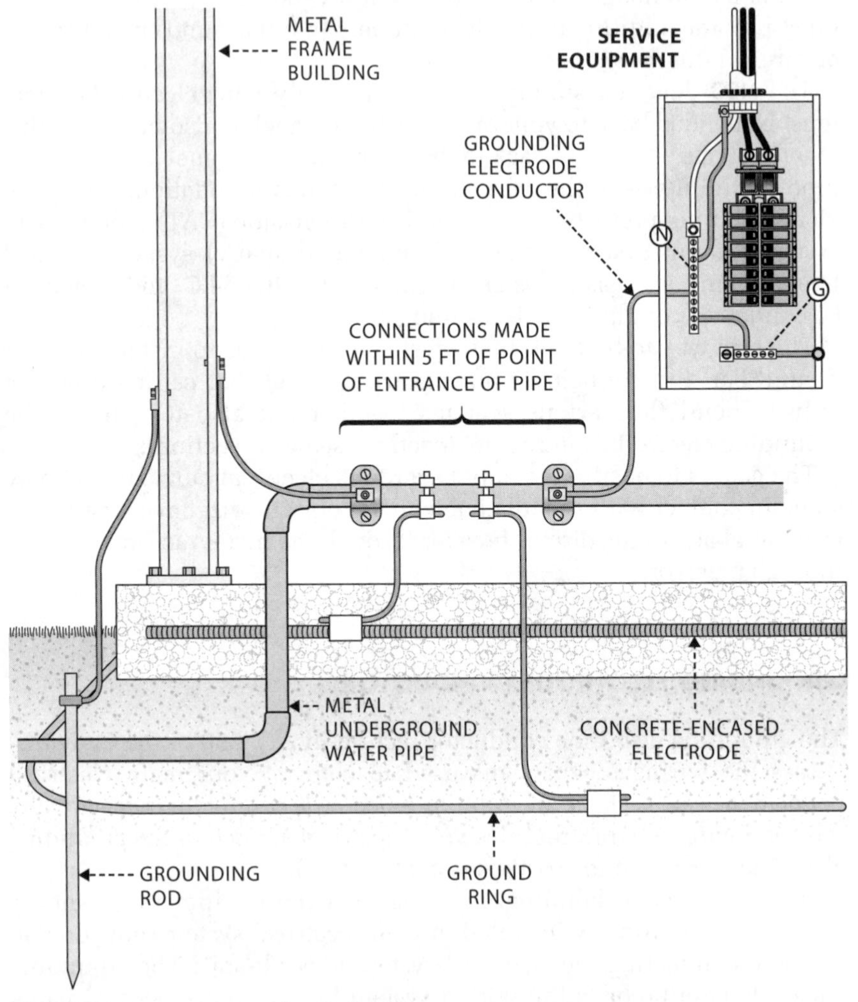

Figure 1.6 Grounding and bonding using copper water pipe.

While the Code may allow use of water pipe and building steel as a grounding electrode conductor, you really should only use a wire-type conductor for this important task. Please see the discussion of building steel and concrete-encased electrodes in the following section and the special note below.

We recommend using a copper ground bar for making all of the necessary grounding electrode connections and all of the required bonding connections (see Fig. 1.7 and section on bonding above).

However you end up connecting your electrical system to your grounding electrodes, the goal is to ensure that the conductor is of sufficient size to handle the fault currents that are likely to occur.

GROUNDING ELECTRODE

The grounding electrode section in Art. 250 of the Code discusses the actual physical connection of a metal object to the earth/soil. Electrodes are the objects that will use the earth/soil to dissipate hazardous electrical energies. In general, the Code only asks that you have two qualified grounding electrodes in contact with the earth/soil. There really are no other requirements, and unfortunately, these minimum standards are really not sufficient for the electrical needs of modern electrical systems (especially commercial and industrial facilities). Figure 1.8 shows a typical grounding electrode system with all of the required bonding for a given structure/building.

The Code recognizes seven types of electrodes: water pipe, building steel, concrete-encased electrodes, ground rings, ground rods/pipes, ground plates, and other listed electrodes. We will not take the time here to discuss the other listed electrodes, as some of these are discussed in Chap. 6 of this book. We also will not waste any time discussing ground plates, other than to say that the sphere of influence is typically so small that they are simply ineffective as electrodes and should not be used.

Sphere of influence The hypothetical volume of soil that will experience the majority of the voltage rise of the ground electrode when that electrode discharges current into the soil. The sphere of influence is equal to the diagonal length of the electrode or electrode system.

The Code allows you to use water pipe as an electrode, if there are is 10-ft of metal pipe in direct contact with the earth. The problem is that most utility (water) companies have replaced old metal water pipes with new nonmetallic ones. And where they do use metal, they have installed a plastic barrier immediately at your building, specifically to stop you

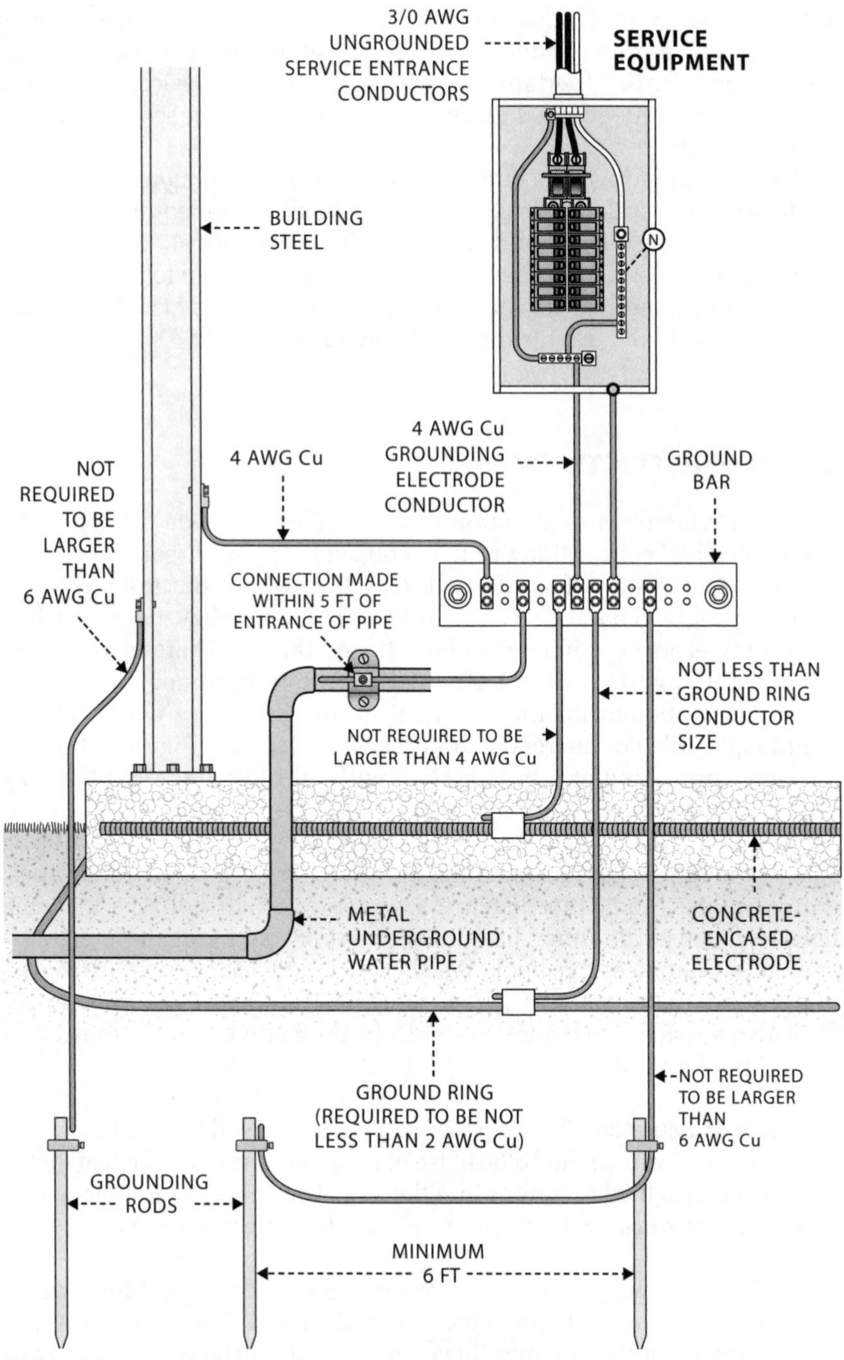

Figure 1.7 Mandatory grounding electrode conductor sizing.

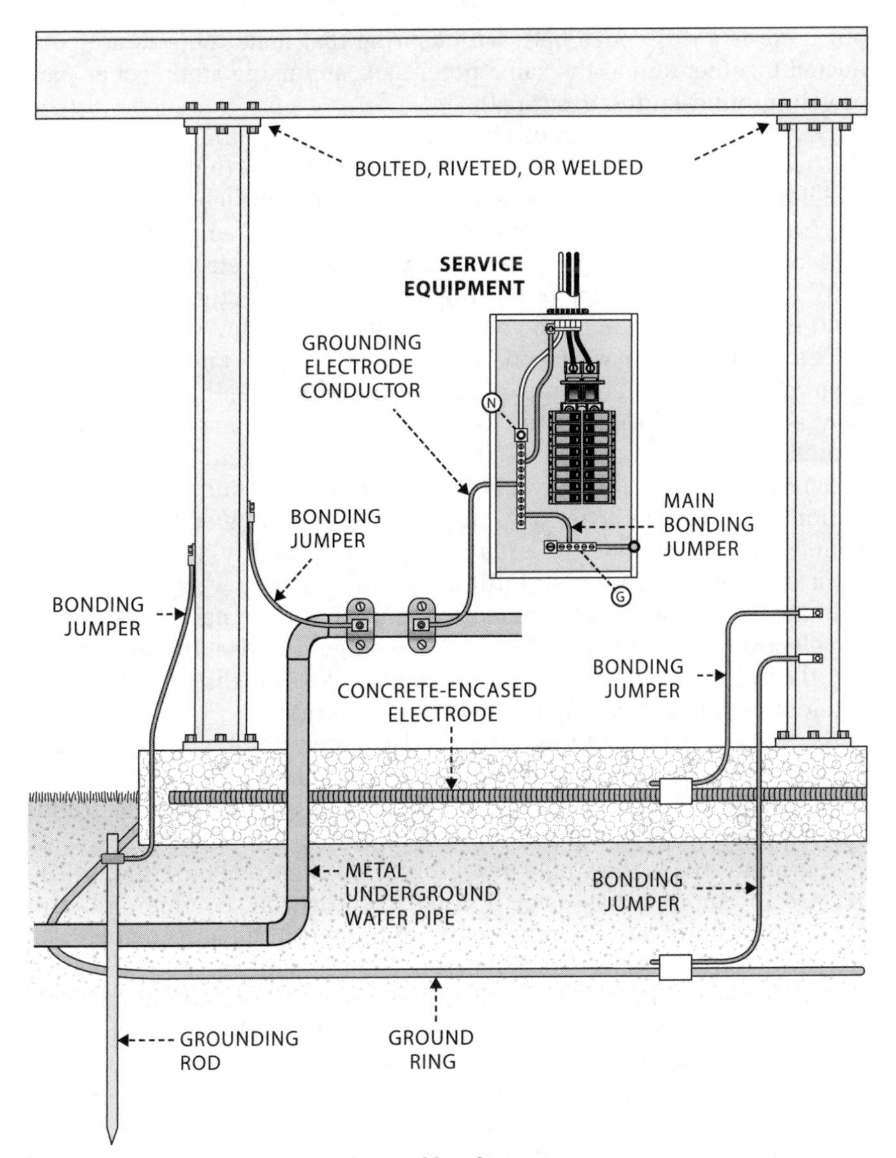

Figure 1.8 Building steel grounding and bonding.

from discharging electrical energy into their piping systems. If your water pipe is plastic, has an isolation system, or is covered in tar or another anticorrosion coating, you may not use water pipe as an electrode.

Using water pipe as an electrode is an old-school process, and the use of water pipe as an electrode should be banned. We recommend that you do not use water pipe as a grounding electrode. That said, you still MUST *bond* to your water pipe (within 5 ft of the building entrance), but there is a difference between ensuring that metal objects are connected together and at the same potentials, and using an object as part of your ground-fault current path.

The Code allows you to use building steel as a grounding electrode, if it is properly bonded to your concrete foundation (with steel rebar) and if the foundation qualifies as a grounding electrode (concrete-encased electrode). This is the most common mistake made when it comes to understanding grounding electrodes; your building structural steel is NOT an electrode. It is a grounding electrode conductor that can only send the electrical energy to your concrete foundation.

The steel rebar in your concrete foundation, also known as a *Ufer ground*, is allowed to be used as a grounding electrode if it has properly bonded, bare steel rebar in the foundation. The rebar may *not* be epoxy coated, as that will insulate the steel and prevent current flow. Also, the concrete must be in direct contact with the earth. If your concrete foundation has a vapor barrier, you may *not* use it as an electrode. And of course, if your concrete foundation does not qualify as a grounding electrode, neither does your building steel.

In any case, it is not recommended to use building steel as a grounding electrode. High-energy electrical discharges can heat the steel, causing the inherent water in the concrete to expand, thus cracking the foundation and destroying its structural integrity.

So, water pipes, building steel, and concrete foundations are poor grounding electrodes and should only be bonded and not used as grounding electrodes: What should you use as a grounding electrode? Easy: ground rings and ground rods.

A ground ring is the single best grounding electrode possible. While its use in America has been limited to industrial, commercial, and technology applications, other parts of the world have recognized its value in reducing electrical hazards. Some countries now mandate ground rings around all structures, including residential homes. This is because the ground ring has the unique ability of being able to bond all the structural steel members of the building to earth, while balancing the resistance to ground across the entire building, thereby dramatically reducing the differences in potential. The ground ring is also great for lightning-protection systems and for both balancing the electrodes and reducing issues caused by the uneven formation of electromagnetic fields. In short, a properly installed ground ring, in conjunction with ground rods, is the best-performing electrode system you can have.

Ground rods are the most common grounding electrode system in the world and have been in use since the time of Benjamin Franklin. Your local electrical inspector will almost certainly insist that your electrical system have at least two ground rods at each meter. There are a number of rules regarding the proper installation of a ground rod, but the short story is to always use 10-ft-long rods, and not the 8-ft-long versions. Please see Chap. 6 for more information regarding ground rods and their proper usage.

In short, the structural and utility systems of your building should not be used as grounding electrodes, even if the Code allows it. A dedicated and purposefully built grounding electrode system should be installed on every building/structure.

Special Note *Regarding Water Pipe, Building Steel, and Concrete-Encased (Rebar in Your Foundation) Electrodes.*

The Code allows you to use these three items as both conductors (grounding electrode conductors and sometimes as equipment-grounding conductors) and as a grounding electrodes. These practices are not recommended for any electrical system. Your grounding conductors (both equipment-grounding conductors and grounding electrode conductors) should be dedicated wire-type conductors installed purposefully for the job.

The water pipe in your building should be used to move water around, building structural steel should be used to hold the roof up over your head, and the rebar in the concrete should be used to hold your foundation together. Maybe these systems are robust enough to also be able to handle high-energy electrical faults ... but then again, maybe they are not. Why risk damaging these critical systems with an electrical fault, when it is simply not needed and will additionally causes problems to your electrical system (i.e., electrical noise)?

CONCLUSION

Article 250 of the National Electrical Code (NEC) requires you to have an electrical system that is contained in an enclosed system that has every element electrically continuous from end to end (bonding). Each branch circuit must have a dedicated conductor from the main electrical panel out to each load in the circuit (equipment-grounding conductor). The main electrical panel must have a conductor system connecting it to the grounding electrodes that is capable of handling the likely fault currents (grounding electrode conductor). The electrical system must be properly connected to earth/soil (grounding electrode). The neutral conductor must be tied to earth/soil in the main electrical panel via a proper neutral-to-ground bond (system bonding).

In Fig. 1-9 we see a complete grounding system for a typical building with all of the required grounding connections discussed in the NEC.

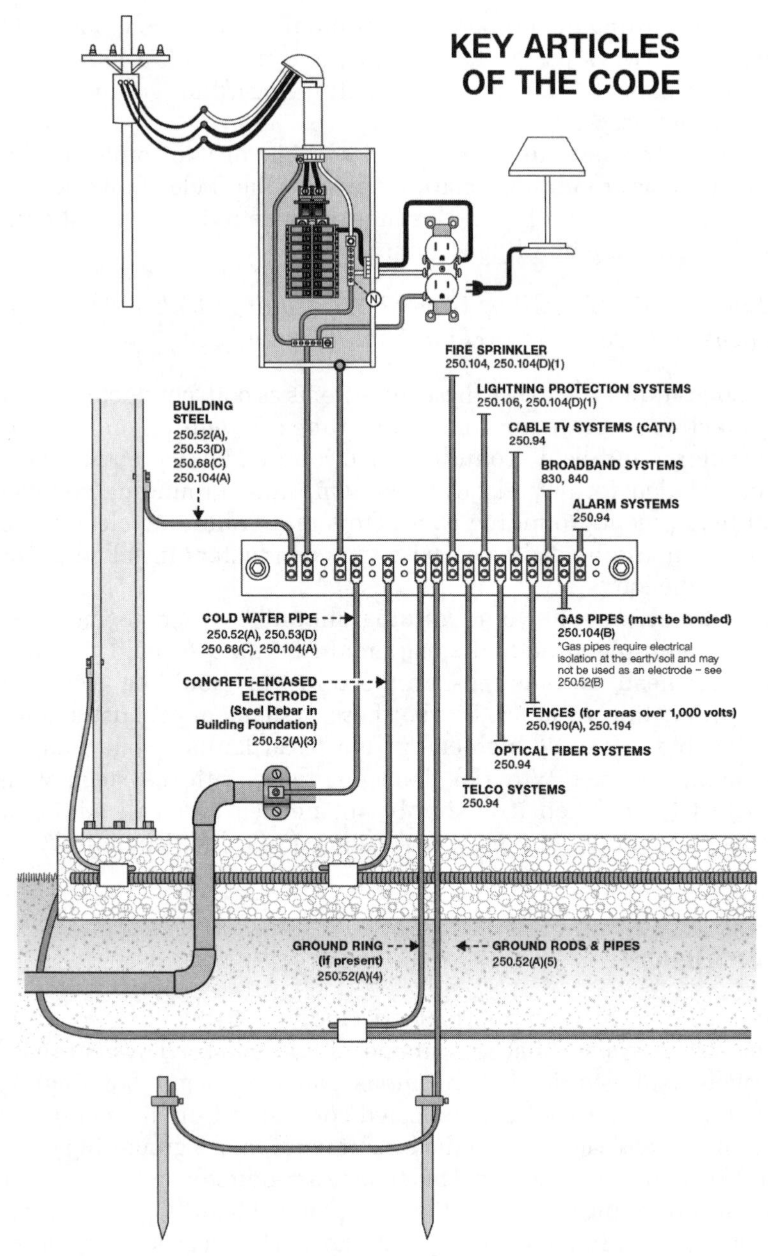

Figure 1.9 Complete grounding and bonding for a typical building.

Including the mandatory connections to alarm systems, gas pipes, broadband systems, fences, cable tv (CATV) systems, optical fiber systems, lightning protection systems (LPS) Telco systems, fire sprinkler systems, and of course the other mandatory grounding electrodes.

Chapter Two

GROUNDING/EARTHING SYSTEMS

INTRODUCTION

The National Electrical Code (NEC) divides grounding into two distinct areas: equipment grounding and system grounding. Equipment grounding is the process of connecting aboveground equipment to the earth. In other words, how to properly bond grounding wires to equipment and route them through conduits, circuit-breaker boxes, and so on. System grounding is the process of intentionally making an electrical connection to the earth itself. This is the actual connection of metal to soil and the minimum standards by which this connection is made. This process is often referred to as *earthing*.

The goal of this chapter is to provide a basic knowledge of system grounding and earthing in an easy-to-read and understandable manner. We will not discuss aboveground wiring issues, except where needed. The topics that will be covered are system grounding, the benefits and features of the available grounding electrodes, the ground potential rise (GPR) hazards of high current discharges, and the effects lightning strikes will have on a grounding system. We will also introduce the principles of proper soil testing, resistance-to-ground (RTG) testing, proper test-well installation, and meter selection.

Both equipment grounding and system grounding are becoming more important as technology rapidly advances. Many of the latest and most advanced systems have stringent grounding requirements. Understanding the available electrical data through proper ground testing enables the electrical engineer to manage grounding systems that will meet specified grounding criteria.

Our goal is to provide the basic knowledge needed to understand and make the right choices when it comes to electrical grounding. Remember: "To protect what's above the ground you need to know what's in the ground."

In the last few decades, much has been learned about the interaction between the grounding electrode and the earth, which is a three-dimensional electrical circuit. Ultimately, it is the soil resistivity (and spatial variations thereof) that determines system design and performance. There are a number of different grounding electrodes in use today. They are the standard driven rod, advanced driven rod, grounding plate, concrete-encased electrode (sometimes called a Ufer ground), water pipes, and the electrolytic electrode.

ZONE (OR SPHERE) OF INFLUENCE

An important concept as to how efficiently grounding electrodes discharge electrons into the earth is called "the zone of influence," which is sometimes referred to as the "sphere of influence" (see Fig. 2.1). The zone of influence is the volume of soil throughout which the electrical potential rises to more than a small percentage of the potential rise of the ground electrode when that electrode discharges current into the soil. The greater the volume compared with the volume of the electrode, the more efficient the electrode. Elongated electrodes, such as ground rods, are the most efficient. The surface area of the electrode determines the ampacity of the device but does not affect the zone of influence. Case in point, the greater the surface area, the greater the contact with the soil and the more electrical energy that can be discharged per unit of time.

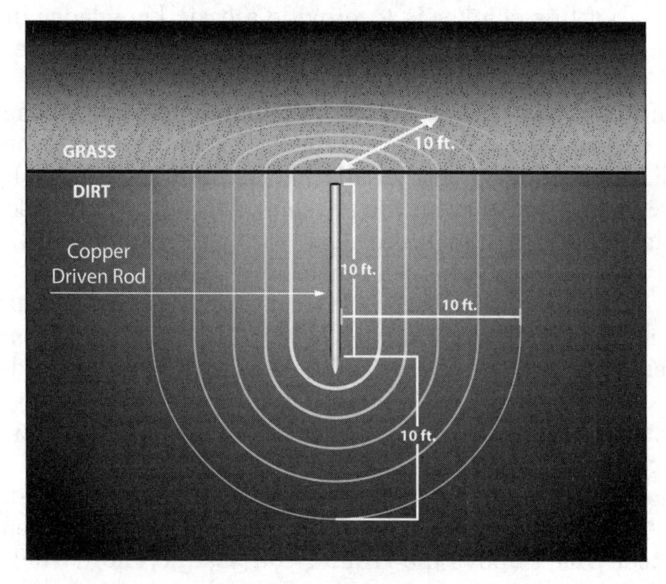

Figure 2.1 Sphere of influence of an earth electrode.

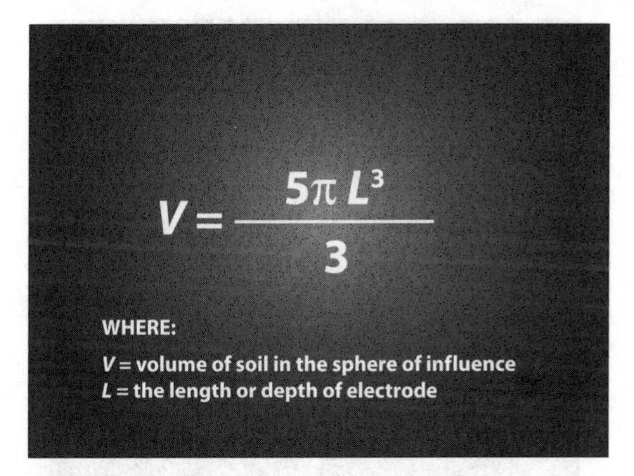

$V = \frac{5\pi L^3}{3}$
WHERE:
V = volume of soil in the sphere of influence
L = the length or depth of electrode

Figure 2.2 Volume of the sphere of influence formula.

The formula for calculating the volume of soil is shown in Fig. 2.2. A simpler version is used when the formula is modified by rounding π (pi) down to 3 and cross-canceling to get the formula:

$$V = 5L^3$$

Thus, a single 10-ft driven rod will utilize 5000 ft^3 of soil, whereas a single 8-ft rod will utilize about half the soil at 2560 ft^3. Going from an 8-ft to 10-ft ground rod can provide a significant reduction in the resistance to ground (RTG), as the sphere of influence will be nearly doubled, given that soil resistivity does not increase with depth.

GROUNDING ELECTRODES

Grounding is the process of electrically connecting any metallic object to the earth by way of an earth electrode system. The **National Electrical Code (NEC)** requires that grounding electrodes be tested to ensure that they are under 25-Ω resistance to ground (earth). It is important to know that aluminum electrodes are not allowed for use in grounding (earthing), as aluminum will rapidly corrode when buried.

Driven Rod. The standard driven rod or copper-clad rod consists of an 8- to 10-ft length of steel with a 5- to 10-mil coating of copper (see Fig. 2.3). This is by far the most common grounding device used today. The driven rod has been in use since the earliest days of electricity, with a history dating as far back as Benjamin Franklin.

Driven rods are relatively inexpensive to purchase; however, ease of installation is dependent upon the type of soil and terrain where the rod is to be installed. The steel used in the manufacture of a standard

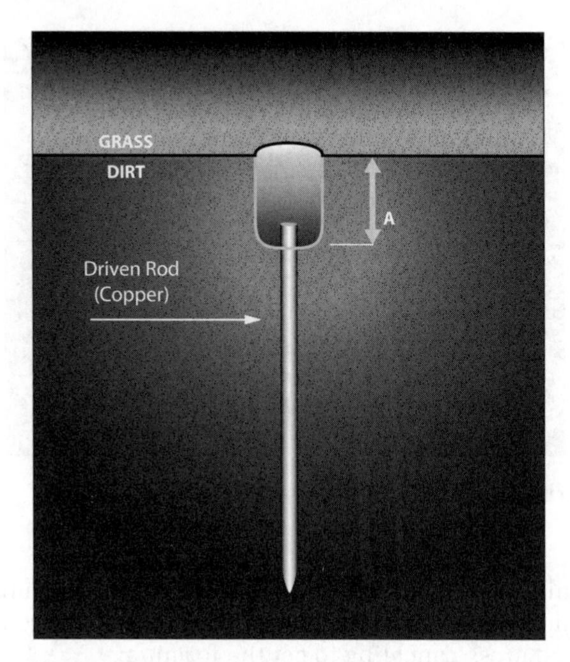

Figure 2.3 Copper-clad driven grounding rod.

driven rod tends to be relatively soft. Mushrooming can occur on both the tip of the rod, as it encounters rocks on its way down, and at the end where force is being applied to drive the rod through the earth. Driving these rods into the earth can be extremely labor intensive when rocky terrain creates problems and the tips of the rods continue to mushroom. Often, these rods will hit a rock and actually turn back around on themselves and pop back up a few feet away from the installation point.

Because driven rods range in length from 8 to 10 ft, a ladder is often required to reach the top of the rod, which can become a safety issue. Many falls have resulted from personnel trying to literally whack these rods into the earth while hanging from a ladder many feet in the air.

The NEC requires that driven rods be a minimum of 8 ft in length and that 8 ft of length must be in direct contact with the soil. To comply with this requirement, the installer will typically use a shovel to dig down into the ground 18 in. before a driven rod is installed, although the most common rods used by commercial and industrial contractors today are rods 10 ft in length, which negates the need for the extra installation process. This can save time as well as meet with the many industrial specifications that also require this length as a minimum.

A common misconception is that the copper coating on a standard driven rod has been applied for electrical reasons. While copper is certainly a conductive material, its real purpose on the rod is to provide corrosion protection for the steel underneath. Many corrosion problems can occur, because copper is not always the best choice in corrosion protection. It should be noted that galvanized driven rods have been developed to address the corrosion concerns copper presents, and in many cases are a better choice for prolonging the life of the grounding rod and grounding systems. Generally speaking, galvanized rods are a better choice in all but high-salt environments.

An additional drawback of the copper-clad driven rod is that copper and steel are two dissimilar metals. When an electrical current is imposed on the rod, electrolysis will occur; also, the act of driving the rod into the soil can further damage the copper cladding, allowing corrosive elements in the soil to attack the bared steel and decrease the life expectancy of the rod. Environment, aging, temperature, and moisture also easily affect driven rods, giving them a typical life expectancy of 5–15 years in good soil conditions. Driven rods also have a very small surface area, which is not always conducive to good contact with the soil. This is especially true in rocky soils, in which the rod will only make contact on the edges of the surrounding rock.

A good example of this is to imagine a driven rod surrounded by large marbles. Actual contact between the marbles and the driven rod will be very small. Because of this small surface contact with the surrounding soil, the rod will increase in RTG, lowering the conductance and limiting its ability to handle high-current faults.

Advanced Driven Rods. Advanced driven rods are specially engineered variations of the standard driven rod, with several key improvements (see Fig. 2.4). Because they present lower physical resistance, advanced rods can now go into terrain where only large drill-rigs could install before and can quickly be installed in less demanding environments. The modular design of these rods can reduce safety-related accidents during installation. Larger surface areas can improve electrical conductance between the soil and the electrode. Of particular interest is that advanced driven rods can easily be installed to depths of 20 ft or more, depending upon soil conditions.

Advanced driven rods are typically driven into the ground with a standard drill hammer. This automation dramatically reduces the time required for installation. The tip of an advanced driven rod is typically made of carbide and works in a similar manner to a masonry drill bit, allowing the rod to bore through rock with relative ease. Advanced driven rods are modular in nature and are designed in 5-ft lengths. They have permanent and irreversible connections that enable an operator to install them safely while standing on the ground. Typically, a shovel is used to dig down into the ground 18 in. before the advanced driven rod is installed. The advanced driven rod falls into the same category as a driven rod and satisfies the same codes and regulations.

In the extreme northern and southern climates of the planet, frost heave is a major concern. As frost sets in every winter, unsecured

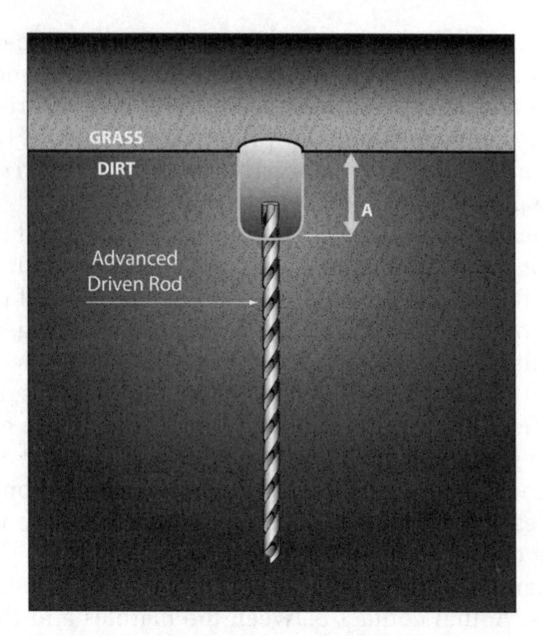

Figure 2.4 Advanced driven grounding rod.

objects buried in the earth tend to be pushed up and out of the ground. Driven grounding rods are particularly susceptible. Anchor plates are sometimes welded to some buried portion of the rods to prevent them from being pushed up and out of the earth by frost heave. However, this requires that a hole be augured into the earth in order to get the anchor plate into the ground, which can dramatically increase installation costs. Advanced driven rods do not suffer from frost-heave issues and can be installed easily in extreme climes.

Grounding Plates. Grounding plates are typically thin copper plates buried in direct contact with the earth. The NEC requires that ground plates have at least 2 ft^2 of surface area exposed to the surrounding soil. Ferrous materials must be at least 0.20-in. thick, while nonferrous materials (copper) need only be 0.060-in. thick. Grounding plates are typically placed under poles or to supplement buried ground rings.

As shown in Fig. 2.5A, grounding plates should be buried at least 30 in. below grade level. While the surface area of grounding plates is greatly increased over that of a driven rod, the zone of influence is relatively small, as shown in Fig. 2.5B. The zone of influence of a grounding plate can be as small as 17 in. This ultra-small zone of influence typically causes grounding plates to have a higher resistance reading than other electrodes of the same mass. Similar environmental conditions, which lead to the failure of the driven rod, also

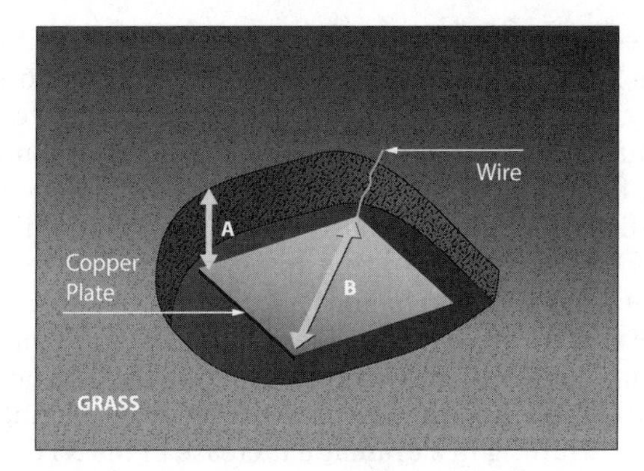

Figure 2.5 Buried copper plate.

plague the grounding plate, such as corrosion, aging, temperature, and moisture.

Ufer Ground or Concrete-encased Electrodes. Originally, Ufer grounds were copper electrodes encased in the concrete surrounding ammunition bunkers. In today's terminology, Ufer grounds consist of any concrete-encased electrode, such as the rebar in a building foundation, when used for grounding, or a wire or wire mesh encased in concrete.

Concrete-encased Electrode. The NEC requires that concrete-encased electrodes use a minimum no. 4 AWG copper wire at least 20 ft in length and encased in at least 2 in. of concrete (see Fig. 2.6). The advantages of concrete-encased electrodes are that they dramatically increase the surface area and degree of contact with the surrounding soil. However, the zone of influence is not increased; therefore, the RTG is typically only slightly lower than it would be for the wire without the concrete.

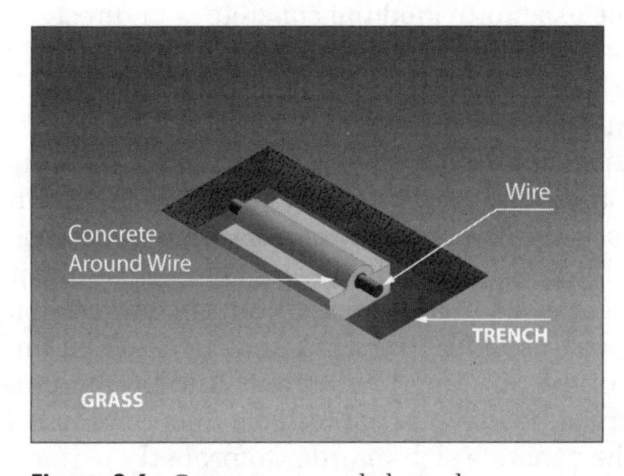

Figure 2.6 Concrete-encased electrode.

Concrete-encased electrodes also have some significant disadvantages. When an electrical fault occurs, the electric current must flow out of the conductor and through the concrete to get to the earth. Concrete, by nature, retains a lot of water that rises in temperature as the electricity flows through the concrete. If the concrete-encased electrode is not sufficient to handle the total current, the boiling point of the water may be reached, resulting in an explosive conversion of water into steam. Many concrete-encased electrodes have been destroyed after receiving relatively small electrical faults. Once the concrete cracks apart and falls away from the conductor, the concrete pieces act as a shield, preventing the copper wire from contacting the surrounding soil and resulting in a dramatic increase in the RTG of the electrode.

There are many new products available that are designed to improve concrete-encased electrodes. The most common are modified concrete products that incorporate conductive materials, usually carbon, into the cement mix. The advantage of these products is that they are fairly effective in reducing the resistivity of the concrete, thus lowering the RTG of the electrode encased. The most significant improvement of these new products is in reducing heat buildup in the concrete during fault conditions, thus lowering the likelihood that steam will destroy the concrete-encased electrode. However, some disadvantages are still evident. Again, these products do not increase the zone of influence, and as such, the RTG of the concrete-encased electrode is only slightly better than that of a bare copper wire or driven rod. Also, a primary concern regarding enhanced grounding concretes is the use of carbon in the mix. Carbon and copper are of different nobilities and will sacrificially corrode each other over time. Many of these products claim to have buffer materials designed to reduce the accelerated corrosion of the copper caused by the addition of carbon into the mix. However, few independent long-term studies are being conducted to test these claims.

Ufer Ground or Building Foundations. Ufer grounds or building foundations may be used, provided the concrete is in direct contact with the earth (no plastic moisture barriers), rebar is at least 0.500 in. in diameter, and there is a direct metallic connection from the service ground to the rebar buried inside the concrete (see Fig. 2.7).

This concept is based on the conductivity of the concrete and the large surface area, which will usually provide a grounding system that can handle very high current loads. The primary drawback occurs during fault conditions, if the fault current is too great compared with the area of the rebar system and moisture in the concrete superheats and rapidly expands, cracking the surrounding concrete and threatening the integrity of the building foundation. Another important drawback to the Ufer ground is that it is not testable under normal circumstances, as isolating the concrete slab in order to properly perform RTG testing, is nearly impossible.

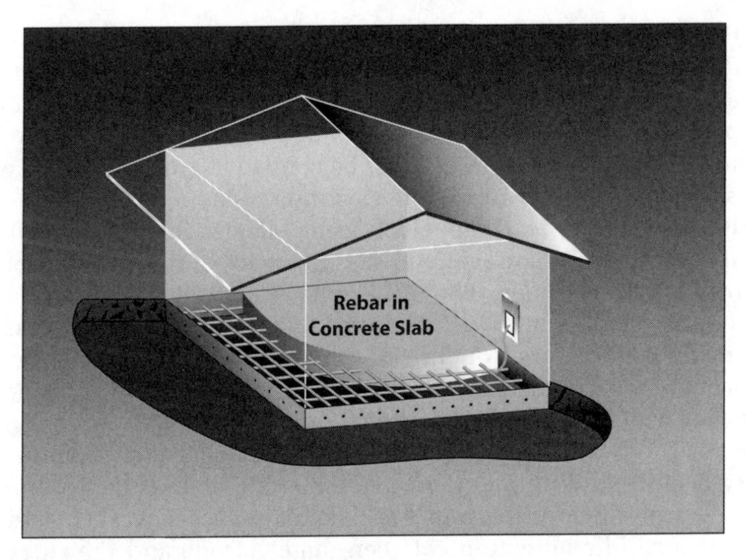

Figure 2.7 Building foundation or Ufer.

The metal frame of a building may also be used as a grounding point, provided the building foundation meets the above requirements and is commonly used in high-rise buildings. It should be noted that many owners of these high-rise buildings are banning this practice and insisting that tenants run ground wires all the way back to the secondary service locations on each floor. The owners will already have run ground wires from the secondary services back to the primary service locations and installed dedicated grounding systems at these service locations. The goal is to avoid the flow of stray currents, which can interfere with the operation of sensitive electronic equipment.

Water Pipes. Water pipes have been used extensively over time as grounding electrodes. Water pipe connections are not testable and are unreliable due to the use of tar coatings and plastic fittings. City water departments have begun to specifically install plastic insulators in the pipelines to prevent the flow of current and reduce the corrosive effects of electrolysis. The NEC requires that at least one additional electrode be installed, when water pipes are used as an electrode. There are several additional requirements, including:

- 10 ft of the water pipe must be in direct contact with the earth.
- Joints must be electrically continuous.
- Water meters may not be relied upon for the grounding path.
- Bonding jumpers must be used around any insulating joints, pipe, or meters.

- Primary connection to the water pipe must be on the street side of the water meter.
- Primary connection to the water pipe must be within 5 ft of the point of entrance to the building.

The NEC requires that water pipes be bonded to ground, even if water pipes are not used as a grounding electrode.

Electrolytic Electrode. The electrolytic electrode was specifically engineered to eliminate many of the drawbacks found in other types of grounding electrodes. The electrolytic electrode consists of a hollow copper shaft filled with salts and desiccants whose hygroscopic nature draws moisture from the air. The moisture mixes with the salts to form an electrolytic solution that continuously seeps into the surrounding backfill material, keeping it moist and high in ionic content. The electrolytic electrode is installed into an augured hole and is typically backfilled with a conductive material, such as bentonite clay. The electrolytic solution and the backfill material work together to provide a solid connection between the electrode and the surrounding soil that is free from the effects of temperature, environment, and corrosion (see Fig. 2.8). The electrolytic electrode is the only grounding electrode that improves with age. All other electrode types will have a rapidly increasing RTG as the seasons change and the years pass. The drawbacks to these electrodes are the cost of installation and the cost of the electrode itself.

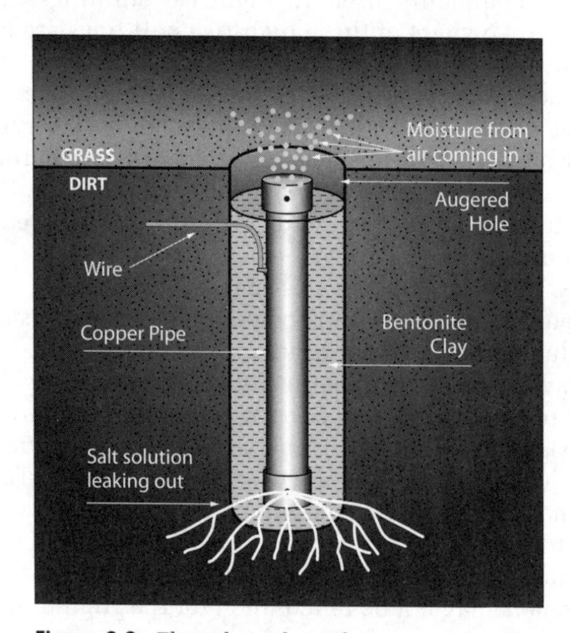

Figure 2.8 Electrolytic electrode.

Various backfill products are available; the primary concern should be whether the product protects the electrode from corrosion and improves its conductivity. Carbon-based products should be avoided, as they will corrode the copper over time.

There are generally two types of electrolytic electrodes that one can install: ones that use sodium chloride (table or rock salt) and those that use magnesium sulfate (Epsom salt). There are advantages and disadvantages for each type. The electrolytic electrodes that use sodium chloride have very long life spans (30–50 years) and as such are often sealed closed, as there is no need to access the tube. The disadvantage is that very little salt actually enters the surrounding soil, so the time it takes to lower the RTG can be very long (years if not decades). The electrolytic electrodes that use magnesium sulfate come with an access cap at the top of the electrode, as the magnesium sulfate will rapidly dissolve away and out of the tube, entering the surrounding soil and thus quickly lowering the RTG. The disadvantages of magnesium sulfate electrodes is that they require annual maintenance to refill the salts in the tube, and if the magnesium sulfate is exposed to high heat, such as from a lightning strike, chemical reactions can occur, resulting in some toxic substances. The material safety data sheet for magnesium sulfate should be consulted prior to use. Some grounding engineers have installed electrolytic electrodes with magnesium sulfate for the first few years of operation so as to rapidly lower the RTG, and then switched over to a sodium chloride and desiccant mix for the long-life and low-maintenance benefits once the surrounding soil has been saturated with conductive materials.

Earth Electrode Comparison Chart. Table 2.1 compares the various types of electrodes with respect to some important characteristics that may prove helpful in selecting proper electrode usage.

Grounding Electrode Test Wells. The purpose of a ground test well is twofold: to be able to quickly and easily measure RTG of the grounding electrode and/or to measure point-to-point continuity (resistance) from one electrode to another. Most grounding test wells are improperly installed preventing accurate RTG measurements of the electrode.

The first part, RTG, is a measurement of the total resistance to the flow of electricity that the earth is providing against the ground system under test. As the three-point fall-of-potential method is not usable for any active grounding system, only the clamp-on or induced-frequency method can be used for testing the RTG of electrodes with a test well. This involves a handheld meter with large jaws that are "clamped" around a grounding conductor under test, thus the need for a test well.

The second part, simple point-to-point continuity (resistance) measurements, can be done at even an improperly installed test well. Point-to-point testing from a ground test well back to a reference point can be

Table 2.1 Earth Electrode Comparison

	Driven Rod	Advanced Driven Rod	Grounding Plate	Concrete-encased Electrode	Building Foundation	Water Pipe	Electrolytic Electrode
RTG	Poor	Average	Poor	Average	Above Average	Poor to Excellent†	Excellent
Corrosion Resistance	Poor	Good	Poor	Good*	Good*	Varies	Excellent
Increase in RTG in Cold Weather	Worsens	Worsens slightly	Worsens	Worsens slightly	Worsens slightly	Minimally affected	Minimally affected
Increase in RTG Over Time	Worsens	Typically unaffected	Worsens	Typically unaffected	Typically unaffected	Typically unaffected	Improves
Electrode Ampacity	Poor	Average	Average	Average*	Above average*	Poor to excellent†	Excellent
Installation Cost	Average	Excellent	Below average	Below average	Average	Average	Poor
Life Expectancy	Poor 5–10 years	Average 15–20 years	Poor 5–10 years	Average* 15–20 years	Above average* 20–30 years	Below average 10–15 years	Excellent 30–50 years

*High-current discharges can damage foundations when water in the concrete is rapidly converted into steam.

†When part of extensive, bare, metallic, electrically continuous water system.

very valuable for understanding the integrity and conductivity of a grounding system. The goal is to install ground test wells capable of allowing both types of measurement.

Improper Installation of a Ground Test. One of the most common ways ground test wells are improperly installed is seen in Fig. 2.9. In this scenario, the installer has simply provided physical access to the ground ring with a ground rod "T-welded" into the loop. This scenario allows us to clamp on to the ground loop conductor above the ground rod itself. However, the signal from the clamp-on ground resistance meter will merely travel around the copper conductor in a loop and is never forced to travel through the earth itself. This test well is improperly installed and will not allow an accurate RTG measurement.

In Fig. 2.10, we see the second most common way test wells are improperly installed, and this is with a twist or "loop" added to the ground conductor. As you can see, this scenario is actually the exact same thing as scenario 1, although it is quite possibly easier to clamp the meter around the conductor. The downside is that this twist in the conductor is a major violation of the rules listed in NFPA® 780 Lightning Standards and the Motorola R56 (and several IEEE standards) in regard to the self-induced coupling effects that high-current, short-duration faults can have on ground systems. When lightning strikes or a major short-circuit fault occurs, a large and powerful magnetic field will form as the current travels through the ground conductor. Any conductors with a "tight radius" are subject to burn-open due to cross-coupling of the magnetic fields. The bottom line is that this test well is improperly installed and not only is it at risk of critical failure under electrical stress, but it will not allow an accurate RTG measurement test.

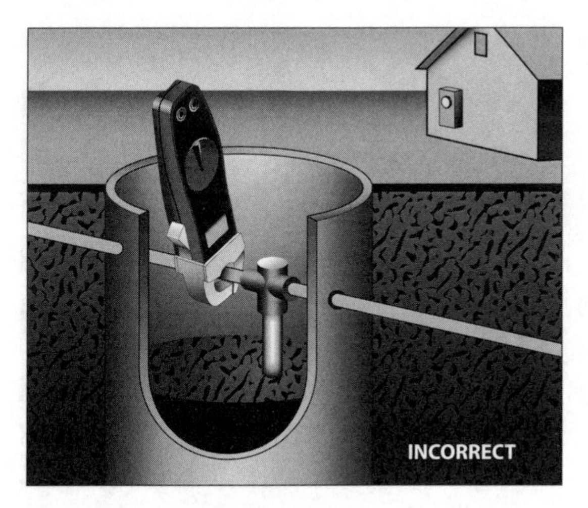

Figure 2.9 Incorrect test well.

Figure 2.10 Incorrect test well with a loop.

Proper Installation of a Ground Test Well. In Fig. 2.11 we can see that installing a grounding electrode in a test well simply requires a short conductor extension or "pigtail" to be added connecting the top of the electrode to the ground loop. While this does require two welds instead of just a single exothermic weld, it is often less labor intensive, as the conductor extension is easier to work with than the ground loop. In this setup, we can see that the clamp-on meter has easy access and the injected signal will be forced down the ground rod and through the earth, enabling an accurate RTG reading of that ground rod. And, of course, accurate point-to-point testing can be conducted as well.

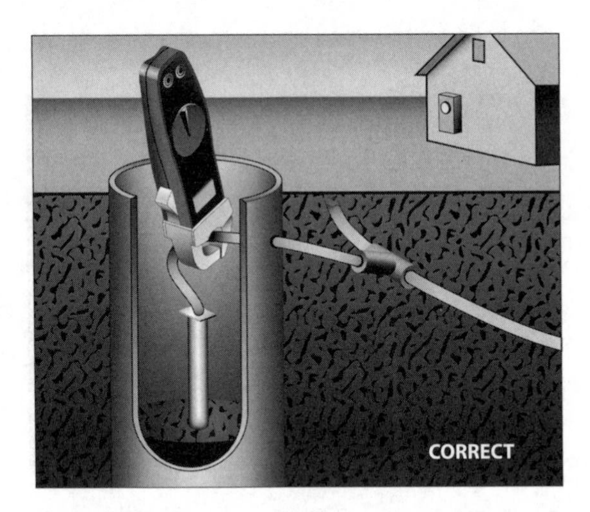

Figure 2.11 Correct test well.

It is important to note that this test well scenario is only valid for testing a single grounding electrode at a time, within the bigger grounding system. It does not provide an RTG of your entire ground system. Depending on the size and requirements of your ground system, it may not even be possible to design a grounding system that can be tested as a whole. This is why multiple ground test wells are often installed at key locations around your ground system; they provide a means of validating the ground system integrity at multiple locations for comparison over time.

SYSTEM DESIGN AND PLANNING

A grounding design starts with a design specification and is often followed by a site analysis, collection of geological data, and a test of soil resistivity of the area. Typically, the site engineer or equipment manufacturers specify a minimum RTG number. The NEC states that the RTG shall not exceed 25 Ω for a single electrode. However, high-technology manufacturers will often specify 3 or 5 Ω, depending upon the requirements of their equipment. For sensitive equipment and under extreme circumstances, a 1-Ω specification may sometimes be required. In designing a ground system, the difficulty and costs increase exponentially as the target RTG approaches the unobtainable goal of 0 Ω.

Data Collection. Once the specifications have been established, data collection begins. Soil resistivity testing, geological surveys, and test borings should provide the basis for all grounding designs. Proper soil resistivity testing using the Wenner four-point method is recommended because of its accuracy. This method will be discussed later in this chapter. Additional data is always helpful and can be collected from existing ground systems located at the site. For example, driven rods at the location can be tested using the three-point fall-of-potential method or an induced-frequency test can be conducted using a clamp-on ground resistance meter.

Data Analysis. Rarely is soil uniform in resistivity from the surface down to depth equal to the whole grounding system's sphere of influence. Soil is typically organized into horizontal layers of largely homogenous materials that have been laid down through the ages, each with different electrical properties. The resistivity of these layers can be inferred with careful measurements and sophisticated computer-modeling techniques. The results are often that the actual soil resistivities from layer to layer can vary by many orders of magnitude. Only by calculating the resistivity and depths of these layers (creating a *soil model*) can you accurately design a grounding system.

Grounding Design. Soil resistivity is the key factor that determines the resistance or performance of a grounding system. It is the starting point of any grounding design. As you can see in Tables 2.2 and 2.3 below,

Table 2.2 Surface Materials versus Resistivity

Type of Surface Material	Average Resistivity in Ohm-meters	
	Dry	Wet
Crusher granite w/fines	140×10^6	1,300
Crusher granite w/fines 1.5"	4,000	1,200
Washed granite: pea gravel	40×10^6	5,000
Washed granite: 0.75"	2×10^6	10,000
Washed granite: 1–2"	1.5×10^6 to 4.5×10^6	5,000
Washed granite: 2–4"	2.6×10^6 to 3×10^6	10,000
Washed limestone	7×10^6	2,000 to 3,000
Asphalt	2×10^6 to 30×10^6	10,000 to 6×10^6
Concrete	1×10^6 to 1×10^9	20 to 100

Table 2.3 Soil Types versus Resistivity

Soil Types or Type of Earth	Average Resistivity in Ohm-meters
Bentonite	2 to 10
Clay	20 to 1,000
Wet organic soils	10 to 100
Moist organic soils	100 to 1,000
Dry organic soils	1,000 to 5,000
Sand and gravel	50 to 1,000
Surface limestone	100 to 10,000
Limestone	5 to 4,000
Shales	5 to 100
Sandstone	20 to 2,000
Granites, basalts, etc.	1,000
Decomposed gneisses	50 to 500
Slates, etc.	10 to 100

soil resistivity varies dramatically throughout the world and is heavily influenced by electrolyte content, moisture, minerals, compactness, and temperature.

SOIL RESISTIVITY TESTING

Soil resistivity testing is the process of measuring a volume of soil to determine the conductivity of the soil. The resulting soil resistivity is expressed in ohm-meters or ohm-centimeters.

Soil resistivity testing is the single most critical factor in electrical grounding design. This is true when discussing simple electrical design, dedicated low-resistance grounding systems, or the far more complex issues involved in ground potential rise (GPR) studies. Good soil models are the basis of all grounding designs, and they are developed from accurate soil resistivity testing.

Wenner Soil Resistivity Test and Other Soil Resistivity Tests. The Wenner four-point method (sometimes called four-pin) is by far the most used test method to measure the resistivity of soil. Other methods do exist, such as the general method and Schlumberger method; however, they are not frequently used for grounding design applications and vary only slightly in how the probes are spaced when compared with the Wenner method.

Electrical resistivity is the measurement of the specific resistance of a given material. It is expressed in ohm-meters and represents the resistance measured between two plates, covering opposite sides of a 1-m cube. This test is commonly performed at raw land sites during the design and planning of grounding systems specific to the tested site. The test spaces four probes out at equal distances to approximate the depth of the soil to be tested. Typical spacings will be in the range of 1, 1.5, 2, 3, 4.5, 7, 10 ft and so on, with each spacing increasing from the preceding one by a factor no greater than 1.5, up to a maximum spacing that is approximately one to three times the maximum diagonal dimension of the grounding system being designed. This results in a maximum distance between the outer current electrodes of three to nine times the maximum diagonal dimension of the future grounding system. This is one "traverse" or set of measurements, and is typically repeated, albeit with shorter maximum spacings, several times around the location at right angles and diagonally to one another to ensure accurate readings.

The basic premise of the test is that probes spaced at 5-ft distances across the surface, will measure the average soil resistivity to an approximate depth of 5 ft. The same is true if you space the probes 40 ft across the earth, you get a weighted average soil resistance from 0 down to 40 ft in depth and at all points in between. This raw data must be processed with computer software to determine the actual resistivity of the soil as a function of depth.

Conducting a Wenner Four-Point Test. The following describes how to take one traverse or set of measurements. As the "four-point" indicates, the test consists of four pins that must be inserted into the earth. The outer two pins are called the *current probes*, C_1 and C_2. These are the probes that inject current into the earth. The inner two probes are the *potential probes*, P_1 and P_2. These are the probes that take the actual soil resistance measurement.

In the test shown in Figs. 2.12 and 2.13 a probe C_1 is driven into the earth at the corner of the area to be measured. Probes P_1, P_2, and C_2 are driven at 5, 10, and 15 ft respectively from rod C_1 in a straight line to

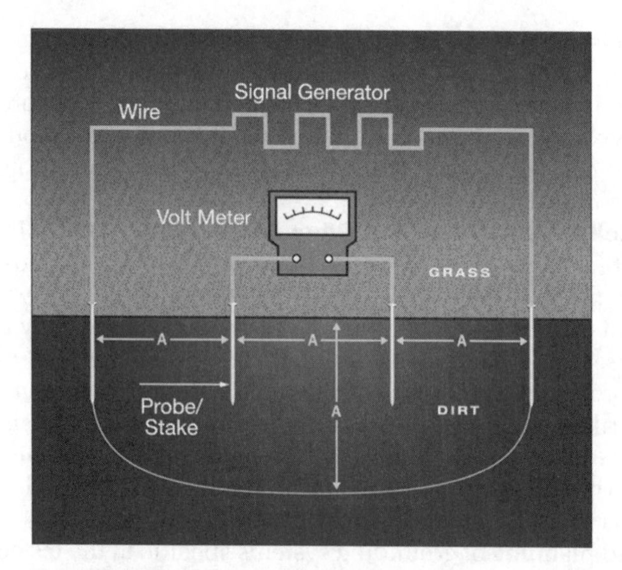

Figure 2.12 A four-point testing pattern.

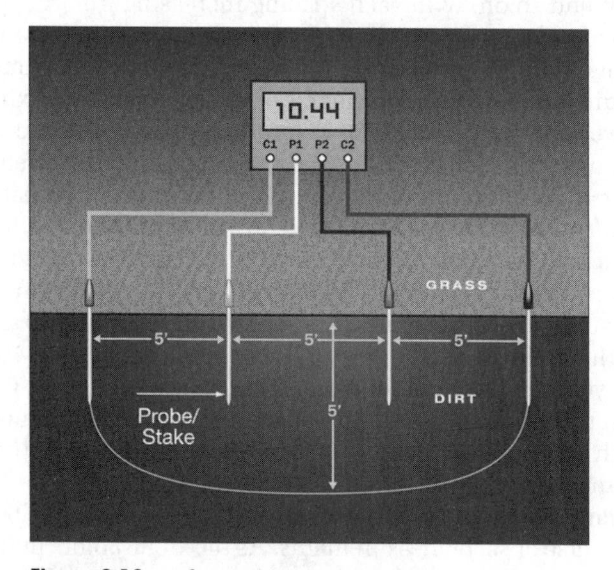

Figure 2.13 A four-point test setup.

measure the soil resistivity from 0 to 5 ft in depth. C_1 and C_2 are the outer probes and P_1 and P_2 are the inner probes. At this point, a known current is applied across probes C_1 and C_2, while the resulting voltage is measured across P_1 and P_2. Ohm's law can then be applied to calculate the measured apparent resistance.

Probes C_2, P_1, and P_2 can then be moved out to 10-, 20-, and 30-ft spacing to measure the resistance of the earth from 0 to 10 ft in depth. You can continue moving the three probes (C_2, P_1, and P_2) away from C_1 at equal intervals to approximate the depth of the soil to be measured. Note that the performance of the electrode can be influenced by soil resistivities at depths that are considerably deeper than the depth of the electrode, particularly for extensive horizontal electrodes, such as water pipes, building foundations, or grounding grids.

Soil Resistance Meters. There are basically two types of soil resistance meters: direct current (DC) and alternating current (AC) models, sometimes referred to as *high-frequency meters*. Both meter types can be used for four-point and three-point testing and can even be used as a standard (2-point) volt meter for measuring common resistances.

Care should always be given when selecting a meter, as the electronics involved in signal filtering are highly specialized. Electrically speaking, the earth can be a noisy place. Overhead power lines, electric substations, railroad tracks, various signal transmitters, and many other sources contribute to signal noise found in any given location. Harmonics, 60-Hz background noise, and magnetic field coupling can distort the measurement signal, resulting in apparent soil resistivity readings that are larger by an order of magnitude, particularly with large spacings. Selecting equipment with electronic packages capable of discriminating between these signals is critical.

Alternating current or high-frequency meters typically use a pulse signal operating at 128 pulses per second or greater. Often these AC meters claim to be using "pulsed DC," which is in reality simply a square-wave AC signal. These AC meters typically suffer from the inability to generate sufficient current and voltage (typically less than 50 mA and under 10 V) to handle long traverses and generally should not be used for probe spacings greater than 100 ft. Furthermore, the high-frequency square-wave signal flowing in the current lead induces a noise voltage in the potential leads that cannot be completely filtered out: this noise becomes greater than the measured signal, as the soil resistivity decreases and the pin spacing increases. High-frequency meters are less expensive than their DC counterparts and are by far the most common meter used in soil resistivity testing.

Direct current meters, which actually generate low-frequency pulses (on the order of 0.5 to 4.0 s/pulse), are the preferred equipment for soil resistivity testing, as they do away with the induction problem from which the high-frequency meters suffer. However they can be very expensive to purchase. Depending upon the equipment's maximum voltage (500 to 2000 mA and 800 V peak to peak), DC meters can take readings with extremely large probe spacings and often many thousands of feet in distance. Typically, the electronics filtering packages offered in DC meters are superior to those found in AC meters. Care should be taken to select a reputable manufacturer.

Data Analysis. Once all the resistance data is collected, the formula shown in Fig. 2.14 can be applied to calculate the apparent soil resistivity in ohm-meters. For example, if an apparent resistance of 4.5 Ω is measured at 40-ft spacing, the soil resistivity in ohm-meters would be 344.7. Figure 2.14 shows the entire soil resistivity formula in detail. One refers to "apparent" resistivity, because this does not correspond to the actual resistivity of the soil. This raw data must be interpreted by suitable methods to determine the actual resistivity of the soil. Also note that the final 1.915 number is calculated by converting meters into feet.

When we describe a soil resistivity test such as the Wenner four-point method, we often correlate the spacings between the probes as a depth or sounding reading. In other words, the distance between the pins in theory equates to the approximate depth being measured. Remember that there are many factors that relate to the actual depth of the measurements read by the meter, so this concept is just a general guideline.

When a resistivity meter takes a measurement, that individual number means nothing in itself; a little math must first be done to determine resistivity. The simplified formula is to take the reading from the meter, multiply it by the probe spacing (in feet), and then multiply it again by 1.915. The result is a resistivity number. When we combine a series of measurements taken at different spacings, we can begin to determine what the characteristics (resistivity) of the earth are like at various depths. The process of comparing numerous individual soil resistivity measurements (at differing spacings) is how one develops a soil model.

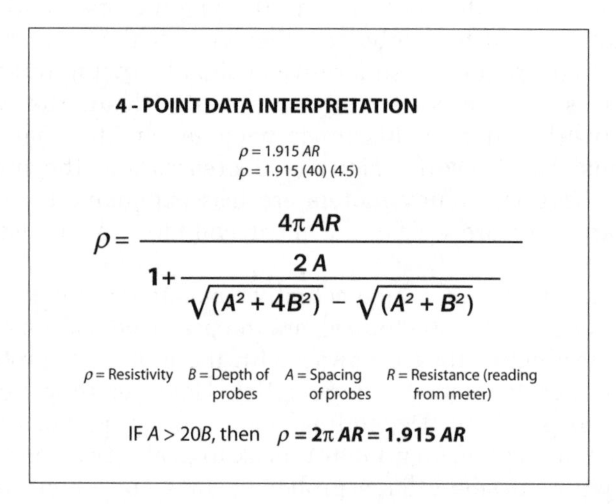

4 - POINT DATA INTERPRETATION
$\rho = 1.915\,AR$
$\rho = 1.915\,(40)\,(4.5)$
$\rho = \dfrac{4\pi AR}{1 + \dfrac{2A}{\sqrt{(A^2 + 4B^2)}} - \sqrt{(A^2 + B^2)}}$
ρ = Resistivity
B = Depth of probes
A = Spacing of probes
R = Resistance (reading from meter)
IF $A > 20B$, then $\rho = 2\pi AR = 1.915\,AR$

Figure 2.14 Soil resistivity calculation.

The soil model will show changes in resistivity of the earth at various depths. What is the resistivity of the soil at 5 ft? What is it at 10 ft? A good soil model will answer these questions. Of course, there are many rules as to how many measurements must be taken and what spacings are required in order to get an accurate model, but that is a different topic. The concern in this case is what happens when a drastic change in resistivity occurs from one layer to the next.

When we conduct a soil resistivity test, we are injecting a test signal (electrical energy) into the surface of the earth, down through the soil to various depths, and recording the loss of energy as a resistance. As the electrical test signal passes from one layer to the next, the test signal will degrade in proportion to the changes in resistance it encounters. This is especially true when the signal must try to move from a very conductive layer of soil to a very resistive layer of soil. The test signal will simply prefer to stay in the most conductive material.

If you have ever seen a submarine war movie, you may have noticed the sub commander will move his submarine below a colder layer of ocean water to avoid being detected by sonar. The cold layer of water will bounce the sonar signal up and away from the submarine, hiding it from the enemy. This is similar to what happens when we conduct soil resistivity tests; the test signal may in fact not penetrate the layers as well as we might hope.

These changes in layer resistivity affect the signal in a predictable way and as such can be calculated and the effects corrected. This is why good engineers prefer soil resistivity models calculated using computer-modeling programs instead of simple hand calculations (good computer-modeling programs perform thousands, if not hundreds of thousands, of calculations). Today's sophisticated algorithms take into account most of the variables and provide vastly superior and more accurate soil resistivity models.

That said, computer algorithms can only help correct the math. A good soil resistivity technician will know how to improve the original signal. The first step is to always use true DC meters. It is recommended to use 800-V p-p DC meters, which require an additional car battery to generate the needed power. The next step is to have many readings starting at 6-in. spacings, with spacing intervals increasing at a factor of no greater than 1.5, with 1.33 preferred. You also need to keep taking readings and increasing your spacings at those intervals until your spacings are at least as great as the depth you are trying to read, preferably two or three times as great. This would certainly mean many dozens of measurements and a lot of work.

Shallow Depth Readings. Shallow depth readings, at as little as 6-in. depths, are exceedingly important for most, if not all, grounding designs. As described above, the deeper soil resistivity readings are actually weighted averages of the soil resistivity from the earth's surface

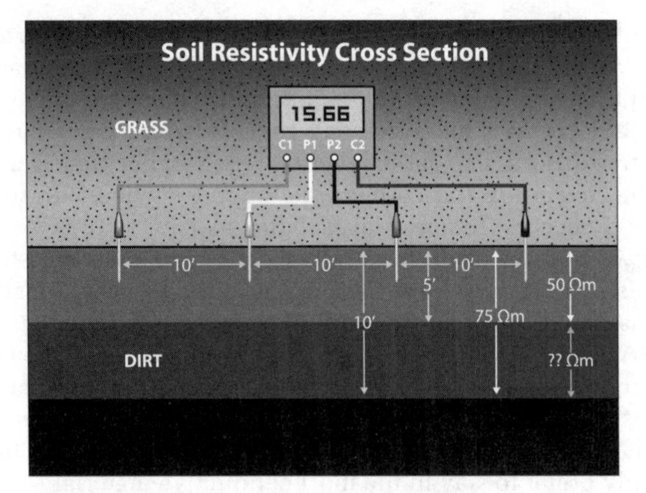

Figure 2.15 Importance of shallow readings.

down to depth and include all the shallow resistance readings above it. The trick in developing the final soil model is to pull out the actual resistance of the soil at depth, and that requires "subtracting" the top layers from the deep readings. Figure 2.15 demonstrates how the shallowest readings impact deeper ones.

As you can see from Fig. 2.15, if you have a 5-ft reading of 50 Ω·m and a 10-ft reading of 75 Ω·m, the actual soil resistance from 5 to 10 ft might be 100 Ω·m (the point here is to illustrate a concept: precomputed curves or computer software are needed to properly interpret the data). The same is true for larger pin spacings. The shallowest readings are used over and over again in determining the actual resistivity at depth.

Shallow depth readings of 6 in. and 1, 1.5, 2, and 2.5 ft are important for grounding design, because grounding conductors are typically buried 1.5 to 2.5 ft below the surface of the earth. To accurately calculate how those conductors will perform at these depths, shallow soil readings must be taken. These shallow readings become even more important when engineers calculate GPR, touch voltages and step voltages.

It is critical that the measurement probes and current probes be inserted into the earth to the proper depth for shallow soil resistivity readings. If the probes are driven too deep, then it can be difficult to resolve the resistivity of the shallow soil. Ideally a 20 to 1 ratio (5 percent) is best; however, when doing very shallow readings that rule cannot always apply. A good rule of thumb when conducting short spacing measurements is that the penetration depth of the potential probes should be no more than 10 percent of the pin spacing, whereas the current probes must not be driven more than 30 percent of the pin spacing.

Deep Readings. The type of meter used often determines the maximum depth or spacing that can be read. A general guideline is that AC soil resistivity meters are good for no more than 100-ft pin spacings, particularly in low-resistivity soils. For greater pin spacings DC soil resistivity meters are required. While DC meters are always the preferred choice, they are the only type of meter that can generate the required voltages needed to push the signal through the soil at long distances, keeping it free of induced voltages and signal noise error from the current injection leads.

Test Location. Soil resistivity testing should be conducted as close as possible to where the proposed grounding system will be located, taking into consideration physical items that may cause erroneous readings. There are two issues that may cause poor quality readings:

1. Electrical interference, causing unwanted signal noise to enter the meter.
2. Metallic objects "short-cutting" the electrical path from probe to probe. The rule of thumb here is that a clearance equal to the pin spacing should be maintained between the measurement traverse and any parallel buried metallic structures.

Testing in the vicinity of the site in question is obviously important; however, it is not always practical. Many electric utility companies have rules regarding how close the soil resistivity test must be in order to be valid. The geology of the area also plays into the equation, as dramatically different soil conditions may exist only a short distance away.

When left with little room or poor conditions in which to conduct a proper soil resistivity test, one should use the closest available open field with as similar geological soil conditions as possible.

TESTING OF EXISTING GROUNDING SYSTEMS

The measurement of ground resistance for an existing earth electrode system is very important. It should be done when the electrode is first installed and then at periodic intervals thereafter. This ensures that the RTG does not increase over time. There are two methods for testing an existing earth electrode system. The first is the three-point or fall-of-potential method (see Fig. 2.16) and the second is the induced-frequency test or clamp-on method. The three-point test requires complete isolation from the power utility. This involves not just power isolation, but also removal of any neutral or other ground connections extending outside the grounding system. This test is the most suitable test for large grounding systems and single isolated grounding electrodes when initially constructed.

The second method is the induced-frequency test and can be performed while power is on. It actually requires the utility to be connected

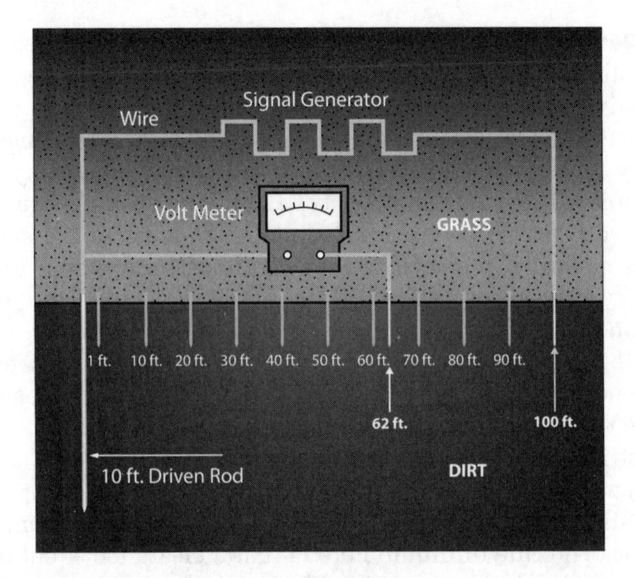

Figure 2.16 A three-point test method.

to the grounding system under test. This test is accurate only for small electrodes, as it uses frequencies in the kilohertz range, which see long conductors as inductive chokes and therefore do not reflect the 60-Hz resistance of the entire grounding system.

Both tests inject a signal into the electrode system; they differ only in the return. The three-point test uses a small probe installed at some distance from the electrode as the signal return; the induced-frequency test uses the utility company's grounding system.

Fall-of-Potential Method or the Three-Point Test. The three-point or fall-of-potential method is used to measure the RTG of existing grounding systems. The two primary requirements to successfully complete this test are the ability to isolate the grounding system from the utility neutral and knowledge of the diagonal length of the grounding system (i.e., a 6-ft × 8-ft grounding ring would have a 10-ft diagonal length). In this test, a short probe, referred to as probe Z, is driven into the earth, at a distance of 10 times the diagonal length of the grounding system (rod X). A second probe (Y) is placed in-line at a distance from rod X equal to the diagonal length of the grounding system.

At this point, a known current is applied across X and Z, while the resulting voltage is measured across X and Y. Ohm's law can then be applied ($R = V/I$) to calculate the measured resistance. Probe Y is then moved out to a distance of two times the diagonal length of the grounding system, in-line with X and Z, to repeat the resistance measurement at the new interval. This will continue, moving probe Y out to three times, four times, ... nine times the diagonal length to complete the three-point test, with a total of nine resistance measurements (see Fig. 2.17).

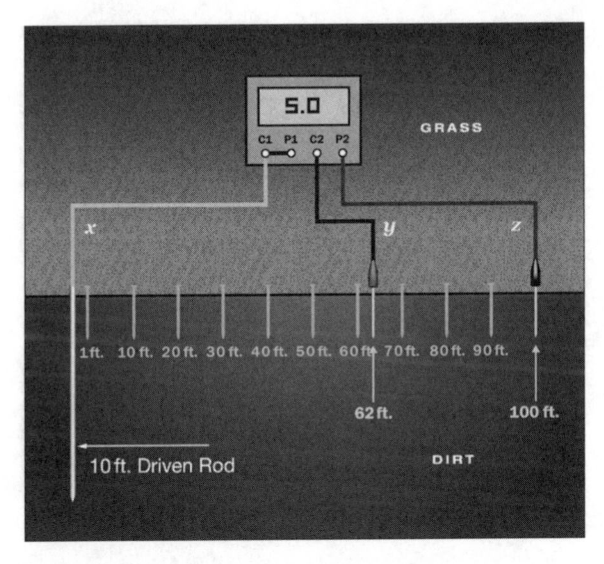

Figure 2.17 A three-point test setup.

Graphing and Evaluation. The three-point test is evaluated by plotting the results as data points with the distance from rod *X* along the *X* axis and the resistance measurements along the *Y* axis, to develop a curve. Roughly midway between the center of the electrode under test and the probe *Z*, a plateau or "flat spot" should be found, as shown in the graph on Fig. 2.18. The resistance of this plateau (actually, the resistance measured at the location 62 percent from the center of the electrode under test, if the soil is perfectly homogeneous) is the RTG of the tested grounding system.

Invalid Tests. If no semblance of a plateau is found and the graph is observed to rise steadily, the test is considered invalid. This can be due to the fact that probe Z was not placed far enough away from rod *X* and can usually indicate that the diagonal length of the grounding system was not determined correctly. If the graph is observed to have a low plateau that extends the entire length and only rises at the last test point, this may also be considered invalid. This is often because the utility or Telco-neutral connection remains connected to the grounding system (see Fig. 2.18).

Induced-frequency Testing or Clamp-on Testing. The induced-frequency test, commonly called the clamp-on test, is one of the newest test methods for measuring the RTG of a grounding system or electrode. This test uses a special transformer to induce an oscillating voltage (often 1.7 kHz) into the grounding system. Unlike the three-point test which requires the grounding system to be completely disconnected and isolated before testing, this method requires that the grounding system under test be connected to the electric utility's ground system (or other large

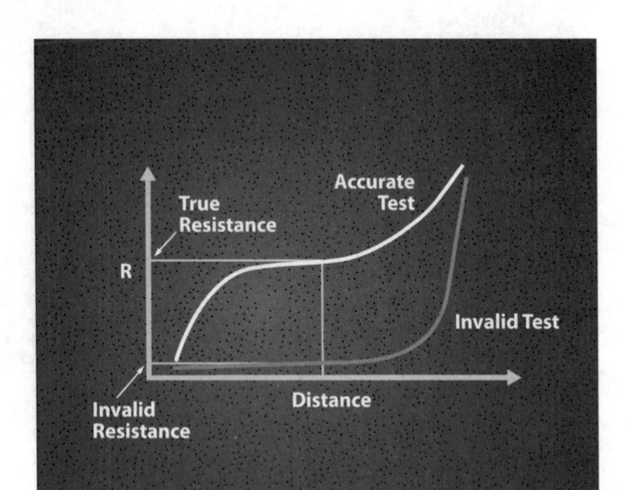

Figure 2.18 Comparison of valid vs. invalid three-point test.

grounding systems, such as one from the telephone companies), with a connecting conductor (typically via the neutral return wire) to provide the return path for the signal. This test is the only test that can be used on live or "hot" systems. However, there are some limitations, primarily:

1. The amount of amperage running through the tested system must be below the equipment manufacturer's limits.
2. The test signal must be injected at the proper location, so the signal is forced through the grounding system and into the earth.
3. The instrument, a "clamp-on meter," actually measures the sum of the resistance of the grounding system under test and the impedance of the utility neutral grounding, including the neutral wiring. Due to the high frequency used, the impedance of the neutral wiring is nonnegligible and can be greater than the ground resistance of a very-low-resistance grounding system under test, which can therefore not be measured accurately (see Fig. 2.19).
4. The ground resistance of a large grounding system at 60 Hz can be significantly lower than at 1.7 kHz.

Many erroneous tests have been conducted in which the technician only measured metallic loops and not the true RTG of the grounding system, and the veracity of the induced-frequency test has been questioned due to testing errors. However, when properly applied to a small to medium-sized, self-standing grounding system, this test is rapid and reasonably accurate.

Test Application. The proper use of this test method requires the utility neutral to be connected to a grounded wye-type transformer. The oscillating voltage is induced into the grounding system at a point where it will be forced into the soil and will return through the utility neutral.

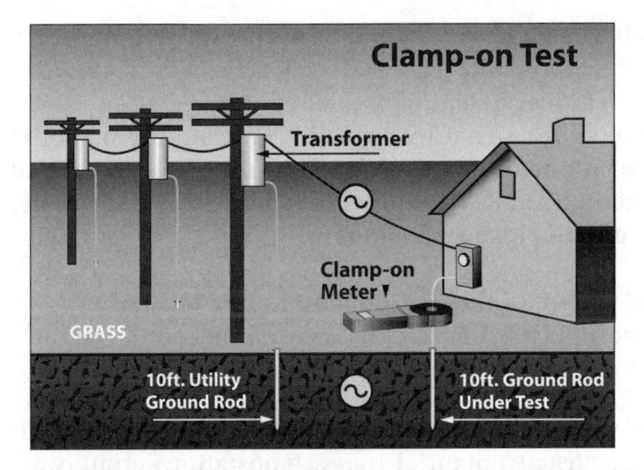

Figure 2.19 Induced-frequency test diagram.

Extreme caution must be taken at this point, as erroneous readings and mistakes are often made. The most common of these occur when clamping on or inducing the oscillating voltage into the grounding system at a point where a continuous metallic path exists back to the point of the test. This can result in a continuity test being performed rather than a ground resistance test. Understanding the proper field application of this test is vital to obtaining accurate results. The induced-frequency test can test grounding systems that are in use and does not require the interruption of service to take measurements.

Ground Resistance Monitoring. Ground resistance monitoring is the process of automated, timed, and/or continuous RTG measurement. These dedicated systems use the induced-frequency test method to continuously monitor the performance of critical grounding systems. Some models may also provide automated data reporting. These new meters can measure RTG and the current that flows on the grounding systems that are in use. Another benefit is that electrical service need not be interrupted for these measurements to be taken.

GROUND POTENTIAL RISE

Ground potential rise is a phenomenon that occurs when large amounts of electricity enter the earth. This typically happens when substations or high-voltage towers fault or when lightning strikes occur. When currents of large magnitude enter the earth from a grounding system, not only will the grounding system rise in electrical potential, but so will the surrounding soil as well.

The voltages produced by a GPR event can be hazardous to both personnel and equipment. As described earlier, soil has resistance that will allow an electrical potential gradient or voltage drop to occur along the path of the fault current in the soil. The resulting potential differences will cause currents to flow into any and all nearby grounded conductive bodies, including concrete, pipes, copper wires, and people.

Ground Potential Rise Definitions

Ground potential rise (as defined in IEEE Std. 367) is the product of a ground electrode impedance, referenced to remote earth, and the current that flows through that electrode impedance.

Ground potential rise (as defined by IEEE Std. 80-2000) is the maximum electrical potential that a (substation) grounding grid may attain, relative to a distant grounding point, assumed to be at potential of remote earth. This voltage, GPR, is equal to the maximum grid current times the grid resistance.

Ground potential rise events are a concern wherever electrical currents of large magnitude flow into the earth. This can be at a substation, high-voltage tower or pole, or a large transformer. In cases in which a GPR event may be likely, grounding precautions are required to ensure personnel and equipment safety.

Electrical potentials in the earth drop abruptly around the perimeter of a grounding system but do not drop to zero. In fact, in a perfectly homogenous soil, soil potentials are inversely proportional to the distance from the center of the grounding system, once one has reached a distance that is a small number of grounding system dimensions away. The formula is as follows:

$$\text{maximum earth potential} = \text{soil resistivity} \times \text{current}/(2 \times \text{pi} \times \text{distance})$$

Where earth potential is in volts, soil resistivity is in ohm-meters, current is the current flowing into the soil from the grounding system in amperes, circle constant pi (unitless), and distance is in meters from source current.

Probably the most commonly noted GPR event involves the death of cows in a field during a lightning strike. Imagine lightning striking the center of an open field where cows are standing. The current injected into the earth flows radially away from the strike point in all directions, creating voltage gradients on the surface of the earth, also in a radial direction. All the cows facing the lightning strike would have their forehooves closer to the strike point than their rear hooves. This would result in a difference of potential between their forelegs and rear legs, causing current to flow through their bodies, including their hearts, and killing the cows. On the other hand, those cows with their flanks turned toward the lightning strike would have a greater chance of surviving, as the distance between their forelegs and therefore the voltage applied between them, would be relatively small, resulting in a lesser current flow.

GPR studies are typically conducted on substations and high-voltage towers. Substations have relatively large grounding areas, especially when compared with high-voltage towers and poles. Towers and poles represent by far the most potentially dangerous and difficult GPR situations to handle and are often not protected, unless they are located in high-exposure areas or have equipment installed at ground level at which service personnel might be required to work.

Ground Potential Rise Analysis. The primary purpose of a GPR study is to determine the level of hazard associated with a given high-voltage location for personnel and/or equipment. As can be seen in Fig. 2.20, electrical faults can occur at a tower, resulting in electrical energy flowing into the earth. When the degree of electrical fault hazard is identified, the appropriate precautions must be made to make the site safe. To do this, the engineer must identify what the minimum grounding system for each location will be. The engineer must also take into consideration all local and federal guidelines, including utility company requirements.

> ***For example*** *Many utility companies require at a minimum that a simple ground ring be installed at least 18 in. below ground and 3 ft from the perimeter of all metal objects. This ground ring is also referred to as a counterpoise.*

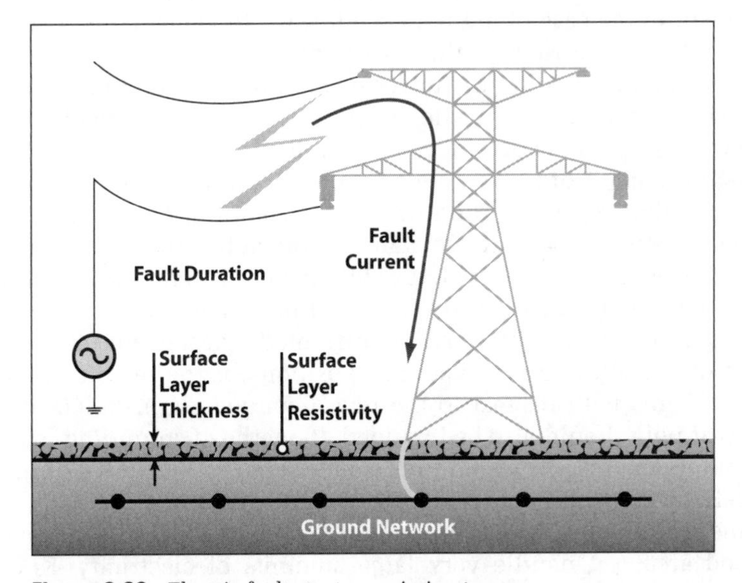

Figure 2.20 Electric fault at a transmission tower.

Once the minimum grounding system is identified, the engineer can run a GPR analysis and identify the extent of any electrical hazards.

Typically, items reported in a GPR study will include the following: the square footage, size, and layout of the proposed grounding grid; RTG of the proposed grounding system; the estimated fault current that would flow into the grounding system; GPR (in volts) at the site; 300-V peak line; the X/R ratio; and the fault clearing time in seconds. Touch and step voltages are usually computed as well, as these are the primary indicators of safety.

The grounding engineer needs three pieces of information to properly conduct a GPR study:

1. Soil resistivity data
2. Site drawings with the proposed construction
3. Electrical data from the power company

Soil Resistivity Data. The soil resistivity data should include apparent resistivity readings at pin spacings ranging from 0.5 or 1 ft to as many as three grounding grid diagonals, if practical. Touch and step voltages represent the primary concern for personnel safety. Understanding the characteristics of the soil at depths ranging from immediately underfoot to one or more grid dimensions is required for a cost-effective and safe grounding system to be designed. See previous section on soil resistivity.

Site Drawings. The proposed site drawings should show the layout of the high-voltage tower or substation and any additional construction for new equipment that may be occurring on the site, including fencing and gate radius. Incoming power and Telco runs should also be included. In the case of high-voltage towers, the height and spacing of the conductors carried on the tower and any overhead ground wires that may be installed on the tower need to be detailed during the survey. This information is needed to properly address all the touch and step voltage concerns that may occur on the site.

Electric Utility Data. The electric utility company needs to provide electrical data regarding the tower or substation under consideration. This data should include the name of the substation or the number of the tower, the voltage level, the subtransient X/R ratio, and the clearing times. In the case of towers, the line names of the substations involved, the amount of current contributed by each substation in the event of a fault, and the type and positions of the overhead ground wires, if any, with respect to the phase conductors installed on each tower or pole should also be included. If overhead grounding wires are present, tower or pole ground resistances along the line are of interest as well, be they measured, average, or design values.

This information is important, as high-voltage towers have small ground area, yet handle very large amounts of electricity. Knowing whether a tower has an overhead ground wire is important, because the overhead wire will carry away a percentage of the current, which will depend on the overhead ground wire type and ground resistances of adjacent towers, to other towers in the run, reducing the GPR event. Additionally, towers with overhead ground wires tend to have shorter clearing times. The same holds for substations: overhead ground wires on transmission lines and neutral wires on distribution lines can significantly reduce the magnitude of fault current that flows into the substation grounding system during fault conditions.

The following information is required from the utility company:

1. Phase-to-ground fault current contributed by each power line circuit
2. Fault clearing time
3. Line voltage
4. Subtransient X/R ratio
5. The make/type/number of overhead ground wires on each tower/pole line and position with respect to the ground wire continuity and bonding configuration back to the tower and substation phase conductors
6. The average distance from tower to tower and tower to substation
7. Typical tower/pole ground resistance: measured or design values

As-built drawings are often acquired and are useful for towers with existing grounding systems. They are also useful in the case of modifications and upgrades to existing substations, which will have extensive grounding systems already installed.

Personnel Safety during Ground Potential Rise Events. The grounding engineer will be required to develop safety systems to protect any personnel working where GPR hazards are known to exist. Federal law mandates that all known hazards must be eliminated from the workplace for the safety of workers. It is the engineer's choice as to which voluntary standards to apply to comply with the law. Title 29 of the Code of Federal Regulations, part 1910.269 (29 CFR 1910.269) specifically states that step and touch potentials must be eliminated on transmission and distribution lines, which include any related communications equipment.

Substations are always considered workplaces, and step and touch potentials (voltages) must always be eliminated. Transmission and distribution towers or poles are not always considered workplaces and therefore are often exempt from these requirements. Take, for example, a lonely tower on a mountainside or in the middle of the desert: these towers are not typically considered workplaces. However, any high-voltage tower or pole becomes a workplace as soon as equipment is installed that is not related to the electric utility company and requires outside vendors to support the new equipment. Cellular telecommunications, environmental monitoring, and microwave relay equipment are good examples of equipment that, when installed on a high-voltage tower, turns the tower into a workplace. This would make the elimination of step and touch potentials required.

Hazardous Voltages. Fibrillation current is the amount of electricity needed to cause cardiac arrest from which recovery will not spontaneously occur in a person and is a value based on statistics. IEEE Std. 80-2000 provides a method to determine the pertinent value of fibrillation current for a safety study, along with a good explanation of how it is derived. Many different methods exist for calculating fibrillation current; however, the 50-kg IEEE method is the one most commonly used in North America. The formula shows that the fibrillation current level is inversely proportional to the square root of the fault duration; however, it must be increased by a correction factor based on the subtransient X/R ratio, which can be quite large for shorter fault durations. If personnel working at a site during fault conditions experience voltages that will cause a current less than the fibrillation current to flow in their bodies, then they are considered safe. If a worker will experience a greater voltage than is acceptable, additional safety precautions must be taken.

The subtransient X/R ratio at the site of the fault is important in calculating the acceptable fibrillation current and determining the maximum allowable step and touch potentials that can occur at any given site.

Fault duration is a required piece of data for properly calculating maximum allowable step and touch potentials. The fault duration is the amount of time it takes for the power company to shut off the current in the event of a fault.

Ultimately, the engineer must determine two things:

1. The site-specific maximum allowable voltage that a person can safely withstand
2. The actual voltages that will be experienced at the site during a fault

Each site will have different levels of voltage for both of the above. Unfortunately, we cannot simply say that a human being can withstand a certain level of current or voltage and use that value all the time, because the maximum safe human voltage threshold is determined by the surface layer resistivity, the fault duration, and the subtransient X/R ratio. Additionally, as each site has different fault durations and different soil conditions, it is critical that calculations be made for each and every possible fault location.

Step Potential. When a fault occurs at a tower or substation, the current will enter the earth. Based on the distribution of varying resistivities in the soil (typically, a horizontally layered soil is assumed), a corresponding voltage distribution will occur. The voltage drop in the soil surrounding the grounding system can present hazards for personnel standing in the vicinity of the grounding system. Personnel "stepping" in the direction of the voltage gradient could be subjected to hazardous voltages (see Fig. 2.21).

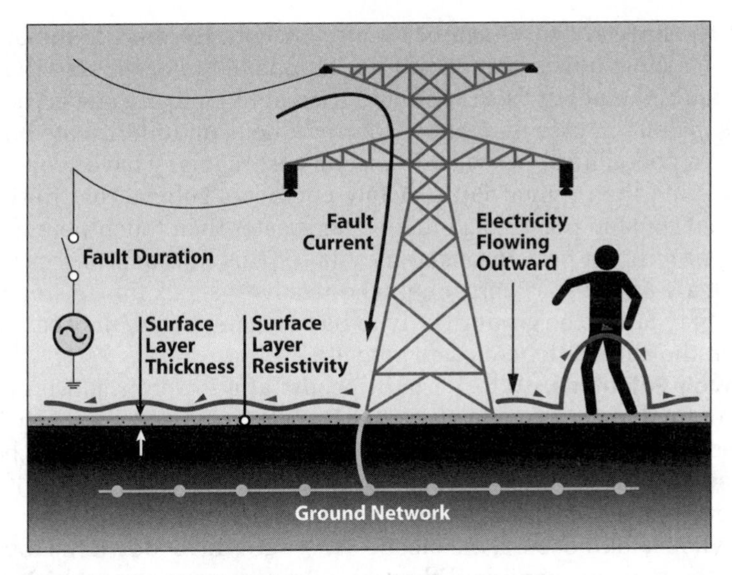

Figure 2.21 Step potential near a transmission tower.

In the case of step potentials, electricity will flow if a difference in potential exists between the two legs of a person. Calculations must be performed that determine how great the tolerable step potentials are and then compare those results with the step voltages expected to occur at the site.

Hazardous step potentials can occur a significant distance away from any given site. The more current that is pumped into the ground, the greater the hazard. Soil resistivity and layering plays a major role in how hazardous a fault occurring on a specific site may be. Low-over-high soil models tends to increase step potentials at farther distances from the ground-fault location. While high-over-low soil models tend to increase step potentials when in close proximity to the ground-fault location.

In other words, a high-resistivity top layer and low-resistivity bottom layer tends to result in the highest step voltages when personnel are in close proximity to the source of the ground fault: the low-resistivity bottom layer draws more current out of the electrode through the high-resistivity layer, resulting in large voltage drops near the electrode.

When personnel are standing farther from the source of the ground fault, the worst-case scenario occurs when the soil has conductive top layers and resistive bottom layers: in this case, the fault current remains in the conductive top layer for much greater distances going away from the electrode.

Fault clearing time is an important factor to consider as well. The more time it takes the electric utility company to clear the fault, the more likely it is for a given level of current to cause the human heart to fibrillate.

It is important to remember that step voltages are life-threatening, because most power companies use automated re-closers. In the event of a fault, the power is shut off and then automatically turned back on. This is done in case the faults occurred due to an unfortunate bird that made a poor choice in where to rest or dust that may have been burned off during the original fault. A few engineers believe that fibrillation current for step potentials must be far greater than touch potentials, as current will not pass through any vital organs in the former case. This is not always true, as personnel who receive a shock due to step potentials may fall to the ground, only to be hit again, before they can get up, when the automatic re-closers activate.

Touch Potentials. When a fault occurs at a tower or substation, the current will pass through any metallic object and enter the earth. Those personnel "touching" an object in the vicinity of the GPR will be subjected to these voltages, which may be hazardous (see Fig. 2.22).

For example, if a person happens to be touching a high-voltage tower leg when a fault occurs, the electricity would travel down the tower leg into the person's hand and through vital organs of the body. It would then continue on its path and exit out through the feet and into the earth. Careful analysis is required to determine the acceptable fibrillation currents that can be withstood by the body were a fault to occur.

Engineering standards use a 1-m (3.28-ft) reach distance for calculating touch potentials. Please note that meters are used instead of feet in this rule as the standards are metric based. A 2-m (6.54-ft) reach distance is used when two or more objects are inside the GPR event area. For example,

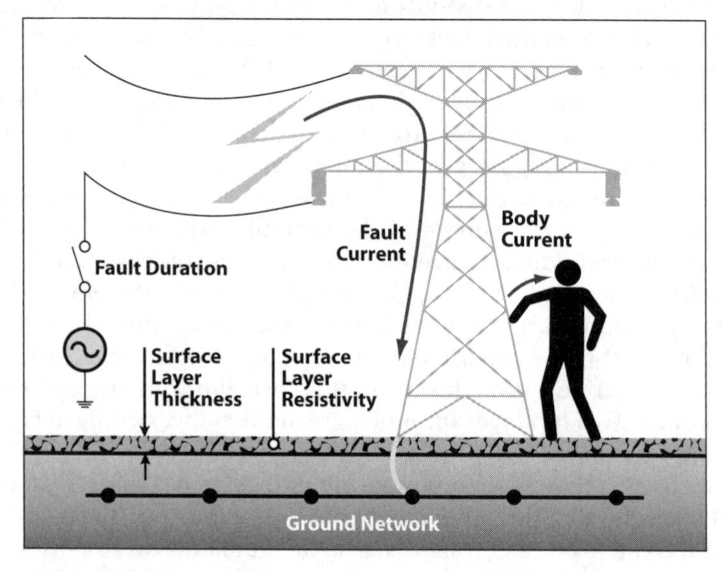

Figure 2.22 Touch potential at a transmission tower.

a person could have both arms stretched out and be touching two objects at once, such as a tower leg and a metal cabinet. Occasionally, engineers will use a 3-m distance to be particularly cautious, as they assume someone may be using a power tool with a power cord 3 m in length.

The selection of where to place the reference points used in the touch potential calculations are critical in getting an accurate understanding of the level of hazard at a given site. The actual calculation of touch potentials uses a specified object (such as a tower leg) as the first reference point. This means that the farther away from the tower the other reference point is located, the greater the difference in potential. If you can imagine a person with incredibly long arms touching the tower leg and yet standing many dozens of feet away, you would have a huge difference in potential between that person's feet and the tower. Obviously, this example is not possible: this is why setting where and how far away the reference points used in the touch calculation is so important and why the 1-m rule has been established.

Mitigating Step and Touch Potential Hazards. Mitigation of step and touch potential hazards is usually accomplished through one or more of the following three main techniques:

1. Reduction in the RTG of the grounding system
2. Proper placement of ground conductors
3. The addition of resistive surface layers

Understanding the proper application of these techniques is the key to reducing and eliminating any GPR hazards. Only through the use of highly sophisticated three-dimensional electrical simulation software, which can model soil structures with multiple layers and finite volumes of different materials, can the engineer accurately model and design a grounding system that will safely handle high-voltage electrical faults.

Reduce the Resistance to Ground Reducing the RTG of the site is often the best way to reduce the negative effects of any GPR event, where practical. The GPR is the product of the fault current flowing into the grounding system times the RTG of the grounding system; in essence, Ohm's law. Thus, reducing the RTG will reduce the GPR to the degree that the fault current flowing into the grounding system does increase in response to the reduced RTG. For example, if the fault current for a high-voltage tower is 5000 A and the RTG of the grounding system is 10 Ω, the GPR will be 50,000 V. If we reduce the RTG of the grounding system down to 5 Ω, and the fault current increases to 7000 A as a result, then the GPR will become 35,000 V.

As seen in the example above, the reduction of the RTG can have the effect of allowing more current to flow into the earth at the site of the fault, but will always result in lower GPR values and touch and step voltages at the fault location. On the other hand, farther away from the fault location, at adjacent facilities not connected to the faulted structure, the increase in current into the earth will result in greater current flow near these adjacent facilities and therefore an increase in the GPR, touch voltages, and step voltages at these facilities. Of course, if these are low to begin with, an increase may not present a problem, but there are cases in which a concern may exist. Reducing the RTG can be achieved by any number of means, as discussed earlier in this chapter.

Proper Placement of Ground Conductors A typical specification for ground conductors at high-voltage towers or substations is to install a ground loop around all metallic objects and connected to the objects; keep in mind that it may be necessary to vary the depth and/or distance that ground loops are buried from the structure in order to provide the necessary protection. Typically, these ground loops require a minimum size of 2/0 AWG bare copper conductor buried in direct contact with the earth and 3 ft from the perimeter of each object and 18 in. below grade. The purpose of the loop is to minimize the voltage between the object and the earth's surface, where a person might be standing while touching the object: that is, to minimize touch potentials.

It is important that all metallic objects in a GPR environment be bonded to the ground system to eliminate any difference in potentials. It is also important that the resistivity of the soil, as a function of depth, be considered in computed touch and step voltages and in determining at what depth to place conductors. For example, in a soil with a dry, high-resistivity surface layer, conductors in this layer will be ineffective; a low-resistivity layer beneath that one would be the best location for grounding conductors. On the other hand, if another high-resistivity layer exists farther down, long ground rods or deep wells extending into this layer will be less effective.

It is sometimes believed that placing horizontal grounding loop conductors very close to the surface results in the greatest reduction in touch potentials. This is not necessarily so, as conductors close to the surface are likely to be in drier soil, with a higher resistivity, thus reducing the effectiveness of these conductors. Furthermore, while touch potentials immediately over the loop may be reduced, touch potentials a short distance away may actually increase, due to the decreased zone of influence of these conductors. Finally, step potentials are likely to increase at these locations: indeed, step potentials can be a concern near conductors that are close to the surface, particularly at the perimeter of a grounding system. It is common to see perimeter conductors around small grounding systems buried to a depth of 3 ft below grade to address this problem.

Resistive Surface Layers One of the simplest methods of reducing step and touch potential hazards is to wear electric hazard shoes. When dry, properly rated electric hazard shoes have millions of ohms of resistance in the soles and are an excellent tool for personnel safety. On the other hand, when these boots are wet and dirty, current may bypass the soles of the boots in the film of material that has accumulated on the sides of the boots. A wet leather boot can have a resistance as low as 100 Ω. Furthermore, it cannot be assumed that the general public, who may have access to the outside perimeter of some sites, will wear such protective gear.

Another technique used in mitigating step and touch potential hazards is the addition of more resistive surface layers. Often a layer of crushed rock is added to a tower or substation to provide a layer of insulation between personnel and the earth. This layer reduces the amount of current that can flow through a given person and into the earth. Weed control is another important factor, as plants become energized during a fault and can conduct hazardous voltages into a person. Asphalt is an excellent alternative, as it is far more resistive than crushed rock, and weed growth is not a problem. The addition of resistive surface layers always improves personnel safety during a GPR event. Please review the previous sections on step and touch potentials.

Telecommunications in High-voltage Environments. When telecommunications lines are needed at a high-voltage site, special precautions are required to protect switching stations from unwanted voltages. Running any copper wire into a substation or tower is going to expose the other end of the wire to hazardous voltages, and certain precautions are required.

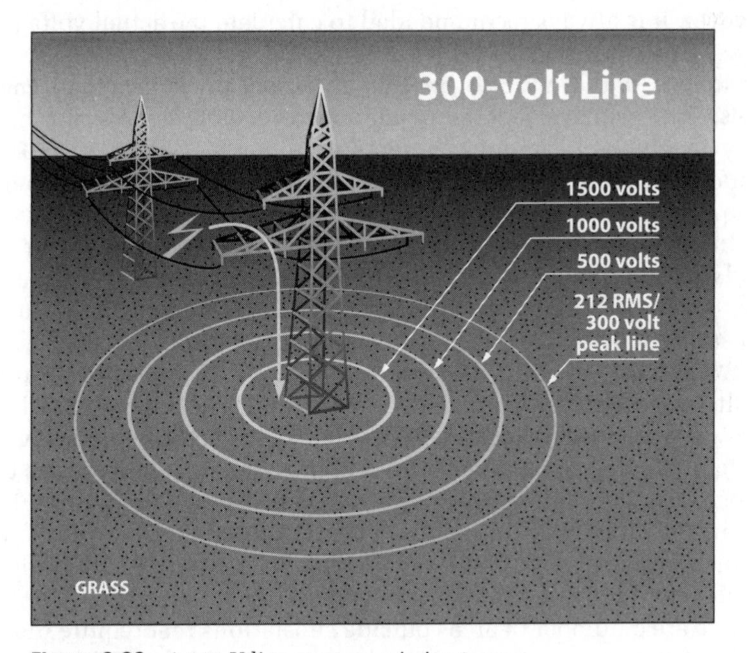

Figure 2.23 A 300-V line at a transmission tower.

Industry standards regarding these precautions and protective requirements are covered in IEEE Standards 387, 487, and 1590. These standards require that a GPR study be conducted, so the 300-volt peak line can be properly calculated.

To protect the telephone switching stations, telecommunication standards require that fiber-optic cables be used instead of copper wires within the 300-V peak line. A copper-to-fiber conversion box must be located outside the GPR event area at a distance in excess of the 300-V peak, which is the same as the 212-V RMS line. This is known in the industry as the *300-V line*. This means that, based on the calculation results, copper wire from the telecommunications company may not come any closer than the 300-V peak distance. This is the distance at which copper wire must be converted over to fiber-optic cable. This can help prevent any unwanted voltages from entering the telephone companies' telecommunications networks.

The current formulae for calculating the 300-V line, as listed in the standards, has led to misinterpretation and divergences of opinion, resulting in order-of-magnitude variations in calculated distances for virtually identical design input data. Furthermore, operating experience has shown that a rigorous application of theory results in unnecessarily large distances. This has caused many compromises within the telecommunications industry. The most noted one is a newer standard, IEEE Std. 1590-2003, which lists a 150-m (~500-ft) mark as a default distance if a GPR study has not been conducted at a given location. However, it is always recommended to calculate the actual voltage and not use this shortcut.

Regulations and Standards for Human Safety in High-voltage Environments. The primary federal regulation and enforcement for human safety in high-voltage environments is under 29 CFR 1910.269. This federal law requires the elimination of known electrical hazards in the workplace, including hazardous step and touch voltages. The law does not inform you how to eliminate these electrical hazards or what standards should be used; it merely requires you do eliminate them. To date, the only known U.S. standard for the elimination of hazardous step and touch voltages is the IEEE Std. 80.

IEEE Std. 80 is the IEEE Guide for Safety in AC Substation Grounding. It was originally published in 1961 and has been updated three times, in 1976, 1986, and finally in 2000. This standard is fairly difficult to understand, not only because of the complex nature of the topic, but because the standard is trying to accomplish two functions. First, it is trying to provide a reasonably safe environment for personnel working in high-voltage substations. Second, it is trying to show how to calculate and mitigate hazardous step and touch voltages.

There are a number of areas outside substations that require the mitigation of hazardous step and touch voltages, and the IEEE Std. 80 is often listed as a requirement for non–substation grounding standards. This has led to a lot of confusion within the grounding industry. When cellular telephone companies install equipment on high-voltage transmission towers, the towers must be made safe for the cellular personnel. The only available standard for mitigating these hazardous voltages is the IEEE Std. 80, which specifically states it is only applicable to substations. As a result, many telephone companies ignore the requirements of IEEE Std. 80, placing their own personnel at risk.

An excellent solution for the IEEE would be if they split Standard 80 into two separate standards, one for the protection of personnel in substations and another for the mitigation of hazardous step and touch voltages.

Computer Modeling for Human Safety in High-voltage Environments. Back in the early 1960s, when the first standards were being developed to mitigate the hazards to human safety caused by step and touch voltages, the best that any engineer could do was go through a series of hand calculations and try to determine the best ways to make a particular high-voltage environment safe. Today, we have the ability to use computer modeling that can take into account far more variables than could ever be reasonably calculated by hand. The goal of this section is to make the argument that computer modeling is the only cost-effective and ethical way to comply with the requirements of federal law found in 29 CFR 1910.269.

There are a number of engineering programs on the market that can be useful in designing safe workplaces in high-voltage environments. These programs generally fall into one of two categories: those that provide a general analysis of the hazards, and those that provide an accurate, detailed analysis. The differences between these two may sound small, but in reality they are quite important. The general analysis programs usually result in overinstallation in grounding in areas that do not need it, and underinstallation in grounding in areas that do need it. The reason for this is that the general analysis programs do not calculate step and touch voltages from specific objects within the computer model, but use an average potential across the entire compound.

The computer programs that provide a detailed analysis of the step and touch voltage hazards at a given site are typically quite expensive and require a great deal of training to master. The simulations conducted by these systems can take hours and hours of computing time using high-speed computers just to run a single simulation. This is in stark contrast to the general analysis systems, which provide near-instantaneous results on even the slowest of computers. Detailed computer simulation programs are also capable of using soil models with three or more layers.

A proper study of the step and touch voltage hazards found at a given site involves a detailed drawing of all the metallic objects at the site, including transformers, towers, switches, concrete foundations, buildings, posts, fences, gates/doors, and any other object that could be touched by a person, along with the grounding system. These objects must be placed in a multilayered soil model. Most sites require soil models with three to five layers in order to accurately model the propagation of electrical energy through the grounding system and into the earth. Single- and two-layer models are almost always inaccurate. Single-layer soils simply do not exist in reality, and two-layer models typically present such a high contrast between the layers as to present calculation errors that are nearly impossible to overcome. The general analysis computer programs are almost always limited to two-layer soil models and often have further limitations as to the available resistivities and depths they can calculate.

Chapter 13 of the 2000 edition of *IEEE Std. 80* deals with the soil structures and the selection of a soil model. The new edition introduced a flawed method for the calculation of a two-layer soil structure and only briefly discusses multilayer structures. This new method uses a uniform soil structure that claims to provide accurate two-layer soil equivalences. In a paper titled *Effects of the Changes in IEEE Std. 80 on the Design and Analysis on Power System Grounding*, by J. Ma, F. Dawalibi, R. Southey, of Safe Engineering Services & Technologies, the authors state that the two-layer soil structure method presented in annex E of the IEEE 80 standard is flawed. The paper shows that in some cases using the IEEE Std. 80 two-layer soil method, when the grounding grid is increased in size, the calculated RTG increased. This is contrary to what will happen in reality, when a grounding grid is increased in size, the RTG of the grid will decrease. In other examples, when the soil resistivity was increased in the two-layer calculation, the resistance of the grounding grid was calculated to decrease. Again, this is contrary to what will occur in real life. The results presented in this paper demonstrate that the calculations found in IEEE Std. 80 are only valid for uniform (single-layer) models and should not be used for multilayer soil configurations. This further demonstrates the need for detailed computer models instead of the suggested hand calculations from annex E of Std. 80.

But even beyond the issue of being able to accurately calculate multilayer soil models, there are still far too many variables to ever calculate by hand. There are the formulas regarding permissible body current that must be analyzed and new requirements for calculating the surface layer derating factor; the calculation of foot resistances is now based on a formula with a rigorous series expression, with each term being a surface integral; and one must also take into account the effects of the DC offset current generated due to the asymmetrical fault current. And these are only a few of the basics. Combining all of this data when trying to determine the touch voltage for a 1-m reach distance at a 45° angle from a specific transformer located at a specific location within a ground grid at a substation on top of a three-layer soil model is nearly impossible to do by hand. And that is just one location around a single transformer that may have dozens of touchable locations around its perimeter! Let alone the thousands, if not hundreds of thousands, of other locations that could be touched within the substation that must also be analyzed.

There are a number of other issues (particularly calculating the touch voltages for the exterior fences) that could be discussed, but by now it should be obvious that hand calculating these variables would be a daunting task. In fact, if you were to print out all the calculations that a detailed computer model performs, it would look like a doctorial thesis. In short, computer modeling is the only viable and ethical way to design a substation grounding grid. In today's computer age and under 29 CFR 1910.269, a jury may consider hand calculations and/or general computer modeling to be criminal negligence. Detailed computer modeling is the only way to accurately conduct the proper IEEE Std. 80 calculations.

THE EFFECTS OF LIGHTNING ON A GROUNDING SYSTEM

Lightning is an atmospheric electrostatic discharge (spark) that can travel at speeds of 220,000 km/h and can reach temperatures approaching 30,000°C, which is hot enough to fuse silica sand into glass. There are some 50–100 lightning strikes occurring somewhere on the planet every second that can carry as much as 200,000 A of electricity and generate more than 1×10^8 V with each strike. The Federal Emergency Management Agency estimates there are 750 severe injuries and 200 deaths in the United States annually from lightning and that it causes an estimated $4–5 billion in damage per year.

There are a number of engineering issues to be concerned with when lightning strikes an object, including a power plant. For this discussion, we will assume that this is a basic power plant/substation with an overhead lightning protection system (LPS) tied into the standard ground grid of the facility. It should be understood that all of the engineering principles mentioned in this article are wholly dependent upon having excellent soil resistivity data and valid soil models. Soil resistivity data is, and always will be, the heart of grounding science. Please refer back to the earlier section on soil resistivity.

Distribution of Current. When lightning strikes an aboveground object, such as an aerial on an LPS, the current will start to divide itself across the LPS proportionally to the impedance it encounters on its way down to the earth. We can imagine the lightning striking the aerial and the current dividing in two as it moves from the aerial into the conductor. As the current flows down the conductor, at each conductor intersection the current will divide again and again, until it finally reaches the earthing electrode system, where it will finally travel into the soil and dissipate. Our primary concern with the distribution of current in a grounding system is whether or not the conductors can handle the current levels, without burning open like a fuse.

Down conductors for an LPS must terminate in a connection to an earthing electrode of some kind. Most typically, it is a single standard ground rod, and on occasion three rods installed in a triangular pattern with conductors tying them together. The effectiveness of the connection to earth of each of these electrodes (or electrode systems) is measurable and is called *resistance-to-ground.* The RTG of any given electrode will vary, given the immediate soil conditions in which the electrode finds itself (moisture content, specific soil chemistry, proximity of nonconductive buried rocks, proximity to other conductive buried objects and/or soils, etc.), and as a result, each electrode will have a specific RTG.

As there will be multiple down conductors for any given LPS, electrodes with a lower RTG will see a proportionally larger percentage of the current. In other words, the electrodes with a better connection to the earth will see more current than the other electrodes, "unbalancing" the LPS. The European Lightning Protection Standard EN 62305 (or BS EN 62305 or IEC 62305) calls for the "balancing" of these electrodes by either supplementing each electrode until all have the same RTG or installing a buried ground ring tying all the electrodes together. The United States has no such requirement.

When lightning enters a conductor, huge magnetic fields are formed as the energy passes through the conductor. These magnetic fields hold huge amounts of inductive energy and will induce currents into nearby metallic objects, including the same conductors (wires) they are currently traveling on. When a conductor is routed in such manner as to enable the magnetic fields from one part of the conductor to induce energy into another part of the same conductor (imagine a tight bend or circle), this is called a *self-induced magnetic coupling.* Self-induced magnetic couplings such as this can quickly lead to a thermal avalanche wherein the two magnetic fields keep cross-coupling into each other, forming a perpetually increasing energy level, thereby increasing the heat in the conductor until it melts and burns open.

All known LPS regulatory codes have detailed instructions on how to properly route conductors so as to prevent these self-induced magnetic couplings (and thus prevent thermal avalanches). Not only must the straight current portion of the lightning strike be considered (impedance of the conductors, current-carrying capacity, etc.), but the magnetic fields that are formed and the subsequent current that will re-enter the system upon the collapse of the magnetic fields, must also be taken into account. Computer modeling along with good design and diligent installation techniques will prevent an overcurrent situation on any one given conductor at any point in the LPS, both above and below grade.

Frequency Spectrum and Time Domain. It is a well known phenomenon that lightning has both an AC component and a DC component at the same time. In fact, lightning will propagate through a structure at many (if not all) possible frequencies. A typical lightning strike will see a range of frequencies from 0 Hz to as much as 10 MHz, and sometimes even higher. However, the distribution of these frequencies is not spread evenly, and certain frequencies will be prominent. This collection of frequencies generated by a lightning strike is called the *frequency spectrum* and is primarily determined by the geometric shape of the structure struck by the lightning.

Just as the length of an antenna determines the best frequency to broadcast/receive radio signals, the same is true for lightning. The lightning will adjust its frequency based on the structure (antenna) it strikes and will resonate due to impedance imbalances between the structure and the earth. In the case of lightning, all structures are antennas. Most structures make up very complex antennas, with the buried portion adding a further complexity, as it will affect the lightning strike, following many of the same principles found in the half-wavelength theory for antennas. All the various variables and nuances of the calculations can become very complex, given all the different types of materials to be found on a typical structure, the erratic pathways of the LPS, and the variations found in the soil; only computer modeling can adequately calculate the expected resonant frequencies for a lightning strike on a given structure.

To calculate the frequency spectrum (or frequency response) of a lightning strike, one must first develop an accurate model of the structure with the LPS (including material types) and develop an accurate soil model for the site. Numerous frequencies must be run individually through the model until an accurate profile can be developed. Mathematical algorithms have been developed to assist in the proper selection of the test frequencies, to reduce run times and improve the statistical accuracy of the profiles. But needless to say, many hundreds if not thousands of individual frequencies must be run through the computer simulation to accurately determine the frequency spectrum. The final result is a graph showing the entire range of expected frequencies on the X axis, with magnitude on the Y axis. Typically, the simulation will demonstrate that two or three frequencies will resonate through the structure during a lightning strike.

In the time domain, you can calculate the length of time it takes the energy from a lightning strike to clear out of your structure. The actual lightning strike itself will start and stop in a very short time frame, typically only a few microseconds. However, as described earlier, magnetic fields will form not only in the LPS, but in the structure itself. The time it takes to generate the magnetic fields to full strength, the reflections of energy through the structure, and the time it takes to collapse the fields (steel is at least 250× more magnetic than copper and therefore holds magnetic fields for far longer) is the time duration of the lightning strike. While the actual lightning stroke may start and stop in microseconds, the time your system will be impacted by the electrical energy from the strike, will almost certainly be many thousands of times longer, as the energy generated by the strike will take time to dissipate out to the earth through the grounding system.

Conducting studies related to frequency spectrum and time domain can have substantial benefits. The primary benefits include:

1. Human safety: Lightning strikes are different from standard electrical utility faults not only in dramatic amounts of current but in the near-limitless voltage potential and the high frequencies that will be generated. The frequency spectrum and time domain are critical for accurately calculating the effects of a lightning strike with regards to human safety studies involving step and touch voltage hazards, GPR studies, and electromagnetic interference studies.
2. The frequency spectrum and time domain can be used in an electrical coordination and short-circuit fault study, to improve the settings of the overcurrent protection devices and reduce downtime due to unintentional power outages caused by lightning strikes. For many facilities, such as military facilities, hospitals, data centers, and power plants, a single power outage can have huge consequences and often is measured in millions of dollars in lost revenue.
3. The frequency spectrum and time domain can be used for improving surge-protection systems. While all sites will want broad-spectrum frequency protection for unwanted surges and transients, the addition of specially tuned surge protectors designed to stop the resonant frequencies determined during the frequency spectrum study can prove especially useful for protecting vital equipment from the impact of lightning strikes.
4. In some countries, these studies are compliance requirements for international lightning protection codes (IEC 62305, BS EN 62305, or EN 62305) without overengineering. These codes have strict rules that in many cases will simply result in overdesigned LPSs for many buildings. A good frequency spectrum and time domain study can prove the effectiveness of an LPS without adding huge installation costs due to overengineering. Or it will at least prove the necessity of those costs.

Additionally, once the resonant frequencies of a lightning strike have been determined, the distribution of current can be better analyzed as impedances can now be properly calculated.

Hazardous Step and Touch Voltages for Human Safety during a Lightning Strike. Human safety is always a paramount concern, and when dealing with lightning strikes, the critical issues can compound quickly. Personnel touching a conductive object or even stepping near a lightning strike can suffer serious injury and even death. Calculating step and touch voltage hazards during a lightning strike is in principle the same as when one calculates the hazard during a line-to-ground fault. However, we must take into account the new frequencies and clearing time that will be generated during the strike, as determined by the frequency spectrum and time domain analysis.

Step and touch voltage hazards are calculated using the strike amperage at the point of contact, the frequency, the X/R ratio, the clearing time (i.e., time domain), and the specific soil resistivity conditions. Only a computer simulation can accurately model the frequency spectrum and time domain of the lightning strike to determine the touch and step voltages that will be experienced during a strike. Once the safety parameters (crushed rock, shoes, etc.) have been applied to the computer model, the overvoltage hazards will appear, and standard mitigation techniques for reducing these hazardous voltages can be applied, thus making the site safe.

Electromagnetic Interference and Ground Potential Rise during a Lightning Strike. Another key factor during a lightning strike is electromagnetic interference. As mentioned earlier in this chapter, large magnetic fields will form in both the LPS and in the buried grounding/earthing system. Additionally, the GPR will cause scalar voltages to form across the surface of the earth with the potential decreasing with distance. These large magnetic fields and GPR effects will transfer voltages and currents (by inductive and capacitive coupling and through earth coupling) into nearby pipelines, railways, communication lines, homes, industrial facilities, farms, and other such utilities, whether buried or aboveground.

The current and voltages that are induced into the nearby utilities can cause great damage. One of the primary concerns is the induced voltages/currents that can form on the data and/or shield lines of communications cables, causing damage, as these transient currents will flow through sensitive electronic equipment on their way to earth. The neutral wires from the utility power company can carry stray currents formed from the electromagnetic fields up and into homes and industrial parks. Nearby gas pipelines can have their protective coatings compromised by the stress voltages caused by the difference in potential between the surrounding earth and the pipe.

Again, using the highest-magnitude frequencies of the lightning strike, as determined by the frequency spectrum analysis, the electromagnetic fields and the GPR can be accurately calculated, and the impact on the surrounding infrastructure can be properly analyzed and potential problems mitigated.

SUMMATION OF LIGHTNING EFFECTS ON GROUNDING SYSTEMS

When calculating the effects of a standard electrical utility fault on a grounding system, we know that the utility fault will have clamped voltages at a very specific frequency (60 Hz). With a lightning strike, the potential voltages are virtually unlimited and will have multiple frequencies based upon the geometric shape of the structure that is struck. The hazards presented by an electrical fault at 60 Hz are very different from the hazards presented by an electrical fault at 60 kHz or even 60 MHz. These order-of-magnitude increases in frequency present unique engineering challenges and are why lightning strikes are so dangerous. The bottom line is that GPR and step and touch voltage hazard studies conducted using power company (utility) fault data are simply not applicable for a lightning strike.

So, in summation, all of the factors presented above need to be taken into account when building an effective LPS, be it for a utility substation, telecommunication site, or simply just a personal home.

STANDARDS FOR LIGHTNING PROTECTION SYSTEMS

There are significant differences between the U.S. standards and the international standards for lightning protection. The international standards, such as the IEC 62305 and its British counterpart, the BS EN 62305, have many additional requirements above and beyond those found in the U.S. standard NFPA 780. The differences between these standards are worth noting for those who are interested in providing lightning protection for their structures.

Lightning Protection Systems. Lightning protection systems are not only one of the most important (and expensive) infrastructure components of a building but also one of the least understood. A poorly designed LPS can add unnecessary expense, add legal liabilities to your construction project, and may not provide the protection you need. A good LPS is engineered and designed specifically for your structure and the needs of your facility. In the United States, most industry and the government facilities are protected by NFPA 780 Standard for the Installation of Lightning Protection Systems. This standard is issued by the National Fire Protection Association® (NFPA), the same group that writes the NEC and provides a guideline for installing a one-size-fits-all LPS.

U.S. Standard NFPA 780. NFPA 780 provides guidelines for how often to place air terminals; spacing for cross- and down conductors; ground rod and loop requirements; surge-protection requirements; and installation of protection for trees, towers, and so on. The standard, however, has two primary shortcomings in that it does not analyze the installed system's ability to handle an actual lightning strike, nor does it take into consideration what the system is protecting. In other words, NFPA 780 has the same requirements for a garage as it does for a billion-dollar computer farm. These shortcomings, along with virtually no legal and/or insurance industry requirements for lightning protection, has led many facilities managers to simply take their chances and forgo protecting their buildings.

International Standard IEC 62305. We can learn a lot about lightning protection by looking at the requirements of IEC 62305, which is significantly more demanding than the U.S. NFPA 780 standard. Here are the basic requirements of IEC 62305:

1. It requires a risk factor assessment that determines the level of required lightning protection. This risk assessment is quite complex, and software is almost always needed to make a proper assessment. The calculation includes human life, public services, cultural heritage, economic risk, and occupancy issues. This risk assessment is both good and bad, in that a garage will have fewer requirements than found in NFPA 780, but billion-dollar computer farms will have greater requirements.
2. The IEC 62305 standard requires an actual assessment of the LPS to insure it is capable of handling a lightning strike. The lightning strike calculations are far more significant for both the time-domain parameter and the actual strike amperages (100–200 kA), than the U.S. industry standard (often only 15 kA). Calculations that are often required include:
   a. Expected amount of current to be carried on individual conductors in DC amps, to ensure that current carrying capacity is not exceeded.
   b. Rolling-ball theory of lightning protection tested against 3D computer models of the structure and surrounding area.
   c. Spark gap and arc-flash calculations from the LPS to adjacent conductive utilities.
   d. Time domain of the lightning strike on the specific structure. This is critical to understanding the amperage carrying capacity of the conductors. Without an actual calculation, the IEC 62305 default time domain could force the unnecessary installation of additional conductors.
   e. Frequency spectrum of the lightning strike on the specific structure. This data is needed for both surge protection and for timing of circuit breakers to prevent power outages
3. In general, IEC 62305 has physical construction and installation requirements that are far greater than NFPA 780.
   a. Down conductors vary from 10-m to 20-m spacings. The NFPA 780 uses one-size-fits-all 30-m spacing.
   b. The zone of protection or rolling-ball theory in the IEC standard varies the angle required based on the risk assessment, which can impact placement of certain types of aerials from 20-m to 60-m heights.
   c. Concrete columns that are used for down conductors must be tested at a 0.2-Ω or less continuity, and rebar must be welded with 20× diameter overlaps. These must be bonded to the floor slab.

- d. Ground rings are required for all nonconductive buildings, buildings housing electronic systems, and certain risk factors. Individual rod installations (without ground rings) must be tested, so each electrode is at the same RTG.
- e. Spark gaps between lightning conductors and other metallic objects must be considered.
- f. Incoming utility services (such as overhead power lines) and adjoining public spaces may also be required to have protection systems installed, based on the risk assessment.
- g. Both internal and external lightning surge-protection systems are mandatory.

4. The IEC 62305 has stringent requirements for annual testing and inspection of the LPS. This, of course, goes along with mandatory maintenance requirements.
5. The IEC 62305 standard gives you three choices when it comes to selecting an interception model; we recommend you use the rolling-sphere model (RSM), sometimes known as the electro-geometric model. The protection angle method and the mesh method should not be used. These are legacy methods that have been left in the standard due to historical reasons; you will be far better served by using the RSM.

Lightning Protection System Recommendation. It is recommended that you look at both the NFPA 780 and the IEC 62305 for guidance. Using a combination of the two, including a reasonable risk assessment and computer modeling, a good engineering firm will be able to maximize human safety and cost effectiveness when designing your LPS. Advanced computer-modeling techniques which include the RSM, frequency spectrum and time-domain analysis, and step and touch voltage analysis will ensure that your facilities' needs are met without over- or underengineering your LPS.

CONCLUSION

We sincerely hope that we have provided you, the reader, with a better overall understanding of system grounding, the benefits and features of the available grounding electrodes, the GPR hazards of high current discharges, and the importance of properly considering soil layering. We were also able to introduce, albeit briefly, the importance and principles of proper soil testing, RTG testing, meter selection, and LPS design. We are confident that with this basic knowledge, you will be able to make the right decision when it comes to grounding choices. Remember: "To protect what's above the ground you need to know what's in the ground."

Chapter Three

ECONOMIC AND LEGAL ANALYSIS

ECONOMIC CONSIDERATIONS

Economics studies the allocation of scarce resources and the trade-offs that entails. If a nearly infinite amount of resources were applied to grounding and otherwise idiot-proofing high-voltage electrical equipment, it would be theoretically possible to reduce the number of deaths due to electrocution from high-voltage electrical equipment to zero. But by so doing, resources would be siphoned from other areas, such as health care or policing, and the overall number of societal deaths would increase.

On the level of a company, if a company does not invest enough in grounding, then it could open itself up to damaging litigation, and if it invests too much, it will no longer be competitive with other firms. So a rational choice must be made that balances potential harm to the firm from an accident against the "harm" of investing in protective measures that do not directly generate revenues. In addition, a company must factor in how much it is willing to spend to maintain its reputation as a trusted service provider and a member of its community. For example, an electrician should not just think about how much he or she would be out of pocket if faulty grounding work caused a fire, but should also consider how that fire might inconvenience others and cause damage to his or her own reputation. And beyond purely economic considerations, no one ever wants to carry the burden of causing harm to others because of substandard or otherwise shoddy grounding work.

Calculating rational investment levels based on potential damage to equipment is the easiest part of the calculation. The traditional method is to multiply the cost of equipment by the likelihood of failure. For example, if one has a $100,000 piece of equipment and a

---

This chapter was written by Christopher Clemmens, M.B.A., M.Acc.

grounding failure could cause a 1 percent chance of failure per year, then the baseline investment in grounding would be $1000 yearly. This assumes that the company has the financial wherewithal to withstand the financial loss and/or that the loss of equipment would not cause a failure in customer service. If the financial loss to the firm would be catastrophic, then a larger investment in grounding and/or insurance would be indicated. But it will be difficult to collect an insurance claim if inadequate investment in grounding leads to negligence on the part of the insured. Almost all insurance policies only cover accidents and have clauses about not covering intentional actions and negligence. This makes sense from the perspective of the insurer, who wants to avoid moral hazard, that is, the taking of unnecessary risks because the insured party knows the insurer will pay. Even if one can get an insurance company to pay for damages in case of negligence, a bad record will lead insurance companies to raise premiums to crippling levels.

Aside from catastrophic damage due to inadequate grounding, there are also slower ways that inadequately grounded equipment is damaged over time, such as corrosion. A key tool in cutting corrosion costs is grounding. In 2002, the U.S. Federal Highway Administration released the first, and so far only, congressionally mandated study on the direct costs associated with metallic corrosion in the economy as a whole. This thorough study took two years and covered nearly every sector of the U.S. economy. Results of the study show that the total annual estimated direct cost of corrosion in the United States is a staggering $276 billion—approximately 3.1 percent of the nation's gross domestic product (GDP) at the time of the study. The study estimated that if indirect costs such as outages, delays, failures, and litigation were factored in, total costs would be about 6 percent of GDP. In today's dollars, direct cost would be about $1500 for every American, and if one included indirect costs, then per capita costs would approach $3000. The study found that 25–30 percent of these costs could be avoided if optimum corrosion prevention methods are used. Further, it showed that corrosion-control methodologies are making slow but steady progress. A Department of Defense study found that corrosion costs over $2.5 billion and over 20 percent of total maintenance costs. It also significantly hurts military readiness, because it sidelines crucial equipment for maintenance and repair.

Of course, the real concern with grounding safety is not loss of equipment but loss of life. Unfortunately, over 400 Americans die from electrocution in the workplace each year, and another 500 die every year in house fires caused by electrical malfunctions. There are different methods for calculating the value of a life. Insurance claims and wrongful-death lawsuits look at present value of lifetime earnings, potential contribution to society, and emotional distress caused to loved ones. To this, courts will often add punitive damages if they find a defendant to be negligent.

Government agencies also calculate the value of life when they are setting safety and other regulatory standards. Then, they calculate whether to mandate a safety feature using a similar calculation that we used above with the cost of equipment. If the value of a life is deemed to be $7 million and a highway guard rail would have a 1 percent chance of saving a life, if the cost of the guard rail were under $70,000, its use would be mandated. If it is over $70,000, its use would not be mandated. This calculation might seem quite grim, but we have to remember that if we mandated a crippling amount of investment in safety, we would have no money to invest in other areas of the economy. Determining the value of life is an inexact science. It varies from agency to agency and across periods of time. It generally increases, but it can also decrease, as when the U.S. Environmental Protection Agency (EPA) dropped from $7.8 million in 2003 to $6.9 in 2008. In the same period that the EPA cut its values, the Department of Transportation raised its values twice. The value of a life also varies by country. According to a 2012 Rosgosstrakh Strategic Research Centre survey, the median value of life in Russia is only $44,700. One can only hope that responsible parties are using their consciences, and not just cold financial calculations, when making grounding decisions in Russia.

The indirect costs for a firm or tradesperson causing injury or death are extensive. Reputational harm might result in having no more customers. A firm involved in an accident may be seen as a dangerous or otherwise undesirable employer. Management may become so distracted by dealing with the fallout that they fall into a reactive management mode rather than focusing on building their business. Worse yet, firms might be dragged down by heavy-handed regulatory agencies and crippling lawsuits.

LEGAL CONSIDERATIONS

Of course it is beyond the scope of this book to offer legal advice nor are the authors qualified to do so. That said, electrical grounding problems can damage equipment and even take lives. This can lead to damaging liabilities that can ruin the finances of tradespeople and companies. Also, when discussing legal issues, one must be aware that regulations vary from country to country and even from jurisdiction to jurisdiction within a country. For example, in some U.S. states, it is possible for tradespeople to discharge liabilities caused by negligence through the process of bankruptcy, while in other states it is not.

Almost all legal actions taken against electrical companies occur in civil, as opposed to criminal, courts. A key distinction is that in a criminal case, the plaintiffs must prove their case "beyond a reasonable doubt," which means the judge or jury is almost certain of the defendant's guilt. In civil courts the plaintiffs must only prove their case by a "preponderance of evidence," which means the defendant is more than 50 percent likely to be liable for whatever damages the plaintiff is claiming. Under normal circumstances, this means that the defendant caused damages and failed to exercise the care that a "reasonable person" would use in similar circumstances. In other words, there were damages, and the damages were likely caused by the defendant's negligence.

> The core idea of negligence is that people should exercise reasonable care when they act by taking account of the potential harm they might foreseeably cause to other people.

Although the "reasonable person" standard is the normal benchmark for determining negligence in civil cases, there are circumstances in which the defendant is held to a higher standard. For example, a certified electrician would be held to a much higher standard of due care than a layperson. Another relevant example would be that a company that deals in ultrahazardous materials, which in many circumstances electricity could be considered to be, is held to a much stricter standard than a company dealing in less inherently dangerous goods.

Issues of electrical grounding span many industries, and there are thus different standards of legally required due care. Most of the early legal precedents for electricity and the law were set by cases involving electric transmission companies. Electric transmission companies are held to the "standard of foreseeability."

There are two elements to this standard. The first is that defendants are held liable for "anticipating all eventualities." This is in contrast to the normal common law standard, which holds plaintiffs only to the standard of taking precautions that a reasonable person would take in similar circumstances and places the burden of proof on the plaintiff to demonstrate that the defendant did not fulfill a burden of due care. The standard of foreseeability shifts the burden of proof from the plaintiff to the defendant, so the defendant must prove that there was no way that the mishap could have been foreseen. Given the long and deep case law associated with the use of electricity and grounding, the burden of proof is heavily stacked against a company arguing that a situation that caused damages was not foreseeable. Occupational Safety and Health Administration's general industry electrical safety standards, published in Title 29 Code of Federal Regulations (CFR), Part 1910.302 through 1910.308—Design Safety Standards for Electrical Systems and 1910.331 through 1910.335—Electrical Safety-Related Work Practices Standards, leave few conceivable areas in which those operating in electrical industries are not responsible.

The second element of the standard of foreseeability is that the defendant must have taken every reasonable action to prevent the damage.

A defense under this standard might be force majeure. Force majeure is Latin for "superior force," but it is often translated as "an act of God." In other words, a force that was completely outside the defendant's hands. For example, if a plane crashed into an electrical line and the electric company had taken every precaution to make sure that electricity is cut off to severed wires, but the live wire still caused a fire or injury, the electric company might escape having to pay damages. Force majeure defenses are anything but an easy out for defendants. Courts scrutinize force majeure claims intensely and do not allow them unless the circumstances were truly unforeseeable and unpreventable. For example, damages caused by a tornado in Oklahoma or an earthquake in California would not qualify as force majeure, because these are common enough occurrences that they should be anticipated.

A related standard applied to companies involved in the electrical industries goes back to at least 1912: *res ipsa loquitur* (Latin for "the thing speaks for itself"). What this standard effectively does is to shift the burden of proof from the plaintiff to the defendant. Or as stated in *Tassin v. Louisiana Power & Light Co.*, "the burden shifts to the defendant to show that the accident was caused by something for which it is not responsible." Under this standard, the plaintiff does not need to prove that the defendant breached a duty or acted unreasonably. Plaintiffs only need to prove they suffered harm and the harm was caused by the defendant. An interesting example of this principle is *Hoffmann v. Wisconsin Electric Power Company*. The plaintiffs were awarded $1.2 million, because they proved that poor grounding work on the part of Wisconsin Electrical Power Company caused a low level of stray voltage (a ground potential rise [GPR] event) over a long period of time on their dairy farm. They claimed that this harmed their dairy farm production. Although Wisconsin Electric Power Company was not in violation of state laws, it had to pay damages nonetheless.

The highest legal standard of liability, strict liability, applies in the cases of stray voltage accidentally discharged into the ground. Strict liability applies in situations that the law deems inherently dangerous. Strict liability is best understood through a classic legal hypothetical. Wild animals, such as lions, are legally considered inherently dangerous. Therefore, an owner of a lion would be liable for the damage that an escaped lion caused regardless of how sturdy the cage in which the lion was kept. Under strict liability the plaintiff need not prove the defendant was negligent. The plaintiff only needs to prove that damage occurred and the damage was caused by the defendant. For example, if a child were harmed due to a grounding problem, the defendant would be prima facie guilty. This standard of law forces defendants to take every possible precaution as opposed to the standard burden of taking "reasonable" precaution. Strict liability applies to cases in which stray electrical voltage is discharged into the ground. Therefore, the only legal defense against damages caused by grounding problems is to ensure the absolutely highest quality of grounding.

In some situations, contributory negligence on the part of the defendant can be used as a defense. Contributory negligence is when plaintiffs, as a result of their own negligence, partially causes their damages. To the extent that plaintiffs contribute to their own damages, compensatory damages, but not punitive damages, will be reduced. A classic example of this would be a person who is electrocuted because he or she is using a hair dryer while in the bath and is electrocuted when the hair dryer falls into the bath. Given that builders now have to install ground-fault circuit interrupters in bathrooms, courts would most likely find the fault to be partially, but not wholly, that of the bather. Contributory negligence will not insure that courts will not impose severe losses upon a defendant. In the famous McDonald's coffee case, *Liebeck v. McDonald's*, the plaintiff was found to be 20 percent responsible for the damages, and this reduced the compensatory damages from $200,000 to $160,000. But that did not stop the jury from awarding her an additional $2.7 million in punitive damages. Although McDonald's ended up settling a lesser undisclosed amount after appeals, the settlement, litigation costs, and reputational damage were significant, even though Ms. Liebeck was found to have contributed to her own damages.

Companies that employ trained professionals are not free from the concerns of legal liability. Professionals working around electrical equipment are expected to practice due care. As mentioned above, over 400 Americans die from electrocution in the workplace each year, and another 500 die every year in house fires caused by electrical malfunctions. But even when an employee is at fault, litigation is expensive, the publicity of a court case can cause significant reputational damage, and juries and judges cannot be relied upon to find for even a reasonably cautious defendant.

Of course not all electrical equipment can be kept in a controlled environment that can only be accessed by professionals. Extreme levels of caution must be taken when the general public might have access to equipment. This is especially true when children are involved, because children are held to a different standard of reasonableness when it comes to assigning contributory negligence. A company might not be found liable if an adult breaches a fence and trespasses onto a substation, but in *Dungee v. Virginia Electric & Power Company*, the defendant had to pay $20 million in damages when a 10-year-old child received severe burns when he sneaked into a substation to retrieve a ball. And aside from the ethical reasons for protecting children, damages are usually greater in cases involving children, because a key variable in value of life calculations is the number of years of life lost, which is clearly higher with children. Further, juries can be swayed by emotion, and they are likely to impose much higher punitive damages when children are involved.

In an action for injury from electricity, the question whether the injured person was guilty of contributory negligence is usually left to the consideration of the jury. A person's intentional conduct in exposing himself or herself to electricity can supersede any alleged negligence or wantonness of the power company, thus precluding liability. The law imposes upon a person sui juris, the obligation to use ordinary care for his or her own protection, and the degree of such care should be commensurate with the danger to be avoided. For example, a reasonable person knows not to hold a fork and stick it into an electrical outlet. Although the perpetrator of this act may win a Darwin Award, it is unlikely that his surviving family members would receive a settlement.

In conclusion, given the engineering complexities, the high costs of copper, and the labor expense of installing extensive below-grade electrical grounding systems, careful considerations must be taken into account when making decisions regarding those systems related to human safety and effective equipment operations, such as earthing and grounding. Some of the issues the company executive must consider are:

- Legal responsibilities of the company
  - The safety of employees and contractors
  - The safety of the public
- The cost of life
- Legal liabilities of the company
  - Negligence: anticipating all eventualities
  - Negligence: standard of foreseeability
  - Strict liability (when dealing with inherently dangerous substances such as electricity)
- Effective equipment operation
  - Costs of equipment failure
  - Costs of increased equipment maintenance
  - Data and communication losses
  - Corrosion

For most executives, the decision to hire third-party engineering firms specializing in grounding and earthing, to design and develop cost-effective and efficient grounding systems is fundamental in mitigating the risks a company and its employees face.

Chapter Four

NEC 250 PART I: GENERAL

250.1. SCOPE

This introductory article for the grounding and bonding section of the National Electrical Code (NEC) is a simple statement covering the scope of the entire section that matches the scope of the NEC. This article includes the fundamental principles of protection contained in International Electro-technical Commission (IEC) Std. 60364-1.

Specifically covered under the code are: public and private buildings, structures, mobile homes, recreational vehicles, floating buildings, yards, lots, parking lots, carnivals, industrial substations, any conductor that connects equipment to the supply of electricity, and any electric utility building (office, warehouse, garage, shops, etc.) that is not an integral part of a generating plant, substation, or control center.

The NEC (and Art. 250) specifically does NOT cover: ships, watercraft (other than floating buildings), railroad cars, aircraft, automotive vehicles (other than mobile homes and recreational vehicles), underground mines (including mining machinery and electrical trailing cables), communications equipment exclusive to the communications utilities (excluding their buildings, offices, garages, etc.), and the power generation, transmission, transformation, or distribution of power associated with both the railways and electrical utility companies. However, the NEC does state clearly that Art. 250 is fully capable of being utilized for electrical installations in most cases, and is in fact generally in use throughout the railway and electrical utility industries.

The specific text of Art. 250.1 tells us that the grounding and bonding requirements found in Art. 250 must be met for all electrical systems that are required, permitted, or not permitted to be grounded (basically all electrical systems regardless of legal or engineering status). This also includes grounded circuit conductors and any location that has grounding connections. The scope will dictate the size of grounding and bonding conductors, the size of electrodes, and the methods of grounding and bonding. When guards, isolation, and insulation are substituted for grounding, Art. 250 will govern under what conditions they may be used.

250.2. DEFINITIONS

The NEC (NFPA 70®) has never been known for providing common-language answers to real-world problems, and its definitions and terminology are often nearly incomprehensible, even to the trained engineer or electrician. Arguably the most notorious section is Art. 250, with the confusing terminology and cluttered definitions that litter its pages. Most of this confusion comes down to a couple of easily corrected terms, the biggest culprit being the word "grounding" itself.

The NEC utilizes a single word for both the above-grade physical wiring called grounding and the below-grade connection to earth, which is also called grounding. What is desperately needed is to clarify the various "grounding" systems that are required in the NEC and other international standards, such as IEC 60364 and BS 7671.

While the NEC only uses the single term grounding, this book will utilize an additional term to signify the difference between above-grade wiring and the below-grade connection to the earth, and that term is *earthing*. Throughout this book, earthing will be utilized to clarify the difference between the above-grade physical wiring and connection requirements found within the code (grounding) and the below-grade connection to earth requirements (earthing).

The terms *isolated grounding* and *equipotential bonding* (swimming pool grounding) cause a great deal of confusion. While both of these topics will be discussed in detail in the following chapters, it is important to have a fundamental understanding of what these two items are really discussing.

Isolated grounding systems with isolated earthing electrodes are often required by various electronic equipment manufacturers and are mistakenly believed to mean that earthing rods are to be installed that do not tie back to the electrical panel. This is not what is intended and is a violation of the NEC and all known electrical standards. The intent of an isolated grounding system and earthing electrode is to provide a ground source that is free from undesirable stray electrical currents (noise, harmonics, transients, etc.). This requires an earthing electrode and a grounding conductor system that does not come in contact with any other metallic objects until the grounding conductor is properly bonded to the first service disconnect. Please see Chapter 10 of this book, which covers Part VII of NEC Art. 250, the chapter covering isolated grounding, for more information.

The last subject we will cover prior to getting into the actual definitions is in regard to swimming pool bonding and grounding, or equipotential bonding. The NEC has requirements under certain circumstances for two different above-grade grounding systems.

The first system is a low-impedance fault-current path back to the electrical power source through a physical conductor and not through the earth itself, to ensure that overcurrent protection devices (OCPDs) such as fuses and circuit breakers will function properly (during a short-circuit condition). This is the primary purpose of the green wire in a typical 120-V outlet.

The second system, called *equipotential bonding*, requires a common bonding grid when installing a swimming pool (or transformer, substation, high-voltage motor, etc.). This common bonding grid is an extra copper grounding path, in addition to the first system, and is used to bond all of the various metallic objects around the pool (the rebar in the concrete, handrails, motor chassis, etc.).

This additional equipotential grounding system is important to understand, as it is not only required under the NEC for swimming pools, but is required for data-communication systems (data centers, computer/server racks, etc.), high-voltage environments (substations, transmission towers, etc.), anticorrosion systems, and many other commercial and industrial applications that fall under a variety of other industrial standards and federal regulations. Equipotential grounding and bonding is sometimes referred to as *chassis ground*, *equipment ground*, *instrumentation ground*, or *safety ground*.

The following is a list of various terms and their definitions found in the NEC and/or grounding and earthing terms currently in common use.

Ampacity The maximum amount of electrical current a conductor or device can carry before sustaining immediate or progressive deterioration. Ampacity is also described as current rating or current-carrying capacity. Ampacity is the RMS electric current that a device or conductor can continuously carry while remaining within its temperature rating.

Bonding (bond, bonded) The permanent joining of metallic parts together to form an electrically conductive path. This path must have the capacity to safely conduct any fault current likely to be imposed on it.

Bonding jumper A reliable conductor sized per Art. 250 to ensure electrical conductivity between metal parts of the electrical installation.

Chassis ground A name for a dedicated grounding path for equipment that is likely to become energized during an electrical fault or lightning strike. *See* Equipotential Bonding.

Circuit or branch circuit The electrical system downstream from the last OCPD. The outlet you plug your computer into would be part of a circuit.

Clearing time The length in time it takes for an OCPD to detect fault currents and de-energize a circuit. Typically expressed in cycles: one cycle = 0.0167 s for 60 Hz and 0.02 s for 50 Hz.

Concrete-encased copper conductor electrode A permitted electrode composed of 20 ft of #4 AWG bare copper conductor trenched into the earth to a depth of at least 30 in. (2.5 ft) and encased in concrete, that is in direct contact with the earth. This is not considered to be a good electrode, due to the cracking of concrete under high-current fault conditions.

Concrete-encased electrode An electrode encased in concrete that is located within a concrete footing or foundation; the electrode, which may consist of reinforcing bars of steel and/or copper conductors, of at least 20 ft in length and encased in 2 in. of concrete, must be in electrical contact with the earth. Often called a Ufer ground.

Dual-fed service An electrical service wherein the utility company uses two different transformers and two separate service feeds to supply power to a premises.

Earthing The below-grade portion of the electrical system that provides the connection to the earth.

Effective ground fault-current path An intentionally constructed, low-impedance, electrically conductive path designed and intended to carry current under ground-fault conditions from the point of a ground fault on a wiring system to the electrical supply source and that facilitates the operation of the OCPD or ground-fault detectors.

Electrical noise (electromagnetic interference) Any electromagnetic disturbance that interrupts, obstructs, or otherwise degrades or limits the effective performance of electronics and electrical equipment.

Electrode A conductor used to establish electrical contact with the earth.

Electrolytic electrode A pipe electrode, commonly made from 2-in. copper piping, that incorporates a desiccant within the hollow portion of the pipe to extract moisture from the air and deliver the water to the base of the electrode, thereby maintaining high levels of moisture to the surrounding backfill material without the need for human assistance. See Art. 250.52(A)(6).

Equipment-bonding jumper or load-side bonding jumper A conductor, screw, or strap that bonds the equipment-grounded conductor (neutral) to the equipment enclosures that are on the load side of the first service disconnect enclosure(s), in accordance with Art. 250.102(D). The term is also used to indicate a conductor, screw, or strap that bonds the equipment-grounding conductor of a circuit to the enclosure or raceway of that circuit, in accordance with Art. 250.146.

Equipment ground A name for a dedicated grounding path for electrical and/or electronic equipment. *See* Equipotential bonding.

Equipment-grounding (bonding) conductor The low-impedance fault-current path that is typically run with or encloses the circuit conductors, used to connect the non-current-carrying metal parts of equipment, raceways, and other enclosures to the grounded (neutral) conductor and equipment-grounding (bonding) conductor at service equipment or at the source of a separately derived system. Often called the *green-wire* or *third-wire* ground conductor.

Equipotential bonding A conductor system designed to reduce earth voltage gradients in the area around a permanently installed pool, transformer, substation, high-voltage motor, or similar installation by the use of a common bonding grid. Equipotential bonding is not intended to provide the primary low-impedance ground fault–current path to the source (which would assist in clearing a ground fault) as required by Art. 250.4(A)(3).

Fault current The electrical energy released by a given electrical system during an unintentional line-to-ground or line-to-line fault.

Feeder circuit All circuit conductors between the first service disconnect and the final OCPD. The circuits after the final OCPD are called the *branch circuits*.

Feeder taps The conductor that is tapped into a feeder circuit.

First-make, last-break A type of electrical connector with pins of staggered lengths. The long pins make contact first during connector mating and break last during unmating.

First service disconnect The very first electrical panel that will disconnect (turn off) the power coming in from the utility company.

Sometimes this term is used to describe the first electrical panel that can turn off power on the secondary or low-voltage side of a transformer.

Frost line the depth to which the groundwater in soil is expected to freeze; also known as *frost depth* or *freezing depth*. Typically, frost lines are legally required by local building codes.

Ground clamp A mechanical device that connects the grounding electrode conductor to a grounding electrode, typically a ground rod.

Grounded conductor (neutral) A system or circuit conductor that is intentionally grounded. This is a current-carrying conductor typically called the *neutral wire*.

Grounded (earthed) Connected to earth/soil.

Grounded, solidly Connected to ground without any resistor or impedance device inserted.

Grounded system An electrical service that has one of the current-carrying conductors intentionally grounded (i.e., the neutral wire). Floating or delta-type electrical systems are not grounded systems, whereas wye-type systems are grounded.

Ground ring An earthing electrode, composed of #2 AWG or larger bare copper wire at least 20 ft in length and buried at least 2.5 ft below grade and in direct contact with the earth, that circumnavigates a structure or building and is reconnected to itself, typically with multiple connections to the building's or structure's steel rebar and/or the building's steel system, and used in conjunction with standard ground rods. It is considered to be the most effective electrode system in use today.

Ground fault An unintentional connection between an energized conductor and earth or metallic parts of enclosures, raceways, or equipment.

Ground fault-current path An electrically conductive path from the point of a ground fault on a wiring system through normally non-current-carrying conductors, equipment, or the earth to the electrical supply source. A ground fault–current path may not use the earth/soil as a conductor or current path, as it is a violation of the NEC Arts. 250.4(A)(5) and 250.4(B)(4). This term is used for both intentional and unintentional current paths.

Grounding (earthing) conductor A conductor used to connect equipment to a grounding (earthing) electrode.

Grounding (earthing) electrode A device that establishes an electrical connection to the earth/soil.

Grounding (earthing) electrode conductor The conductor used to connect the grounding (earthing) electrode(s) to the equipment-grounding (bonding) conductor, to the neutral (grounded) conductor, or to both in accordance with Arts. 250.66 and 250.142.

Grounding (earthing) electrode system Two or more grounding electrodes that are bonded together to form a single system. For a typical building/structure, a grounding electrode system should consist of: two ground rods, water pipe, steel frame of the building, the steel rebar in the concrete foundation, and any ground rings.

Grounding electrode conductor system The network of above-grade conductors used to connect building steel, cold-water pipe, concrete-encased electrodes (steel rebar in the foundation), ground rings, ground rods, and so on to the first service disconnect, per Art. 250.66.

Ground rod An earthing electrode composed of stainless steel, copper-coated steel, or zinc-coated steel of at least 5/8 in. (15.87 mm) in diameter with at least 8 ft (2.44 m) of length in direct contact with the earth below the permanent moisture level. This is the most common grounding electrode in use today.

Impedance grounded neutral system An electrical system wherein a grounding impedance, usually a resistor or an inductor, is added between the neutral point (XO terminal) of the transformer and the grounding electrode to limit the amperage of fault currents.

Inductive choke An electrical phenomenon that occurs when alternating current (AC) is passed through an insulated nonferrous (copper or aluminum) conductor in proximity to a ferrite material (steel/iron), causing two different electrical fields to form and resulting in a dramatic increase in impedance. The two materials will form differing magnetic fields that will resist changes to the AC passing through them (which reverses direction rapidly), resulting in an impedance increase with frequency. This typically occurs when a single conductor is routed through a metallic raceway without bonding the two together; high-frequency/high-current events, such as lightning or electrical faults, will cause the raceway to act as an inductor, thus increasing the impedance of the current path by as much as 97 percent, resulting in catastrophic failure of the conductor.

Instrumentation ground A name for a dedicated grounding path for sensitive electronic equipment. *See* Equipotential bonding.

Isolated building or structure A building or structure that has no continuous metallic path to any other building or structure. A continuous metallic path between buildings/structures could be formed from metal conduit, a green-wire grounding conductor, metal pipes (water/gas), rebar in concrete, coaxial lines, data and communication shield wires, and so on. Isolated buildings/structures should be tested using a resistance measurement method.

Listed or listed item An item, product, or device that has been specifically developed for a particular function and has been independently evaluated by a product-testing organization, such as the Underwriters Laboratory (UL) in North America or the Conformité Européenne (CE) if you are in Europe.

Load The electrical energy consumed by a component, circuit, device, piece of equipment, or system that is connected to a source of electric power to perform its function. This term often refers to all electrical components and equipment that consume electrical energy downstream from the first service disconnect. It can also be called an *electric load.*

Low-impedance fault-current path *See* Effective ground fault–current path.

Luminaire Complete lighting unit, consisting of one or more lamps (bulbs or tubes that emit light), along with the socket and other parts that hold the lamp in place and protect it, wiring that connects the lamp to a power source, and a reflector that helps direct and distribute the light.

Main Bonding Jumper A conductor, screw, or strap that bonds the neutral (grounded) conductor to a grounding conductor of some type. The term *main bonding jumper* is confusingly used in the Code in various places to indicate both a *supply-side bonding jumper* and a *system bonding jumper* (load side). For more details, see Arts. 250.24(A)(4), 250.24(B), 250.28, and 408.3(C).

Multigrounded/multiple-grounded system A system of conductors in which a neutral conductor is intentionally grounded solidly at specified intervals. A multigrounded or multiple-grounded system, one of the most predominant electrical distribution systems used in the world, is one in which one transformer provides neutral (grounded conductors) wires to multiple electrical services. Consider a single pole-mounted transformer feeding several homes as an example.

Neutral-to-ground connection (bond) is where a grounded current-carrying conductor, the neutral, is intentionally bonded to earth/ground. In general, this should only occur at the transformer and at the first service disconnect(s). The Code uses several names for neutral-to-ground connections: system bonding jumper, main bonding jumper, and supply-side bonding jumper.

Not-permitted electrode system A metal structure in direct contact with the earth that may not be used as an electrode to carry objectionable currents to the earth. Gas pipes and anything aluminum are not-permitted electrodes.

Not separately derived system(s) A premises wiring system whose power is derived from a source of electrical energy other than a utility service. Examples include solar panels, generators, and wind turbines that have a direct connection to another electrical source. Specifically the neutral conductor is solidly bonded to the neutrals of another source.

Objectionable current Electrical current on grounding and bonding paths that occurs when an improper neutral-to-ground bond (or neutral-to-case) creates a parallel path for neutral current to return to the power supply via metal parts of the electrical system in violation of Art. 250.142. This would be better defined as *objectionable neutral current.*

Objectionable direct current (DC) Unwanted DC electrical current on grounding and bonding paths that occurs when cathodic protection systems impress DC currents onto the grounding system. *See* Objectionable current.

Objectionable neutral current *See* Objectionable current.

Overcurrent device Usually a fuse or circuit breaker, a device that limits the maximum amount of current that can flow in a circuit. Sometimes called an *overcurrent protection device* or *OCPD.*

Permitted electrode An intentional connection to the earth of a copper and/or protected steel system of sufficient size/length that is capable of handling likely objectionable currents from lightning strikes, electrical faults, and the like, safely into the earth. These electrodes can be metal structures in direct contact with the earth that may be used as an electrode to carry objectionable currents to the earth.

Permanent moisture level Undefined within the NEC but commonly linked to either the frost line or the permanent wilting point of the site. Permanent moisture level is also often thought of as being either 18 in. (1.5 ft) or 30 in. (2.5 ft) in depth, due to various interpretations of several codes, regulations, and industry standards.

Permanent wilting point The minimal point of soil moisture a plant requires not to wilt; typically water content of soil at any point lower than –1500 J/kg of suction pressure.

Power ground The ground system the supply of electrical power is connected to, such as an electrical panel or first service disconnect.

Raceway An enclosed channel of metal or nonmetallic material designed expressly for holding wires, cables, bus bars, and so on. Raceways include all types of conduit, flexible tubing, floor raceways, busways, and so on.

Safety ground A name for a dedicated grounding path for equipment and/or chemicals sensitive to static charges. *See* Equipotential bonding.

Separately derived system(s) A premises wiring system whose power is derived from a source of electrical energy other than a utility service. Examples include solar panels, generators, and wind turbines that have no direct connection to another electrical source. Specifically, the neutral conductor must not be solidly bonded to the neutrals of other sources. This term is also used in regard to isolation and/or step-down transformers that generate a distinct power source separate from the utility.

Service The conductors and equipment for delivering electric energy from the serving utility to the wiring system of the premises.

Service conductor The conductors from the service point to the service-disconnecting means. This is a broad term that may include service drops, service laterals, and/or service entrance conductors. Generally considered to be the utility company's wires that run from the pole to the masthead of buildings.

Service conductor, grounded The current-carrying service conductor coming off the low side of the transformer that is to be, or has been, bonded to the ground (grounded). In other words, the neutral wire coming from the utility into a main electrical panel.

Service entrance conductor The service conductors that bring power to the meter and enter the building, up to the first service disconnect.

Service equipment The necessary equipment, usually consisting of circuit breaker(s), switch(es), and/or fuse(s) and their accessories, connected to the load end of service (utility) conductors to a premises and intended to constitute the main control and cutoff of the supply.

Single-point ground A technique of bonding different grounding systems together at only one place to minimize the transfer of electrical noise from one grounding system to the other.

Solidly grounded The intentional electrical connection of the neutral terminal (grounded conductor) to the equipment-grounding (bonding) conductor per Art. 250.30(A)(1).

Sphere of influence The hypothetical volume of soil that will experience the majority of the voltage rise of the ground electrode when that electrode discharges current into the soil. The sphere of influence is equal to the diagonal length of the electrode or electrode system.

Strike-termination devices A component of a lightning protection system designed to intercept lightning strikes that is in turn connected to a conductive path to ground. Strike-termination devices include lightning rods, finials, air terminals, masts, catenary systems, and other permanent metal structures that qualify.

Structural metal grounding electrode A structural metal member that establishes an electrical connection to the earth/soil as defined in Art. 250.52(A)(2). Specifically, the metal frame of a building or structure that is properly bonded to the steel rebar in the building's or structure's foundation (concrete-encased electrode without a vapor barrier), as required in Art. 250.52(A)(3).

Supply-side bonding jumper A conductor installed on the supply (utility) side of a service, or within a service equipment enclosure(s), or for a separately derived system, that ensures the required electrical conductivity between metal parts required to be electrically connected; sometimes called a *main bonding jumper*. A conductor, screw, or strap that bonds the neutral (grounded) conductor to the service equipment enclosures on the supply side. The term *supply-side bonding jumper* is used to distinguish the bond at the supply (utility) side of the service from the *system bonding jumper*, which is used on the load side of the service (at the first service disconnect). For more details, see Arts. 250.24(A)(4), 250.24(B), 250.28, 250.28(D)(2), and 408.3(C).

Note *The* NEC *occasionally uses this term to indicate an equipment-grounding conductor (see Art. 250.30).*

Supply station, DC A battery or a series of batteries, generator, power supply, solar panel(s), or other device that provides DC electrical power to a circuit.

Switchboard One or more panels accommodating control switches, indicators, and other apparatus for operating electric circuits.

System bonding jumper A conductor, screw, or strap that bonds the equipment-grounding conductor to the neutral (grounded) conductor; sometimes called the *main bonding jumper.* The term *system bonding jumper* is used to distinguish the bond at the load side of the service (at the first service disconnect) from the *supply-side bonding jumper,* which is the bonding used on the utility or supply side of the service. For more details, see Arts. 250.24(A)(4), 250.24(B), 250.28, and 408.3(C).

Tapped grounding electrode conductor A grounding electrode conductor sized in accordance with Art. 250.66 that is used to bond multiple transformers (separately derived systems) located in a single building to the grounding electrode system.

Transfer switch An automatic or nonautomatic device for transferring one or more load conductor connections from one power source to another. See Art. 250.30.

Ufer ground A concrete-encased electrode system integral to a building's structural system, such as the steel rebar in a concrete foundation. Copper conductors added to increase the conductivity of the steel rebar would also be a Ufer ground. Ufer grounds are named after Herbert G. Ufer, who developed the system in World War II to ground bomb storage vaults. *See* Concrete-encased electrode.

Ungrounded system An electrical service that does not have any of the current-carrying conductors intentionally grounded (i.e., no neutral wire). Delta-type electrical systems and some standby generators are examples of ungrounded systems.

Ungrounded conductor A current-carrying conductor that has not been grounded. A phase conductor is an ungrounded conductor.

Unintentional neutral current path Where an inappropriately applied neutral-to-ground connection has established a parallel path for neutral conductor currents to travel back to the source on the ground conductors and/or metal enclosures of an electrical system.

Yoke or contact yoke A set of circular metal tabs found on both ends of a standard three-wire, 120-V outlet, often called "Mickey Mouse ears" due to their similarity in shape. These metal tabs provide a metal-to-metal contact between the circuit's (receptacles) equipment-grounding conductor and the receptacle's enclosure via an internal connection integral to the receptacle, as required under Arts. 250.146(A) and (B). An isolated grounding receptacle, does NOT have the internal connection from the yoke to the equipment-grounding conductor and requires two equipment-grounding conductors, one wire type for the grounding terminal on the receptacle per Art. 250.146(D), and one for the receptacle's metal enclosure (box) per Art. 250.118.

Zipse's law "In order to have and maintain a safe electrical installation: All continuous flowing current shall be contained within an insulated conductor or if a bare conductor, the conductor shall be installed on insulators, insulated from earth, except at one place within the system, and only one place, can the neutral be connected to earth."

250.3. APPLICATION OF OTHER ARTICLES

There is a full-page table in the NEC labeled Table 250.3. This is a list of other sections (articles) within the NEC that have grounding and bonding requirements linked to Art. 250. For example, swimming pools have many grounding and bonding requirements unique to swimming pools that are discussed in Art. 680. But other areas, such as agricultural buildings (Art. 547), communications circuits (Art. 800), electrolytic cells (Art. 668), elevators (Art. 620), floating buildings (Art. 553), radio and television equipment (Art. 810), and x rays (Art. 660), to name only a few, have some specific requirements in addition to, or separate from, the general grounding rules. Please refer to Table 250.3 in your NEC for further information.

250.4. GENERAL REQUIREMENTS FOR GROUNDING AND BONDING

Article 250.4 itself is actually quite small and only covers a few pages. It is broken down into two sections: (A) Grounded Systems and (B) Ungrounded Systems. In this case, the NEC is discussing the difference in grounding/bonding requirements between electrical service provided by wye-type transformers (grounded systems) and electrical service provided by Delta-type transformers (ungrounded systems).

250.4(A). Grounded Systems. A grounded electrical system is one in which the electrical service has the three single-phase transformers connected to a common point called the neutral. This configuration is called a *wye*, because in an electrical drawing it looks like the English alphabet's letter "Y." The center point where all three phases meet will have a voltage of zero and is typically referred to as the *XO terminal*, as it is commonly labeled as such in the transformers. The XO point in a wye-type transformer is bonded to earth or grounded. This is where we get the terms *grounded electrical system, "solidly-grounded"*, or *grounded system*.

Electrical engineers ground electrical systems for many reasons: to limit the voltages caused by lightning or by accidental contact with supply conductors; to stabilize line voltages under normal operation relative to ground; to provide a reference point in an electrical circuit from which other voltages are measured; to provide a common return path for electric current; to connect exposed metal parts to the common return path (ground), which prevents user contact with dangerous voltage if electrical insulation fails; and to limit the buildup of static electricity. This can be especially important when dealing with electrostatic-sensitive devices (computers, televisions, stereos, etc.) and flammable products. Please see Fig. 4.1 for a schematic of a typical electrical installation showing a solidly-grounded wye-type transformer providing power to an electrical panel.

250.4(A)(1). This portion of the Code requires you to connect the XO terminal of the wye-type transformer to earth in a proper manner. The intent is to limit the voltage imposed on the electrical system by lightning strikes, line surges, or unintentional contact with higher-voltage lines.

This means the common 0-V point (the XO terminal) must be bonded to both earth and to a common above-grade grounding system of your electrical service. There are some very important specifics about how these bonds must occur that will be discussed in other parts of this book. But the primary issue when bonding the XO terminal of the transformer to earth is to ensure that the grounding electrode conductor and/or the electrode itself does not accidentally come in contact with other metallic objects. This could form an unintentional return path for neutral currents through the non-current-carrying ground system.

The Code further informs you that it is important to consider the length and routing of your grounding electrode conductor; this is the wire that runs from the ground bar in your primary electrical cabinet (first service disconnect) to your grounding/earthing electrodes. The manner in which you route this critical conductor (and its length) will

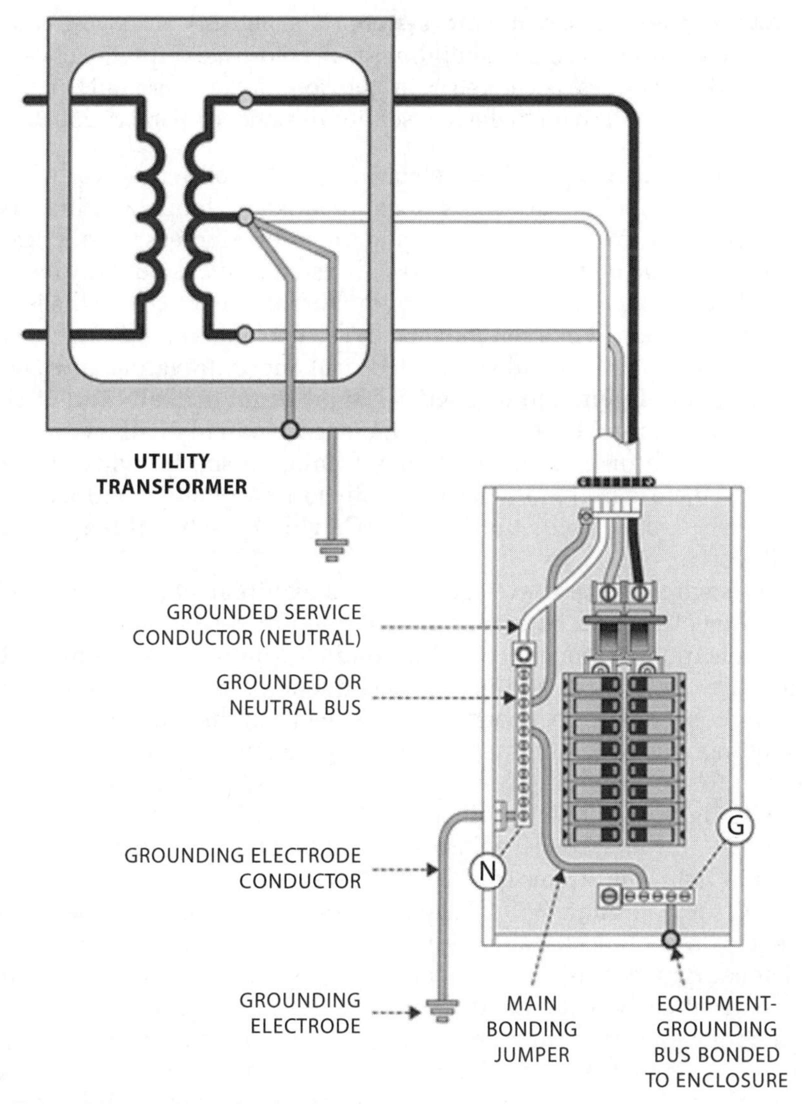

Figure 4.1 Typical single-phase wiring.

dictate how effectively this critical conductor will be able to limit the imposed voltages on the electrical system.

Grounding (earthing) electrode conductor The conductor used to connect the grounding (earthing) electrode(s) to the equipment-grounding (bonding) conductor, to the neutral (grounded) conductor, or to both in accordance with Arts. 250.66 and 250.142.

Grounding electrode conductor system The network of above-grade conductors used to connect building steel, cold-water pipe, concrete-encased electrodes (steel rebar in the foundation), ground rings, ground rods, and so on to the first service disconnect, per Art. 250.66.

Put more simply, a grounding electrode conductor that is overly long and routed in such a manner as to add bends and loops will increase the impedance of the conductor and therefore increase the overall imposed voltage on the entire electrical system. This is not only due to the added impedance of the extra conductor length but also to the issue of *self-inductance*. While self-inductance can get quite complicated, what is important to understand is that electrical energy traveling down any conductor will generate a magnetic field around itself, and if the conductor is coiled back on itself, the magnetic fields will try to form back and forth on each of the loops, forming a self-inductive choke that can be measured in the form of inductance or henrys. Inductance, as you will recall, limits the flow of AC, which is a bad thing in electrical safety.

This is why you may have come across a requirement to have conduit grounding clamps on both ends of your grounding electrode conductor when it is routed through a metallic conduit. Routing this conductor in a metallic conduit can add impedance, due to an inductive choke that can form between the poorly bonded steel conduit and the copper grounding electrode conductor. This requirement to bond a metallic conduit to the grounding conductor is especially pertinent when dealing with lightning down conductors, as the effects are especially prominent due to the extreme voltages, currents, and frequencies that are found in lightning strikes (see NFPA® 780 Annex C).

250.4(A)(2) through (5). This portion of the Code requires the metallic enclosures of electrical equipment and any other conductive material or equipment (cases, boxes, conduit, chassis, etc.) that are normally non-current-carrying be bonded to earth. Furthermore, they should be bonded together to form an effective ground fault–current path. What this means is that the metallic enclosures surrounding and protecting current-carrying conductors must not only be bonded to earth, but must be bonded together to form a continuous low-impedance metallic path capable of handling the likely fault currents imposed on it. It is important that fault currents have a low impedance path back to the appropriate circuit breakers, fuses, and any ground-fault detectors, so that the electrical fault can be quickly terminated by the overcurrent protection devices (OCPD).

Please note that this requirement has nothing to do with the green-wire ground requirement. The green wire routed inside the conduit is an additional grounding system required by the Code, above and beyond the requirement to bond the chassis of electrical enclosures together.

Effective ground fault-current path An intentionally constructed, low-impedance, electrically conductive path designed and intended to carry current under ground-fault conditions from the point of a ground fault on a wiring system to the electrical supply source and that facilitates the operation of the OCPD or ground-fault detectors.

250.4(A)(5). One of the most important parts of the entire Code is found in Art. 250.4(A)(5). This section outlines where the Code prohibits the use of the earth as a ground fault–current path: *"The earth shall not be considered as an effective ground fault–current path."*

What this states is that you must have a metallic path back to your OCPD, such as circuit breakers and fuses, for ALL electrical services in your system, including services with isolated grounding. This is the primary purpose of the green-wire ground, to ensure the proper operation of OCPDs (which need short durations of maximum current in order to operate).

This is the part of the Code that gets many an electrician and electrical engineer confused when installing/designing isolated or dedicated grounding/earthing. They mistakenly believe that isolated ground systems are not tied back to the common grounding system.

This is, of course, NOT how isolated grounding/earthing systems work. While isolated grounding is covered more thoroughly in Chapter 10—in this book, it is worth a quick mention here.

Isolated grounding is mistakenly thought to mean that electrical service is to be provided to a piece of equipment without any grounding conductors (the green-wire ground and/or the chassis ground-fault path) and should have a grounding rod installed near the equipment in complete isolation. You have undoubtedly seen instances in which someone has simply driven a ground rod near a piece of equipment and bonded it without installing a ground wire back to the electrical panel. This is a clear violation of **NEC** Art. 250.4(A)(5) and places personnel and equipment in serious jeopardy.

The reason this is illegal (and a hazardous practice) is that if an electrical fault were to occur in such circumstances, how would the fault current travel back to the circuit breaker? Remember, circuit breakers and fuses work by having a short duration of high-current. If the only available fault-current path is through the earth, there is simply no way to insure that enough current will travel through the resistive soils and back to the circuit breaker to cause it to trip and shut off power to the equipment.

Isolated or dedicated grounding should be more accurately thought of as single-point grounding. Truthfully, we would be better served to use that term and eliminate the use of the other terminology.

250.4(B). Ungrounded Systems. An ungrounded electrical system is one in which the electrical service has the windings of three single-phase transformers connected in series with one another to form a closed circuit. This configuration derives its name from its appearance in an electrical drawing—it looks like the Greek symbol Δ (triangle) for the letter "delta."

For delta transformers, there is no center point where the voltage is zero, and therefore no XO terminal that can bonded to earth or grounded. This is where we get the term *ungrounded electrical systems* or *ungrounded systems*. This also means there is no neutral wire for the system.

Delta power systems present some very special issues with regard to the detection of ground faults and the ability of OCPDs to operate properly and shut off the power. While the reasons for this get quite technical, the thing to remember is that delta power systems can remain operational even during fault conditions. This means a serious electrical fault could be occurring where an entire phase (leg) is faulting to ground, and yet the entire system keeps operating as if nothing were wrong at all! When dealing with delta power systems, grounding and bonding become more important, as proper detection via ground-fault monitors (zero-sequence unbalance detection) will heavily rely on the equipotential ground plane as a reference source for its measurements.

The NEC requirement for ungrounded electrical systems is very similar to the requirements it has for grounded electrical systems, with the very important exception that there is no XO terminal on the transformer to bond to earth. There is also, of course, no neutral. But beyond this, the grounding requirements are very similar. You still must bond the electrical enclosures together and ground them to earth, just as in any other system.

250.4(B)(1) through (4). This portion of the Code requires that the metallic enclosures of electrical equipment and any other conductive material or equipment (cases, boxes, conduit, chassis, etc.) that are normally non-current carrying, shall be bonded to earth. Furthermore, they shall be bonded together to form an effective ground fault–current path.

This means that the metallic enclosures surrounding and protecting current-carrying conductors, must not only be bonded to earth, but must be bonded together to form a continuous low-impedance metallic path capable of handling the likely fault currents imposed on it. It is important that fault currents have an effective path back to the appropriate circuit breakers, fuses, and/or to operate any ground-fault detectors, so that the electrical fault can be quickly terminated.

250.4(B)(4). One of the most important parts of the entire Code is found in Art. 250.4(B)(4). This section is where the Code prohibits the use of the earth as a ground fault–current path: *"The earth shall not be considered as an effective ground fault–current path."* This means that you must have a metallic path back to your OCPD and ground-fault detectors, for ALL electrical services in your system, including services with isolated grounding. This is the primary purpose of the green-wire ground, to ensure the proper operation of OCPDs and/or the detection of ground faults.

250.6. OBJECTIONABLE CURRENT

Article 250.6 can be summed up in one simple sentence: *"Neutral currents shall not use the grounding system as a return path."* With that said, it is one of the more confusing sections in the NEC and has led to some very serious and hazardous problems. The goal of this section is to ensure that neutral currents are not on the grounding system during normal operations and to allow some minor changes to the typical grounding protocols, under certain circumstances, if corrective action is needed to stop objectionable neutral currents.

There are two circumstances that are the primary concern of Art. 250.6. They are for removing unintended neutral currents from the grounding system and/or removing DC that is being impressed on them from cathodic protection systems (cathodic protection systems are commonly used in industrial facilities that have extensive piping systems, as the impressed DC currents help to protect the pipes from corrosion).

Objectionable current Electrical current on grounding and bonding paths that occurs when an improper neutral-to-ground bond (or neutral-to-case) creates a parallel path for neutral current to return to the power supply via metal parts of the electrical system in violation of Art. 250.142. This would be better defined as *objectionable neutral current.*

Electrical noise (electromagnetic interference) Any electromagnetic disturbance that interrupts, obstructs, or otherwise degrades or limits the effective performance of electronics and electrical equipment.

Zipse's law "In order to have and maintain a safe electrical installation: All continuous flowing current shall be contained within an insulated conductor or if a bare conductor, the conductor shall be installed on insulators, insulated from earth, except at one place within the system, and only one place, can the neutral be connected to earth."

The confusion in this section of the Code comes from the lack of a definition for objectionable current in the NEC. The intent of this section of the Code is to handle special issues that may arise in regard to neutral currents using the grounding system as a return path. This would be better defined as *objectionable neutral current.* However, the lack of a proper definition has led some inexperienced individuals to believe that the objectionable current discussed in Art. 250.6 is electrical noise (electromagnetic interference) that can cause data errors in electronic equipment. Article 250.6 specifically does not cover ways to correct problems involving electrical noise; those are discussed under Arts. 250.96(B) and 250.146(D).

This section of the Code mandates that the engineer design and the electrician construct an electrical system that does not use the grounding system as a return path for neutral currents. This section of the Code additionally gives you the ability to take logical corrective actions to remove objectionable neutral currents that may have unintentionally occurred.

250.5(A)(5) and 250.4(B)(4) The earth shall not be considered as an effective ground fault–current path.

What the Code makes very clear in Arts. 250.6(B) and 250.6(D) is that no matter what the conditions may be, you must have a grounding conductor (third-wire or green-wire ground) leading back to where the OCPDs are located.

250.6(A). Arrangement to Prevent Objectionable Current. This section simply states that electrical systems, including circuit conductors, surge arresters, surge-protection devices, and non-current-carrying metal parts of equipment (enclosures, cabinets, raceways, conduits, etc.) shall be installed in a manner to prevent objectionable current.

250.6(B). Alterations to Stop Objectionable Current. This section states that if you do find that neutral currents have somehow gotten on to the grounding system, you may use one of four types of changes to the standard grounding protocols. However, you must ensure that the earth (soil) is not being used as a ground fault–current path. The four standard grounding protocols that may be modified are:

1. *You may discontinue one or more but not all of such grounding connections.* This is the part that allows you to remove bonding tabs and other connections that typically bond metal components together.
2. *You may change the locations of the grounding connections.* In certain circumstances, you may need to ground to the chassis directly and/or install ground wires from other sources, for example, instrumentation grounding directly to grounding terminals.
3. *You may interrupt the continuity of the conductor or conductive path causing the objectionable current.* This is the part that allows you to electrically isolate sensitive electronic systems from the conduit and raceways.
4. *You may take other remedial and approved action.* In certain cases, unique methods may be required that must be approved by the regulatory authority governing your electrical installation.

250.6(C). Temporary Currents Not Classified as Objectionable Currents. This section tells us that a full ground fault (where you get a phase-to-ground or phase-to-case fault) is not considered an objectionable current. This may sound odd, as a phase-to-ground fault would seem to be one of the worst objectionable currents. However, the point of this section is to state clearly that the requirements in Arts. 250.6 (A) and (B), which prohibit neutral currents from using the grounding system as a return path, are required for the normal operation of equipment. These requirements are not about protecting the electrical system during short-term (temporary) abnormal conditions (i.e., ground faults and lightning strikes that may place electrical current temporarily on the grounding system), but are intended to mandate that neutral currents shall not use the grounding system as a return path.

250.6(D). Limitations to Permissible Alterations. This section has been specifically placed in the NEC to let you know that Art. 250.6 is NOT discussing electrical noise (electromagnetic interference) that may cause data errors in sensitive electrical equipment. Even if you have sensitive electronic equipment and your client is requesting isolated grounding, your electrical system must be in compliance with Art. 250.6. See Arts. 250.96(B) and 250.146(D) for the proper grounding of sensitive electronic equipment.

Again, Art. 250.6 is only referring to neutral currents using the grounding system as a return path, and the isolation of direct currents being imposed on the grounding system by cathodic protection devices.

Electrical noise (electromagnetic interference) Any electromagnetic disturbance that interrupts, obstructs, or otherwise degrades or limits the effective performance of electronics and electrical equipment.

250.6(E). Isolation of Objectionable Direct-Current Ground Currents. When you are dealing with systems that have cathodic protection (cathodic protection systems are commonly used in industrial facilities that have extensive piping systems, as the impressed DC currents help to protect the pipes from corrosion), objectionable DC currents may form on the grounding system. Article 250.6(E) allows you to install approved/listed AC-coupling/DC-isolating devices in the grounding conductor path. These systems will block DC from entering the grounding system, but will allow AC to have an effective ground fault–current path. These devices must be listed for this function and evaluated by a product-testing organization.

250.8. CONNECTION OF GROUNDING AND BONDING EQUIPMENT

This section permits the use of machine screws and thread-forming machine screws as an acceptable method of bonding in a grounding system. It also allows the use of listed twist-on wire connectors (wire nuts) and listed pressure connectors, even if those connectors are not colored green. These devices must be listed for this function and evaluated by a product-testing organization, such as the UL or CE.

Wood screws, sheet-metal screws, drywall screws, and any other form of screw that is not a machine screw are prohibited (not allowed).

Additionally, fittings that depend solely on solder are not allowed. However, silver solder is generally considered acceptable by independent product-testing organizations due to its higher melt temperatures.

250.8(A). Permitted Methods. This section of the Code states that equipment-grounding conductors, grounding electrode conductors, and bonding jumpers must be connected via approved methods. Those methods are:

1. Listed pressure connectors
2. Terminal bars
3. Pressure connectors listed for the grounding and bonding of equipment
4. Exothermic welding process
5. Machine screws with at least two threads engaged or secured with a nut
6. Thread-forming screws with at least two threads engaged in the enclosure
7. Connections that are part of a listed assembly
8. Other listed connections

Equipment-grounding (bonding) conductor The low-impedance fault-current path that is typically run with or encloses the circuit conductors, used to connect the non-current-carrying metal parts of equipment, raceways, and other enclosures to the grounded (neutral) conductor and equipment-grounding (bonding) conductor at service equipment or at the source of a separately derived system. Often called the green-wire or third-wire ground conductor.

Grounding (earthing) electrode conductor The conductor used to connect the grounding (earthing) electrode(s) to the equipment-grounding (bonding) conductor, to the neutral (grounded) conductor, or to both in accordance with Arts. 250.66 and 250.142.

Bonding jumper A reliable conductor sized per Art. 250 to ensure electrical conductivity between metal parts of the electrical installation.

Listed or listed item An item, product, or device that has been specifically developed for a particular function and has been independently evaluated by a product-testing organization, such as the UL or CE.

250.8(B). Methods Not Permitted. This section states that devices, fittings, and/or connections that use only solder to secure the connection(s) are NOT allowed. This is because most solder is what is called *soft solder*, which has a melting range between 180 and 190°C (360 and 370°F), but could be as low as 90°C (190°F). The concern is that during electrical fault conditions, the connections experience excessive current flow and could heat to temperatures that could melt the solder, thereby compromising the connection.

Solder is a fusible metal alloy used to join together metal pieces and has a melting point below that of the piece(s). It is commonly used in electronics, plumbing, and the assembly of sheet-metal parts. Soldering performed using alloys with a melting point above 450°C (840°F), called *hard soldering*, *silver soldering*, or *brazing*, is generally accepted by independent product-testing organizations.

250.10. PROTECTION OF GROUND CLAMPS AND FITTINGS

This simple requirement mandates that all clamps and fittings (mechanical connections) in your grounding system that are subject to physical damage must be enclosed in a metal, wood, or another such protective covering.

Listed or listed item An item, product, or device that has been specifically developed for a particular function and has been independently evaluated by a product-testing organization, such as the UL or CE.

All of these devices, from the clamps and fittings to the protective enclosures, must be listed for this function and evaluated by a product-testing organization.

The 2014 version of the Code simplified the verbiage in this section, but did not make any major changes.

250.12. CLEAN SURFACES

This is another very simple statement that mandates the removal of paint, lacquer, enamel, and other types of coatings from the threads and bonding points where you are making your grounding connections. This is to ensure that a solid electrical connection that has a low impedance and good electrical continuity is established. Paint is well known for preventing a good bond when grounding to the chassis of electrical enclosures. In fact, there are numerous ANSI, EIA, ITA, IEEE, MIL-SPEC, and other international standards that mandate the use of special tri-lobed screws and the use of antioxidant joint compounds to ensure proper connectivity between a grounding conductor and the metallic chassis to which it is bonded.

Chapter Five

NEC 250 PART II: SYSTEM GROUNDING

This section of Art. 250 explains which electrical systems must be grounded. Electrical systems that must be grounded include alternating current (AC) systems, impedance-grounded neutral (grounded conductor) systems, vehicle-mounted and portable generators, permanently installed generators, buildings and structures supplied by feeders or branch circuits. This article also details which circuits do not need to be grounded. Keep in mind that the Code uses the term *grounded* to mean a variety of things, such as connecting an electrical circuit to the earth (earthing), connecting metallic components together to ensure there are no hazardous differences in potential (bonding), and to provide effective current paths for electrical faults to travel on back to the first service disconnect, to ensure the proper operation of overcurrent protection devices (OCPDs), such as circuit breakers and fuses (equipment-grounding conductor or green wire grounding). The single use of the word "grounded" can often be confusing, as the Code rarely clarifies why and for what purpose it requires items to be grounded.

Zipse's law "In order to have and maintain a safe electrical installation: All continuous flowing current shall be contained within an insulated conductor or if a bare conductor, the conductor shall be installed on insulators, insulated from earth, except at one place within the system, and only one place, can the neutral be connected to earth."

This particular section instructs us to bond (ground) one of the current-carrying conductors, typically the neutral (grounded conductor), to ground. It tells us that this should occur only at the transformer and at the first service disconnect panel and in no other place.

It is important to remember that grounding and earthing serve multiple purposes in regard to human safety, the proper operation of electrical gear, and the quality of electrical power. One of the primary purposes of grounding is to limit the voltage impressed on the system due to line surges and lightning.

250.20. ALTERNATING-CURRENT SYSTEMS TO BE GROUNDED

This article tells us that there are grounding requirements for all AC systems. What this part of the Code is specifically talking about is the requirement to intentionally ground one of the conductors in the electrical system, typically the neutral (grounded conductor) or white wire.

This includes power from wye-type transformers, delta transformers, and corner-grounded delta transformers. There is also a blanket statement that "other systems shall be permitted to be grounded." This catchall statement in the Code tells us that you may have grounding requirements for metallic items the Code does not specifically mention. So, if the NEC does not specifically mention your exact electrical system by name, you may still be required to ground it and to ground it in compliance with the NEC.

250.20(A). Alternating-current Systems of Less Than 50 Volts. You must bond (ground) one of the current-carrying conductors coming off the low side of the transformer to ground on an AC system operating with less than 50 V: if it is supplied by transformers that have more than 150 V to ground, or if the electrical system supplying power to the transformer itself is ungrounded, located outdoors, or providing electricity via outdoor overhead conductors. Figure 5.1 shows the proper neutral-to-ground bonding arrangement for the most common types of transformers.

In other words, if you have an indoor AC transformer for which the supply power (high side) is less than 150 V to ground and properly grounded, and the low side is less than 50-V AC, then you do not need to bond one of the current-carrying conductors coming off the low side of the transformer to ground.

250.20(B). Alternating-current Systems of 50 Volts to 1000 Volts. This section tells us that you must bond (ground) one of the current-carrying conductors coming off the low side of the transformer to ground for all electrical systems that provide AC power between 50 to 1000 V to any premises where, for the following conditions:

1. The system can be grounded so that the phase conductors (ungrounded conductors) do not exceed 150 V to ground.
2. In a three-phase, four-wire wye-type system, the neutral conductor (grounded conductor) is one of the circuit conductors.
3. In a three-phase, four-wire delta-type system, the midpoint of one of the phase windings is used as a circuit conductor.

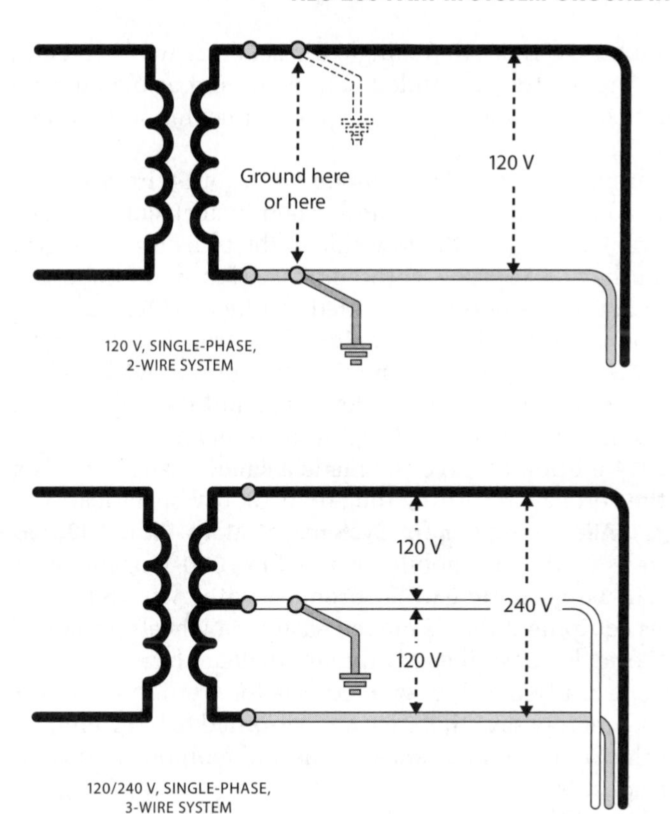

Figure 5.1 Transformer grounding locations.

This includes any system that can be grounded so the maximum voltage to ground on the ungrounded conductors does not exceed 150 V. This means all of the 120-V AC systems that are most familiar. It also includes all three-phase, four-wire delta- and wye-type transformers in which the neutral (grounded) conductor is used as a circuit conductor.

Wye-type transformers are to have the center tap of the windings (XO terminal) tied to ground. This is also where the neutral (grounded) conductor is connected. There is a further requirement to bond the neutral (grounded conductor) bar in the service panel to ground.

Delta-type transformers that are to have a "stinger" used as a neutral are to have the midpoint of one phase winding tied to ground. This is also where the neutral (grounded) conductor is connected. There is a further requirement to bond the neutral (grounded conductor) bar in the service panel to ground.

For example:

- If you have a 120-V, single-phase, two-wire system, you must bond one of the current-carrying conductors coming off the low side of the transformer to ground.

- If you have a 120/240-V, single-phase, three-wire system, you must bond the neutral (grounded conductor and also a current-carrying conductor) coming off the midpoint or center of the transformer to ground.
- If you have a 120/208-V, three-phase, four-wire wye system, you must bond the neutral (grounded conductor) (current-carrying conductors) coming off the low side of the transformer to ground.
- If you have a 120/208-V, three-phase, four-wire delta system, you must bond the neutral (grounded conductor) (current-carrying conductors) coming off the low side of the transformer to ground.

In Fig. 5.2 we can see a schematic of the proper wiring all the way to the first service disconnect for both wye and delta type transformers, including the proper neutral-to-ground connection.

The 2014 edition of the Code made a small change in this section to clarify the verbiage regarding the 50- to 1000-V specification.

250.20(C). Alternating-current Systems of More Than 1000 Volts. This section tells us that all mobile or portable 1000-V or higher AC equipment shall be grounded in accordance with Art. 250.188. "Mobile" describes equipment that is moved easily via wheels, treads, skids, and so on. "Portable" describes equipment that can be easily carried.

The Code really does not give criteria for grounding of systems over 1000 V, but merely says that they are permitted to be grounded. Remember that the Code excludes substations and temporary substations from its requirements.

The 2014 edition of the Code made a small change in this section to clarify the verbiage regarding the 50- to 1000-V specification.

250.20(D). Impedance-grounded Neutral (Grounded Conductor) Systems. Impedance-grounded neutral (grounded conductor) systems are electrical systems wherein a grounding impedance, usually a resistor or inductor, is installed between the grounding electrode conductor and the transformer's neutral (grounded conductor) point. This is done to limit the ground-fault current to a low value.

Grounding (earthing) electrode conductor The conductor used to connect the grounding (earthing) electrode(s) to the equipment-grounding (bonding) conductor, to the neutral (grounded) conductor, or to both in accordance with Arts. 250.66 and 250.142.

This part of the Code tells us that impedance-grounded neutral (grounded conductor) systems shall bond (ground) one of the current-carrying conductors coming off the low side of the transformer to ground in accordance with Art. 250.36 or Art. 250.186. Please see these sections for additional information.

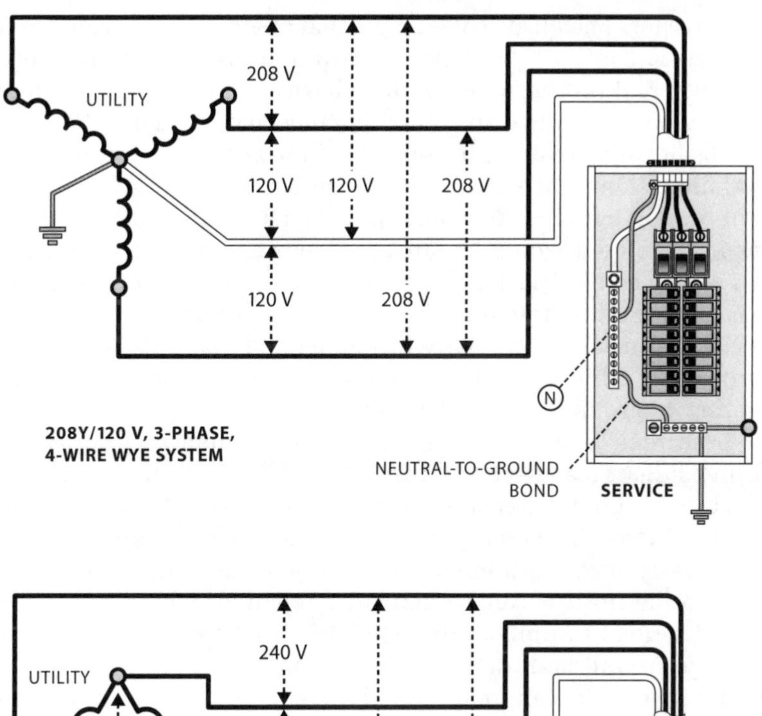

Figure 5.2 Typical wiring configurations for wye-type and delta-type transformers.

250.21. ALTERNATING-CURRENT SYSTEMS OF 50 VOLTS TO 1000 VOLTS NOT REQUIRED TO BE GROUNDED

This section tells us that certain electrical systems operating at 50 to 1000 V have the option of bonding (grounding) one of the current-carrying conductors coming off the low side of the transformer to ground, or not. The electrical systems that qualify for this rule are ones in which the continuity of power is required for the life safety of persons or where it is otherwise unsafe to have an uncontrolled shutdown.

It has been argued that in certain circumstances, such as the surgery room of a hospital, that it is better and safer to continue electrical service to certain systems than to allow the unplanned loss of power due to circuit-breaker tripping. You can also imagine a crane lifting very heavy items in the air—it would be less dangerous to have a controlled shutdown during an electrical fault than to have power shut off in midlift.

It is clear that the intent of the Code is to only allow the use of this rule as a last resort. It is highly recommended that all systems should be grounded. This is because there are serious disadvantages when operating ungrounded electrical systems, including increased problems from high transient voltages, impressed voltages from lightning and line surges, faster rates of insulation deterioration, and of course, the inability of circuit breakers to operate. This means that underground-fault conditions, in which the hot wire comes into contact with the metal chassis of the equipment, or in a line-to-line fault, the condition will continue until it can be manually repaired, which can result in fires and serious equipment damage. This is why ungrounded systems must have ground-fault detectors installed.

Before you design an ungrounded system, you should first put into place all other measures: backup systems, redundant circuits, uninterruptable power supplies, and so on. In particular, high-impedance neutral grounded systems have many of the advantages of ungrounded systems, without the downsides. Please see the ANSI/IEEE Std. 142-1982 (Green Book) for more information.

250.21 (A). General. This section tells us that certain 50- to 1000-V AC systems are permitted, but not required, to be grounded. In other words, having a neutral (grounded) conductor is optional, based on your system's specific needs.

The following electrical systems have the option of having grounded (neutral) systems or not:

1. Electrical systems used exclusively for industrial electric furnaces that melt, refine, temper, and so on.
2. Separately derived systems that exclusively provide power for rectifiers (direct current) used for adjustable-speed industrial drives.
3. Separately derived transformers of 1000 V or less that are used exclusively for:
   a. Control circuits
   b. Places where only qualified maintenance personnel service the equipment
   c. Places where continuity of control power is required
4. Other electrical systems that are not required to be grounded in accordance with Art. 250.20:

Separately derived system(s) A premises wiring system whose power is derived from a source of electrical energy other than a (utility) service. Examples include solar panels, generators, and wind turbines that have no direct connection to another electrical source. Specifically, the neutral conductor must not be solidly bonded to the neutrals of other sources. This term is also used in regard to isolation and/or step-down transformers that generate a distinct power source separate from the utility.

Service The conductors and equipment for delivering electric energy from the serving utility to the wiring system of the premises.

The 2014 edition of the Code made a small change in this section to clarify the verbiage regarding the 50- to 1000-V specification.

250.21(B). Ground Detectors. If you have an ungrounded AC system, you must install ground detectors (ground-fault detectors) to monitor the system. A *ground detector* is an electrical/electronic device that senses when a ground fault occurs and provides a visible and/or audible signal to alert personnel. The intent is to allow personnel to make an immediate but orderly shutdown, so repairs to the electrical system can be made. The installation must be made in accordance with the following:

1. If your electrical system is an AC ungrounded system operating between 120 V and 1000 V, as permitted in Art. 250.21(A)(1) through (A)(4) above, you must install ground detectors on your system.
2. The ground detectors must be installed as close as practicable to the power supply of your system.

The 2014 edition of the Code made a small change in this section to clarify the verbiage regarding the 50- to 1000-V specification.

250.21(C). Marking. If you have an ungrounded AC system, you must durably and legibly mark the source and/or first disconnect as:

"Caution Ungrounded System Operating _____ Volts Between Conductors."

The 2014 edition of the Code made a change in this section regarding the required verbiage to be marked on ungrounded systems.

250.22. CIRCUITS NOT TO BE GROUNDED

Circuits are the part of an electrical system downstream from the transformer, the disconnect panels, and the last OCPD (circuit breaker). A 120-V outlet would be part of a circuit. The Code uses the word "circuit" in this section; however, when you go through the accompanying articles, they often mean that the entire electrical system shall not be grounded.

The five areas are as follows:

1. Electric cranes operating over combustible fibers (NEC Art. 503.155)
2. Health-care facilities where anesthesia is used (NEC Arts. 517.61 and 517.160)
3. Working zones for electrolytic cells (NEC Art. 668)
4. Secondary lighting systems operating under 30 V (NEC Art. 411.5 (A))
5. Certain Underwriters Laboratory (UL)-listed transformers for underwater luminaries (NEC Art. 680.23(A)(2))

250.24. GROUNDING SERVICE-SUPPLIED ALTERNATING-CURRENT SYSTEMS

This section tells us how and where to bond the neutral (grounded conductor) wire to ground what is coming from the electrical utility company (utility side of your electrical meter). This includes outdoor transformers, dual-fed systems, power coming from raceways, delta-connected services, overhead and underground services, and high-impedance services. Typically, only the utility company (power company) will deal with the issues in this section.

250.24(A). System Grounding Connections. The power for your building or premises is either separately derived (comes from solar panels, wind turbines, generators, etc.) or is supplied by the utility company. It is also almost certainly a grounded service (for ungrounded power service see Arts. 250.21 and 250.22). This section of the Code tells us that you must have at each service a grounding electrode conductor connected to the neutral (grounded) service conductor. In other words, you must bond the neutral (grounded conductor) wire coming in from the utility to ground for each service. This is sometimes called a neutral (grounded conductor)-to-ground bond.

Grounding (earthing) electrode conductor The conductor used to connect the grounding (earthing) electrode(s) to the equipment-grounding (bonding) conductor, to the neutral (grounded) conductor, or to both in accordance with Arts. 250.66 and 250.142.

Service conductor, grounded The current-carrying service conductor coming off the low side of the transformer that is to be, or has been, bonded to the ground (grounded). In other words, the neutral wire coming from the utility into a main electrical panel.

Grounded conductor (neutral) A system or circuit conductor that is intentionally grounded. This is a current-carrying conductor typically called the *neutral wire*.

250.24(A)(1). General. The grounding electrode conductor must be made accessible to at the point where overhead service conductors, service drops, underground service conductors, and/or service lateral(s) meet the first service disconnect.

This section tells us that the neutral (grounded conductor) wire coming in from the utility must be bonded to ground at the first service disconnect. This neutral-to-ground connection almost always occurs directly in the panel at the neutral (grounded conductor) bar. The main neutral (grounded conductor) wire from the utility will be bonded to the neutral (grounded conductor) bar in the electrical cabinet. Typically, a screw or jumper wire bonds the chassis of the electrical panel directly to the neutral (grounded conductor) bar. There must also be a conductor (the grounding electrode conductor) bonding the neutral (grounded conductor) bar directly to the earthing electrode.

First service disconnect The very first electrical panel that will disconnect (turn off) the power coming in from the utility company. Sometimes this term is used to describe the first electrical panel that can turn off power on the secondary or low-voltage side of a transformer.

This part of the Code also allows some flexibility in where this neutral (grounded conductor)-to-ground bond can occur. While it is highly unlikely and not recommended, the bond could occur in the utility power meter enclosure or at the top of the masthead (service drop).

The 2014 edition of the Code added overhead and underground service conductors to the requirements found in this section.

250.24(A)(2). Outdoor Transformer. The outdoor transformer that supplies power to your premises, must have a grounding electrode with a wire (a grounding electrode conductor) bonded to the neutral (grounded conductor) or XO terminal of the transformer. This is true for single-phase and three-phase transformers, wye or delta type (see Fig. 5.3).

Exception no. 1: High-impedance grounded neutral (grounded conductor) systems that have an additional ground impedance, usually a resistor or inductor, do not need to meet this rule. These systems must be in compliance with Art. 250.36. Please see Art. 250.36 for more information.

250.24(A)(3). Dual-fed Services. When the power coming into your premises is from two (dual) separate transformers (feeds), a single grounding electrode conductor may be used. This is true if the feeds go into a common enclosure or are grouped together in separate enclosures with a secondary tie.

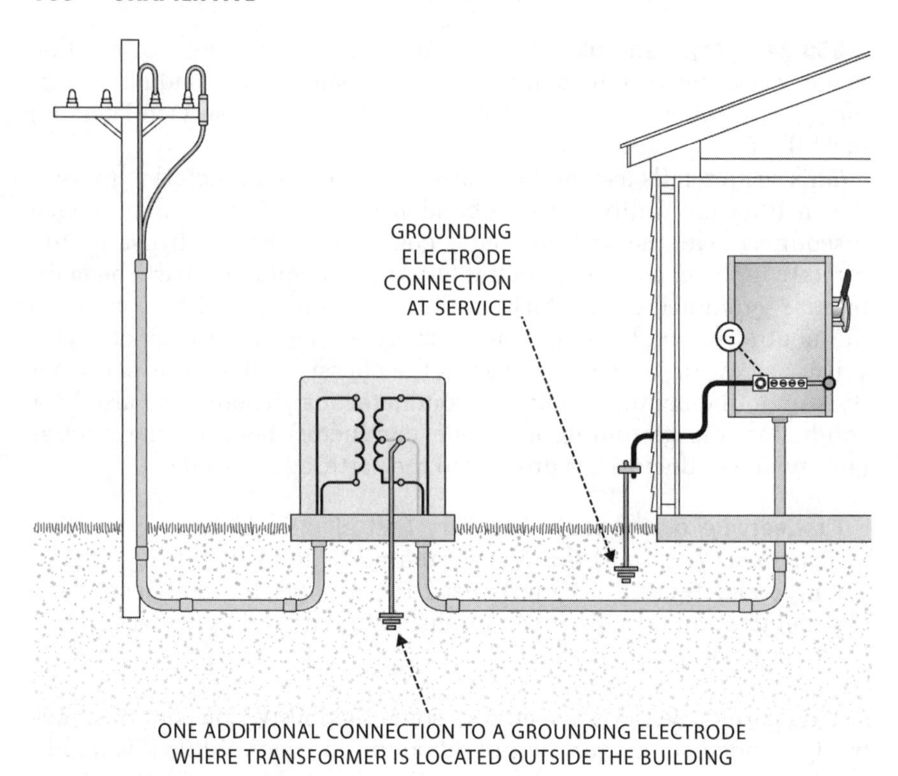

Figure 5.3 Required transformer and service disconnect grounding.

Dual-fed service An electrical service wherein the utility company uses two different transformers and two separate service feeds to supply power to a premises.

Grounding (earthing) electrode conductor The conductor used to connect the grounding (earthing) electrode(s) to the equipment-grounding (bonding) conductor, to the neutral (grounded) conductor, or to both in accordance with Arts. 250.66 and 250.142.

In other words, if the power company uses two transformers and brings two sets of wires into your electrical panel, you only need one grounding electrode conductor from the neutral (grounded conductor) bar to the earthing electrode. If your primary electrical panel has a series of cabinets with bus bars that take each phase and the neutral (grounded conductor) across all the cabinets (as in an industrial setting), you only need one grounding electrode conductor from the neutral (grounded conductor) bar to the earthing electrode, even if the power company brings in two sets of wires from two different transformers.

250.24(A)(4). Main Bonding Jumper as Wire or Bus Bar. This section simply tells us that if the neutral (grounded conductor) bar in your electrical cabinet is bonded to the ground bar (via a main bonding jumper), you may connect the grounding electrode conductor to the ground bar instead of to the neutral (grounded conductor) bar.

Main bonding jumper A conductor, screw, or strap that bonds the neutral (grounded) conductor to a grounding conductor of some type. The term *main bonding jumper* is confusingly used in the Code in various places to indicate both a *supply-side bonding jumper* and a *system bonding jumper* (load side). For more details, see Arts. 250.24(A)(4), 250.24(B), 250.28, and 408.3(C).

Grounding (earthing) electrode conductor The conductor used to connect the grounding (earthing) electrode(s) to the equipment-grounding (bonding) conductor, to the neutral (grounded) conductor, or to both in accordance with Arts. 250.66 and 250.142.

250.24(A)(5). Load-side Grounding Connections. This is one of the most important sections of the Code. This section and Art. 250.142 prohibit you from connecting the neutral (grounded conductor) wire to ground at any other place in the entire electrical system, other than the main bonding jumper. See Art. 250.24(B) below. This means that you may not reground the neutral (grounded conductor) wire anywhere else or use the neutral (grounded conductor) wire to ground an item downstream (the load side) of the first service disconnect. You must bond the neutral (grounded conductor) wire to ground at the transformer and in the first service disconnect panel, but in no other place. The goal is to prevent neutral (grounded conductor) currents from traveling on the chassis of equipment and/or through the green wire ground.

Neutral-to-ground connection (bond) is where a grounded current-carrying conductor, the neutral, is intentionally bonded to earth/ground. In general, this should only occur at the transformer and at the first service disconnect(s). The Code uses several names for neutral-to-ground connections: system bonding jumper, main bonding jumper, and supply-side bonding jumper.

Main bonding jumper A conductor, screw, or strap that bonds the neutral (grounded) conductor to a grounding conductor of some type. The term *main bonding jumper* is confusingly used in the Code in various places to indicate both a *supply-side bonding jumper* and a *system bonding jumper* (load side). For more details, see Arts. 250.24(A)(4), 250.24(B), 250.28, and 408.3(C).

While the entire neutral (grounded conductor)-to-ground bond scenario can be very confusing, it is extremely important that it is done correctly. Imagine a small automobile repair shop with a single-phase, 240-V main electrical panel for the entire building and one secondary electrical panel for the garage. The garage panel is fed from the main electrical panel via two hot wires, one neutral (grounded conductor) wire and a ground wire. The main electrical panel has the neutral (grounded conductor)-to-ground bond located inside.

If an additional neutral (grounded conductor)-to-ground bond were to be erroneously installed in the garage panel, neutral (grounded conductor) currents would travel back to the main electrical panel through both the neutral (grounded conductor) wire and the green ground wire, and of course, through the chassis of electrical panels and via metal conduit. Ideally, we want all of the neutral (grounded conductor) currents to travel back to the main electrical panel via the neutral (grounded conductor) wire only.

Improper neutral-to-ground connections are a very serious safety issue. Not only can they place neutral (grounded conductor) currents on to the chassis of electrical equipment, raceways, and conduits, where they can electrocute people, but they can cause heating and fire. Additionally, they will generate objectionable currents and electrical noise that can result in electronic data loss, and cause ground-fault circuit interrupters (GFCIs) to fail to function properly (trip to open for no apparent reason). GFCI breakers will sense the difference in current between the neutral (grounded conductor) wire and the hot wires (only 10 mA or less) and will immediately disconnect the power.

See Art. 250.142 for more additional information on this subject.

250.24(B). Main Bonding Jumpers. This section tells us that if you have multiple disconnecting systems at your premises, you must have a neutral (grounded conductor)-to-ground bond in each of the disconnects.

In other words, if you have a small apartment building with four apartments and four power meters being fed by a single common bus system (the power company only brought in a single set of wires for all four electrical meters), you must have a neutral (grounded conductor)-to-ground bond in each of the four first service disconnects. You may not make the neutral (grounded conductor)-to-ground bond above the four meters. A shortcut some electricians have made in the past is to make a single neutral-to-ground connection at the bust system above the meter to prevent the need for the connection in each of the four boxes; a practice that is not allowed according to Art. 250.24(B).

Main bonding jumper A conductor, screw, or strap that bonds the neutral (grounded) conductor to a grounding conductor of some type. The term *main bonding jumper* is confusingly used in the Code in various places to indicate both a *supply-side bonding jumper* and a *system bonding jumper* (load side). For more details, see Arts. 250.24(A)(4), 250.24(B), 250.28, and 408.3(C).

Neutral-to-ground connection (bond) is where a grounded current-carrying conductor, the neutral, is intentionally bonded to earth/ground. In general, this should only occur at the transformer and at the first service disconnect(s). The Code uses several names for neutral-to-ground connections: system bonding jumper, main bonding jumper, and supply-side bonding jumper.

250.24(C). Grounded Conductor Brought to Service Equipment. This section of the Code applies to grounded systems (wye-type) of 1000 V or less, and simply tells us that the utility must run the neutral (grounded conductor) wire from the transformer to the first service disconnect, even if you do not plan to use the neutral (grounded conductor) for the load. For example, in a three-phase wye-type service in which you will only have three-phase loads, you still must have a neutral (grounded conductor) wire routed in the conduit from the transformer to the panel, and that neutral (grounded conductor) must have a neutral (grounded conductor)-to-ground bond (be bonded to the metal chassis of the cabinet and to the earthing electrode).

In other words, if your transformer is three-phase, four-wire, you must make your system three-phase, four-wire throughout, even if you are only using three-phase loads. For example, if your service is three-phase, four-wire, 208-V, and you will NOT have a 120-V circuit at all, you must still run a neutral (grounded conductor) in the conduit from the transformer to the panel. This neutral (grounded conductor) must be bonded to the chassis and to a grounding electrode, just like all other neutrals (grounded conductors) in first service disconnect cabinets.

This is extremely important, as the neutral (grounded conductor) provides the fault-current path should an electrical fault occur. Imagine for a moment, a simple maintenance building with nothing but three-phase, 208-V loads in it. Located a few feet outside the building is the utility transformer, which is connected to the building via an underground PVC conduit. Consider what would happen during an electrical fault if the utility were to mistakenly run only three conductors, one for each phase, into the building, and forgot to run the neutral (grounded conductor). If one of the three-phase machines in the shop were to fault, the electrical current would travel back to the panel via the metal conduit and/or the green ground wire. Once the fault current is in the panel, it would need to travel down the grounding electrode, through the earth (soil) itself, back up the transformer's XO terminal ground rod, to the phase coil, and back to the panel in the shop, in order to trip the breaker. As we learned in Arts. 250.4(A)(5) and 250.4(B)(4), you may not use the earth as a conductor for fault currents. The earth (soil) is simply not an effective enough conductor of electricity to ensure that enough current will pass through it to cause the breaker to trip.

In fact, in the scenario above, you could consider the earth as a resistor, and theoretically conduct a Wenner four-point soil resistivity test to calculate the value of that resistor. With that resistor in the circuit, would the fault current that reaches the circuit breaker be enough to ensure that it trips and shuts off the power?

The problem is solved, of course, by installing the neutral (grounded conductor) wire, which is the point of Art. 250.24(C).

This part of the Code also allows you to run a single neutral (grounded conductor) conductor for two or more loads, with a few requirements. Primarily, the voltage of the AC service must be less than 1000 V, and the first service disconnects for each of the loads must be located in a single assembly. Of course, as we learned in Art. 250.24(B), each of the disconnects must have a main bonding jumper. There are some additional requirements that will be discussed below in Art. 250.21(C)(1) to (4).

First service disconnect The very first electrical panel that will disconnect (turn off) the power coming in from the utility company. Sometimes this term is used to describe the first electrical panel that can turn off power on the secondary or low-voltage side of a transformer.

Main bonding jumper A conductor, screw, or strap that bonds the neutral (grounded) conductor to a grounding conductor of some type. The term *main bonding jumper* is confusingly used in the Code in various places to indicate both a *supply-side bonding jumper* and a *system bonding jumper* (load side). For more details, see Arts. 250.24(A)(4), 250.24(B), 250.28, and 408.3(C).

Unintentional neutral current path Where an inappropriately applied neutral-to-ground connection has established a parallel path for neutral conductor currents to travel back to the source on the ground conductors and/or metal enclosures of an electrical system.

The 2014 edition of the Code made a small change in this section to clarify the verbiage regarding the 1000-V or less specification.

250.24(C)(1). Sizing for a Single Raceway. This is a very simple section of the Code that simply states the neutral (grounded conductor) wire will be sized in accordance with Table 250.102(C) (see Appendix B) when dealing with a single raceway.

The notes for Table 250.102(C) (see Appendix B) state that the neutral (grounded conductor) does not need to be larger than the largest phase conductor (ungrounded conductor) for the service. In addition, if your phase conductors are larger than 1100 kcmil copper or 1750 kcmil aluminum, the neutral (grounded conductor) shall be at least 12½ percent of the circular mil area of the largest phase conductor (ungrounded conductor). In other words, the neutral (grounded conductor) wire must be sized to meet code.

> **Raceway** An enclosed channel of metal or nonmetallic material designed expressly for holding wires, cables, bus bars, and so on. Raceways include all types of conduit, flexible tubing, floor raceways, busways, and so on.

The 2014 edition of the Code changed the reference to the new table: Table 250.102(C) (see Appendix B).

250.24(C)(2). Parallel Conductors in Two or More Raceways. This section of the Code is very similar to Art. 250.24(C)(1), in that it has the same requirements, with the exception that when dealing with multiple raceways, the neutral (grounded conductor) cannot be less than 1/0 AWG for each raceway and must be installed in parallel with the ungrounded conductors (the phase conductors). See NEC Art. 310.10 (H) for more information. Figure 5.4 illustrates the proper neutral-to-ground bond for parallel conductors feeding multiple disconnects.

Keep in mind that when the NEC talks about *parallel conductors*, it is referring to when two or more conductors are electrically joined at both ends, so the multiple conductors are in essence treated as a single conductor.

You will typically see this for a large industrial electrical service in which the utility needs to provide a 1000-A, 208-V, three-phase, four-wire service. To do this, you will need at least three large conduits running from the transformer into the building's first service disconnect. so you can route three 500-kcmil cables for each phase and the neutral into the building. Each 500-kcmil cable can handle 380 A, so three of these conductors will give you 1140 A of capacity, easily meeting the 1000 A required. This section of the Code tells us that you must group the neutral wire in the same conduit as the phase wires. You may not route all the neutrals in one conduit and all the phase wires in the other conduits. In fact, you must group one L1, L2, L3, and the neutral together, and route this grouping in each of the conduits.

It is very important to include the neutral wire in the conduit, particularly in single-phase systems, as failing to do so can actually cause inductive-phase heating of the metallic conduits. The neutral wire helps to cross-cancel the magnetic fields that form in the hot wires as the current flows through the conductors. Without the neutral being routed in the metal conduit, you can heat the metal conduit to the point that it will actually glow red-hot! Routing the neutral separately from the hot wires is illegal under this section of the Code, but it is also illegal in other areas as well. Please see NEC Arts. 215.4(B), 300.3(B), and 300.20(A).

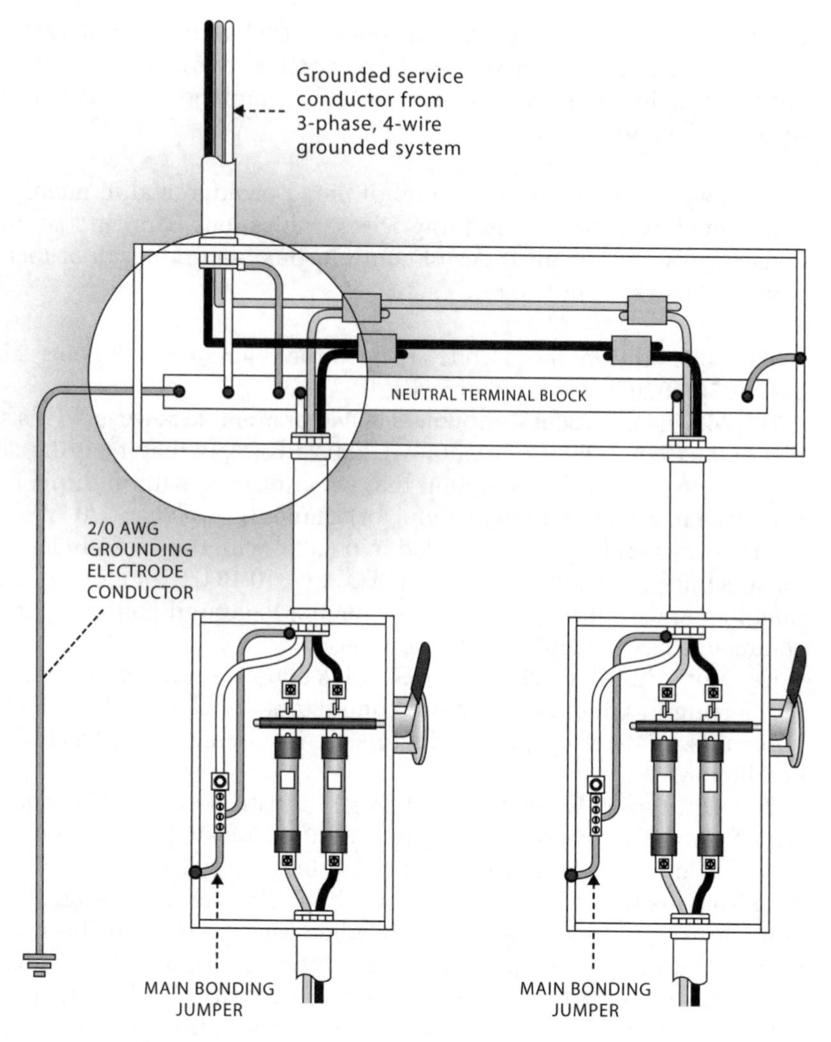

Figure 5.4 Bonding jumper requirements for multiple enclosures.

The bottom line is that the neutral wire must be grouped with the hot wires. Not only will this reduce inductive heating of adjacent metal objects, but it also reduces the generation of electrical noise, reduces accidental power outages, and of course, prevents fires.

250.24(C)(3). Delta-connected Service. With a delta-type transformer, the neutral wire (grounded conductor) must be same as the phase conductors (ungrounded conductors).

250.24(C)(4). High Impedance. When dealing with a high-impedance grounded neutral system, you must be in accordance with Art. 250.36.

Basically, the neutral (grounded conductor) must be sized to handle the new current levels that can be seen due to the impedance that has been added. Remember, high-impedance, grounded, neutral electrical systems add an impedance, usually a resistor, between the XO terminal of the transformer and the grounding electrode to limit the maximum possible ground-fault current that can be seen on the neutral conductor (grounded conductor). This means that the neutral conductor only needs to be sized to match the maximum likely fault current, which will be greatly reduced due to the added impedance. However, the neutral may not be less than 8 AWG copper or 6 AWG aluminum.

250.24(D). Grounding Electrode Conductor. This section simply tells us that you shall have a grounding electrode conductor connect the equipment-grounding conductors, the service equipment enclosures (chassis of cabinets, panels, etc.), and the neutral (grounded service conductor), where applicable. This must be done in compliance with Part III of Art. 250. And of course, impedance-grounded neutral systems must be constructed as covered by Art. 250.36.

Grounding (earthing) electrode conductor The conductor used to connect the grounding (earthing) electrode(s) to the equipment-grounding (bonding) conductor, to the neutral (grounded) conductor, or to both in accordance with Arts. 250.66 and 250.142.

Equipment-grounding (bonding) conductor The low-impedance fault-current path that is typically run with or encloses the circuit conductors, used to connect the non-current-carrying metal parts of equipment, raceways, and other enclosures to the grounded (neutral) conductor and equipment-grounding (bonding) conductor at service equipment or at the source of a separately derived system. Often called the *green-wire* or *third-wire* ground conductor.

The 2014 edition of the Code removed a reference to Art. 250.24(A).

250.24(E). Ungrounded System Grounding Connections. This section of the Code tells us that even for ungrounded systems (such as an ungrounded delta transformer and/or as discussed in Art. 250.21 or 250.22), you must have a grounding electrode conductor and a grounding electrode installed in compliance with Part III of Art. 250. You must connect the grounding electrode conductor to the metal enclosure of the service conductors on the load side of the service drop, service lateral, underground service conductor, and/or overhead service conductor.

In other words, you still must bond the metal chassis of ungrounded electrical systems at the first service disconnect. Again, in the NEC, an ungrounded system is an electrical service that does not use a neutral (grounded conductor) as part of the power supply system.

Grounded system An electrical service that has one of the current-carrying conductors intentionally grounded (i.e., the neutral wire).

Floating or delta-type electrical systems are not grounded systems, whereas wye-type systems are grounded.

Ungrounded system An electrical service that does not have any of the current-carrying conductors intentionally grounded (i.e., no neutral wire). Delta-type electrical systems and some standby generators are examples of ungrounded systems.

The 2014 edition of the Code added the overhead and underground service conductors to the text in this section.

250.26. CONDUCTOR TO BE GROUNDED—ALTERNATING-CURRENT SYSTEMS

This section tells us what conductor is to be grounded or which current-carrying conductor is to be the neutral. This section works in conjunction with Art. 250.20(B).

This is a pretty common-sense section and will seldom be needed, as the manufacturer of the transformer supplying power will almost certainly dictate which of the current-carrying conductors is to become your neutral (grounded conductor).

Grounded conductor (neutral) A system or circuit conductor that is intentionally grounded. This is a current-carrying conductor typically called the *neutral wire*.

Here is a list of which conductor is to be grounded as determined by the electrical system type:

1. Single-phase, two-wire: one conductor is to be grounded
2. Single-phase, three-wire: the neutral conductor is to be grounded
3. Three-phase, four-wire, wye-type: the neutral conductor is to be grounded
4. Three-phase, four-wire, delta-type: one phase conductor is to be grounded
5. Multiphase system, where one phase is used as a neutral: the neutral conductor is to be grounded

The 2014 edition of the Code changed the term *common conductor* to *neutral conductor*.

250.28. MAIN BONDING JUMPER AND SYSTEM BONDING JUMPER

This section of the Code is one of the longer and more complex parts of the NEC. Confusion resulting from a poor understanding of how the one or two mandatory neutral-to-ground bonds must occur in the system is arguably one of the mistakes electricians and engineers make in designing/building an electrical service. We will try to remove some of that confusion while still detailing the Code requirements.

To start with, you really need to understand that the Code uses two terms to discuss the same basic engineering function, which is to tie the neutral conductor to ground at the transformer, and when there is no equipment-grounding conductor between the transformer and the service, a neutral-to-ground connection must occur at the first service disconnect as well.

This is often called the *neutral-to-ground connection*, a term NOT used in the NEC, but commonly used by electricians and engineers. The two terms used by the NEC are *main bonding jumper* and *system bonding jumper*, which are the same thing, other than one occurs at the transformer and the other occurs at the first service disconnect, respectively.

We should note that the Code also breaks out the system bonding jumper into an additional term, the *supply-side bonding jumper* in Art. 250.28(D)(2). This is to help distinguish an additional neutral-to-ground bonding requirement when dealing with multiple disconnects being fed from a single electrical source. See Art. 250.28(D)(2) for more information.

Main bonding jumper A conductor, screw, or strap that bonds the neutral (grounded) conductor to a grounding conductor of some type. The term *main bonding jumper* is confusingly used in the Code in various places to indicate both a *supply-side bonding jumper* and a *system bonding jumper* (load side). For more details, see Arts. 250.24(A)(4), 250.24(B), 250.28, and 408.3(C).

System bonding jumper A conductor, screw, or strap that bonds the equipment-grounding conductor to the neutral (grounded) conductor; sometimes called the *main bonding jumper*. The term *system bonding jumper* is used to distinguish the bond at the load side of the service (at the first service disconnect) from the *supply-side bonding jumper*, which is the bonding used on the utility or supply side of the service. For more details, see Arts. 250.24(A)(4), 250.24(B), 250.28, and 408.3(C).

Supply-side bonding jumper A conductor installed on the supply (utility) side of a service, or within a service equipment enclosure(s), or for a separately derived system, that ensures the required electrical conductivity between metal parts required to be electrically connected; sometimes called a *main bonding jumper*. A conductor, screw, or strap that bonds the neutral (grounded) conductor to the service equipment enclosures on the supply side. The term *supply-side bonding jumper* is used to distinguish the bond at the supply (utility) side of the service from the *system bonding jumper*, which is used on the load side of the service (at the first service disconnect). For more details, see Arts. 250.24(A)(4), 250.24(B), 250.28, 250.28(D)(2), and 408.3(C).

Note *The NEC occasionally uses this term to indicate an equipment-grounding conductor (see Art. 250.30).*

Neutral-to-ground connection (bond) is where a grounded current-carrying conductor, the neutral, is intentionally bonded to earth/ground. In general, this should only occur at the transformer and at the first service disconnect(s). The Code uses several names for neutral-to-ground connections: system bonding jumper, main bonding jumper, and supply-side bonding jumper.

Equipment-grounding (bonding) conductor The low-impedance fault-current path that is typically run with or encloses the circuit conductors, used to connect the non-current-carrying metal parts of equipment, raceways, and other enclosures to the grounded (neutral) conductor and equipment-grounding (bonding) conductor at service equipment or at the source of a separately derived system. Often called the *green-wire* or *third-wire* ground conductor.

Article 250.8 is going to tell us the types of materials we can use, the wire sizes, and attachment methods, and where to make the bond in a number of different scenarios. The rule of thumb regarding the neutral-to-ground connection is that there must be a neutral-to-ground bond at the transformer (main bonding jumper) and another one at each of the first service disconnects that are supplied by the transformer (system bonding jumper). The remaining portions of this section will simply get into a few details regarding the proper way of how to do this.

Article 250.28 states that for grounded electrical systems, you shall install a main bonding jumper and a system bonding jumper, per this section.

Grounded system An electrical service that has one of the current-carrying conductors intentionally grounded (i.e., the neutral wire). Floating or delta-type electrical systems are not grounded systems, whereas wye-type systems are grounded.

Ungrounded system An electrical service that does not have any of the current-carrying conductors intentionally grounded (i.e., no neutral wire). Delta-type electrical systems and some standby generators are examples of ungrounded systems.

250.28(A). Material. This section says that the bonding jumpers (both the main bonding jumper and the system bonding jumper) shall be made of copper or some other corrosion-resistant material. It also states that these jumpers shall be a wire, a bus, a screw, or similar suitable conductor.

> **Note** *Aluminum, lead, tin, zinc, magnesium, iron and steel are generally considered NOT to be corrosion resistant. Copper, brass, nickel (passive), stainless steel (all types), brass (all types), titanium, molybdenum, platinum, silver, and gold ARE considered to be corrosion resistant.*

250.28(B). Construction. This section states that if the bonding jumper (main bonding jumper and/or system bonding jumper) is a screw, the screw must be identified with a green finish that is visible when the screw is installed. The green screw is a favorite among electricians. However, if the green screw is lost, you must replace it with another green screw. You cannot use just any standard machine screw.

250.28(C). Attachment. This section says that you must make the two bonding jumpers (both the main bonding jumper and the system bonding jumper) in accordance with Art. 250.8.

Article 250.8 tells us that you may use machine screws and thread-forming machine screws. But you may not use drywall screws, wood screws, sheet-metal screws, or regular solder (silver solder is generally considered acceptable due to its higher melt point). These devices must be listed for this function and evaluated by a product-testing organization, such as the UL.

250.28(D). Size. Bonding jumpers (main bonding jumpers and system bonding jumpers) should be sized according to the rules outlined in the following three sections.

250.28(D)(1). General. This section of the Code simply states that the bonding jumpers (main bonding jumpers and system bonding jumpers) will be sized in accordance with Table 250.102(C) (see Appendix B). Remember that Table 250.102(C) (see Appendix B) says that when the supply conductors (phase conductors) are larger than 1100 kcmil copper or 1750 kcmil aluminum, the bonding jumpers (main bonding jumpers and system bonding jumpers) shall be at least 12½ percent of the circular mil area of the largest phase conductor (ungrounded conductor).

In a grounded electrical system (as most are), the bonding jumpers (main bonding jumpers and system bonding jumpers) must be able to handle the full ground-fault current. If you experience a phase-to-ground fault, in which the full electrical energy from a phase wire hits the chassis of a metal enclosure or conduit, the bonding jumpers (main bonding jumpers and system bonding jumpers) will be the path the electrical-fault energy must take in order to reach the circuit breakers or fuses. This means that in very-large-capacity electrical services, the neutral-to-ground bond (main bonding jumpers and system bonding jumpers) may actually be larger than the grounding electrode conductor.

Unlike the grounding electrode conductor, which must be sized to conduct current to the earth, the bonding jumpers (main bonding jumpers and system bonding jumpers) are directly in the ground fault–current path, so they must be able to handle the fault currents.

> **Grounding (earthing) electrode conductor** The conductor used to connect the grounding (earthing) electrode(s) to the equipment-grounding (bonding) conductor, to the neutral (grounded) conductor, or to both in accordance with Arts. 250.66 and 250.142.

The 2014 edition of the Code changed the reference to the new table: Table 250.102(C) (see Appendix B).

250.28(D)(2). Main Bonding Jumper for Service with More Than One Enclosure. When you have an electrical service (a grounded electrical service) that feeds more than one first service disconnect (or enclosure), you must have bonding jumpers (main bonding jumpers and system bonding jumpers) in each of the enclosures. This section of the Code tells us that the bonding jumpers (main bonding jumpers and system bonding jumpers) must be sized to match the phase wires that feed each enclosure.

In other words, if the utility is feeding your building with phase wires that are 500-kcmil conductors to a main bus system that in turn feeds several disconnect enclosures (first service disconnects), the supply-side bonding jumper (service equipment enclosure on the supply side of the disconnects) must be sized according to the 500 kcmil, which would be a 2/0 AWG conductor (see Table 250.66 in Appendix B). The system bonding jumpers (first service disconnect side) must be sized according to the largest phase wire in each enclosure. So, if the first of the disconnects is fed with 4/0 AWG wire, the system bonding jumper must be 4/0 AWG. If the remaining disconnects are fed by 2/0 AWG wires, then the system bonding jumpers must be 2/0 AWG. In Fig. 5.5 we see how the grounding electrode conductor must be sized according to the incoming phase conductor for each individual service.

The proper sizing of all these conductors can be found in Table 250.66 (see Appendix B).

Of note is that even in the NEC *Handbook*, the various terms for the name *bonding jumpers* is often confused. This can be seen in the illustrations, where system bonding jumpers are labeled as main bonding jumpers or vice versa. The name of these neutral-to-ground bonds is not important; what is important is that each equipment enclosure on the supply side (enclosures above the first service disconnects and the first service disconnects themselves) must have neutral-to-ground bonds that are sized to match the largest phase conductor in that enclosure.

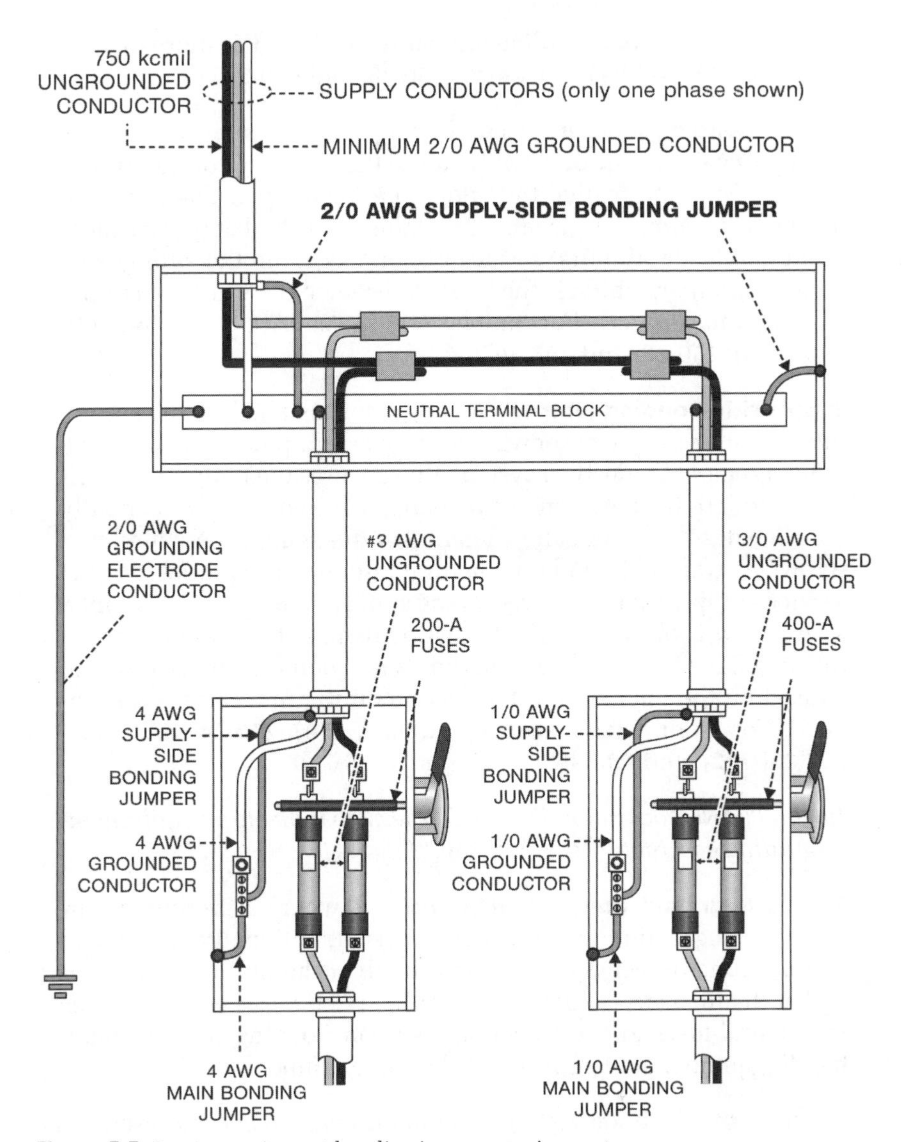

Figure 5.5 Service equipment-bonding jumper requirements.

First service disconnect The very first electrical panel that will disconnect (turn off) the power coming in from the utility company. Sometimes this term is used to describe the first electrical panel that can turn off power on the secondary or low-voltage side of a transformer.

Main bonding jumper A conductor, screw, or strap that bonds the neutral (grounded) conductor to a grounding conductor of some type. The term *main bonding jumper* is confusingly used in the Code in various places to indicate both a *supply-side bonding* *jumper* and a *system bonding jumper* (load side). For more details, see Arts. 250.24(A)(4), 250.24(B), 250.28, and 408.3(C).

System bonding jumper A conductor, screw, or strap that bonds the equipment-grounding conductor to the neutral (grounded) conductor; sometimes called the *main bonding jumper.* The term *system bonding jumper* is used to distinguish the bond at the load side of the service (at the first service disconnect) from the *supply-side bonding jumper*, which is the bonding used on the utility or supply side of the service. For more details, see Arts. 250.24(A)(4), 250.24(B), 250.28, and 408.3(C).

Supply-side bonding jumper A conductor installed on the supply (utility) side of a service, or within a service equipment enclosure(s), or for a separately derived system, that ensures the required electrical conductivity between metal parts required to be electrically connected; sometimes called a *main bonding jumper.* A conductor, screw, or strap that bonds the neutral (grounded) conductor to the service equipment enclosures on the supply side. The term *supply-side bonding jumper* is used to distinguish the bond at the supply (utility) side of the service from the *system bonding jumper*, which is used on the load side of the service (at the first service disconnect). For more details, see Arts. 250.24(A)(4), 250.24(B), 250.28, 250.28(D)(2), and 408.3(C).

Note *The NEC occasionally uses this term to indicate an equipment-grounding conductor (see Art. 250.30).*

Neutral-to-ground connection (bond) is where a grounded current-carrying conductor, the neutral, is intentionally bonded to earth/ground. In general, this should only occur at the transformer and at the first service disconnect(s). The Code uses several names for neutral-to-ground connections: system bonding jumper, main bonding jumper, and supply-side bonding jumper.

Just to clarify, the Code uses at least three different names to describe the various neutral-to-ground connections (bonds): main bonding jumper, system bonding jumper, and supply-side bonding jumper. All three are just different names for neutral-to-ground bonds that must occur at different locations in your electrical system. Frankly, one can argue what the proper name of a given neutral-to-ground bond should be, and to many engineers and electricians the Code appears to use these proper names differently at alternating points throughout the Code. However, the proper name is irrelevant. What is important is that the neutral-to-ground bonds be installed in the right locations and sized properly to handle the fault currents. See Art. 250.28 for more information.

An additional requirement of the Code under this section (which is not mentioned in the Code itself, but is discussed at great length in the NEC *Handbook*) is that the metallic conduits containing the phase conductors (service conductors) coming from the transformer must be bonded to ground (via grounding bushings). In NEC terminology, this is the metal conduit that contains the service entrance conductors, sometimes called the *service laterals*. This bonding jumper is also called a supply-side bonding jumper and must be sized per Table 250.102(C) (see Appendix B), which appears only in the 2014 edition of the Code. You will note that the sizing in the last line of the table is now determined by the table notes in the 2014 code. If you are referencing from an earlier version of the Code, you must size this supply-side bonding jumper in accordance with Table 250.66 (see Appendix B) via a reference from Art. 250.102(C).

Additionally, the number of allowable disconnects under any one service is limited by the requirements of Art. 230.71(A), which basically says that you must be able to turn all the power off to any premises by throwing no more than six switches. See Art. 230.71(A) for more information.

250.28(D)(3). Separately Derived System with More Than One Enclosure. This section of the Code tells us that electrical service coming from separately derived systems (generators, solar panels, etc.), must comply with Art. 250.28(D)(1). Now, you would think that the Code would simply state that separately derived systems with more than one enclosure must comply with Art. 250.28(D)(2), which would be simple and easy, but no. Just to make this more confusing, the Code decided to give you two choices. You, of course, may comply with Art. 250.28(D)(2), or you can simply provide a single bonding jumper (a main bonding jumper) at the source.

Separately derived system(s) A premises wiring system whose power is derived from a source of electrical energy other than a (utility) service. Examples include solar panels, generators, and wind turbines that have no direct connection to another electrical source. Specifically, the neutral conductor must not be solidly bonded to the neutrals of other sources. This term is also used in regard to isolation and/or step-down transformers that generate a distinct power source separate from the utility.

In order to use the single bonding jumper option, you must evaluate the largest phase conductor that yields the largest collective circular mil area, in order to figure out the size of the single neutral-to-ground bonding jumper. This is determined by the largest sum of the areas of the corresponding conductors of each set and compared to Table 250.66 (see Appendix B).

In other words, for the first option, you simply compare the largest phase conductor coming from the separately derived system against

Table 250.66 (see Appendix B). If your largest phase conductor is a 3/0 AWG copper wire, you will find from Table 250.66 (see Appendix B) that you need a #4 AWG copper or a #2 AWG aluminum. But this means that you must install an individual main bonding jumper (#4 AWG copper or a #2 AWG aluminum) for each enclosure between the first service disconnects and the separately derived source.

The second method is to add the area of all the phase conductors (not the neutral) together and then refer to Table 250.66 (see Appendix B). For example, if the separately derived system is supplying three-phase, four-wire power via four 3/0 AWG copper conductors, you simply add the circular mils (cross-sectional area) of the three-phase wires. So, a single 3/0 conductor has a circular mil of 167,800 (see NEC, Chap. 9, Table 9-8). You have three phase wires, so 167,800 × 3 = 503,400 circular mils, or 503 kcmil. If you look at Table 250.66 (see Appendix B), you will see that 503 kcmil require a 1/0 AWG copper or a 3/0 AWG aluminum main bonding jumper.

So, in the first option, you must install a #4 AWG copper bonding jumper in every enclosure between the separately derived source and the first service disconnect. In the second option, you only need to install a single 1/0 AWG copper main service bonding jumper at the neutral terminal of the transformer (XO terminal) of the separately derived source.

250.30. GROUNDING SEPARATELY DERIVED ALTERNATING-CURRENT SYSTEMS

Separately derived grounded electrical systems must comply with Art. 250.30(A) and ungrounded systems must comply with Art. 250.30(B). Additionally, separately derived systems must comply with Art. 250.20, 250.21, 250.22, or 250.26, as applicable. This is also true for systems with connected multiple, parallel, separately derived systems .

This section of the Code is one of the longest and most complex of Art. 250. The primary purpose of this part of the Code is to instruct you on how to properly install the neutral conductors, neutral-to-ground bonds, and the grounding systems (equipment-grounding conductors) required for backup generators (separately derived AC system).

To start with, you have a choice in how your backup generator is to be installed. You may either install it as *separately derived system* (generator) or as a *not separately derived system* (generator). The difference in the two systems is all in how the neutral is handled at the transfer switch.

Transfer switch An automatic or nonautomatic device for transferring one or more load conductor connections from one power source to another. See Art. 250.30.

In a separately derived generator, the transfer switch is a four-pole system, in that it not only transfers the three-phase conductors from one power source to the other (from the utility power to the generator), but it also transfers the neutral (grounded conductor). This means that you must have a neutral-to-ground bond at the XO terminal of the generator with a grounding electrode installed nearby. In Fig 5.6 we the complete wiring with grounding for an electrical system with an emergency generator connected via a 4-pole transfer switch.

In a not separately derived generator, the transfer switch is a three-pole system, in that it only transfers the three-phase conductors from one power source to the other and has a neutral bar to solidly bond both neutrals together (the service utility neutral and the generator's neutral are both tied to a neutral bar in the cabinet). This means that you may NOT have a neutral-to-ground bond at the generator. The neutral-to-ground bond for a not separately derived generator is located in the service equipment enclosure. A second neutral-to-ground bond (at the not separately derived generator) would allow neutral currents during normal operation to return through the equipment-grounding conductor and the metal conduit of the system. This is sometimes called an *unintentional neutral current path*.

In Fig 5.7 we the complete wiring with grounding for an electrical system with an emergency generator connected via a 3-pole transfer switch.

Unintentional neutral current path Where an inappropriately applied neutral-to-ground connection has established a parallel path for neutral conductor currents to travel back to the source on the ground conductors and/or metal enclosures of an electrical system.

Separately derived system(s) A premises wiring system whose power is derived from a source of electrical energy other than a (utility) service. Examples include solar panels, generators, and wind turbines that have no direct connection to another electrical source. Specifically, the neutral conductor must not be solidly bonded to the neutrals of other sources. This term is also used in regard to isolation and/or step-down transformers that generate a distinct power source separate from the utility.

Not separately derived system(s) A premises wiring system whose power is derived from a source of electrical energy other than a (utility) service. Examples include solar panels, generators, and wind turbines that have a direct connection to another electrical source. Specifically the neutral conductor is solidly bonded to the neutrals of another source.

Neutral-to-ground connection (bond) is where a grounded current-carrying conductor, the neutral, is intentionally bonded to earth/ground. In general, this should only occur at the transformer and at

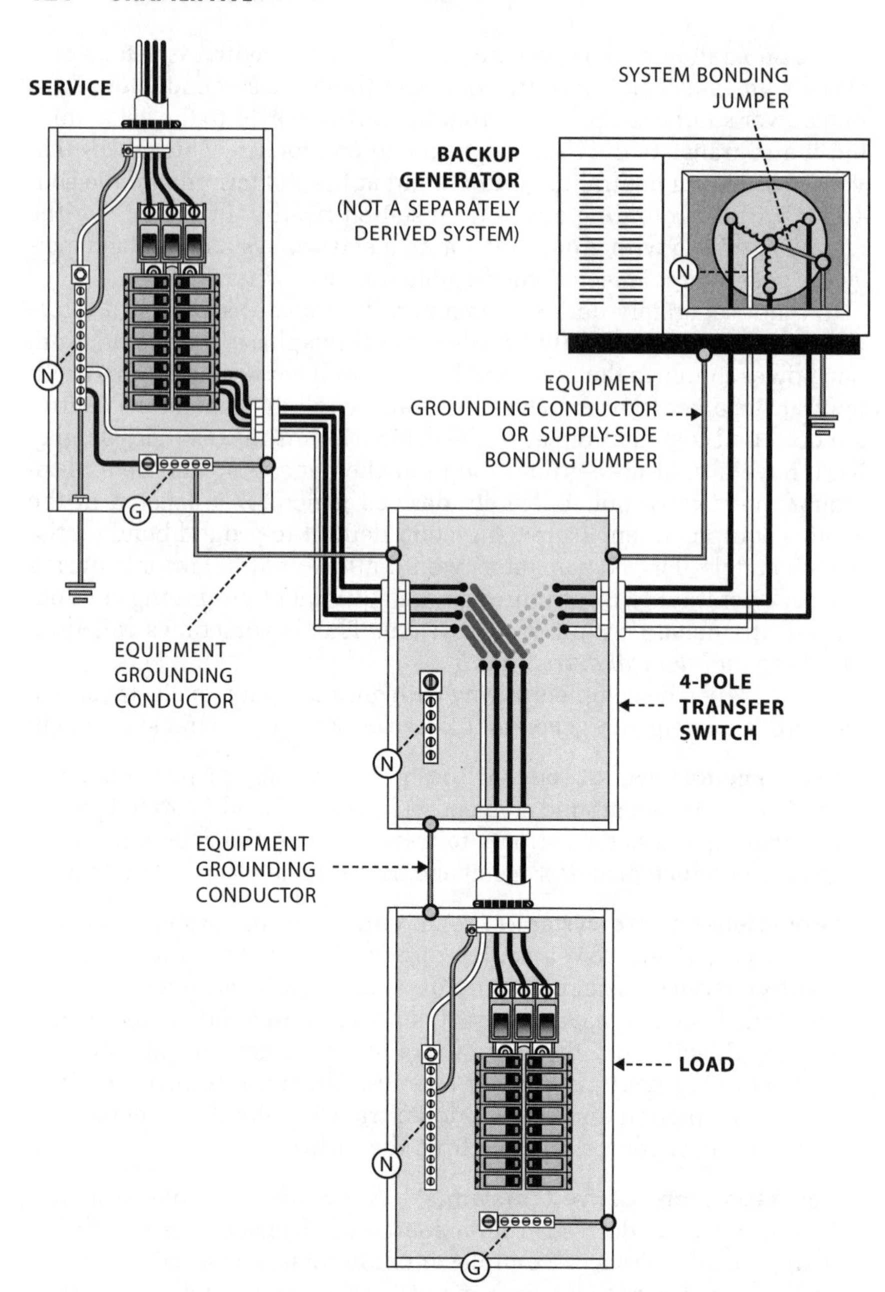

Figure 5.6 Proper wiring for a 4-pole transfer switch.

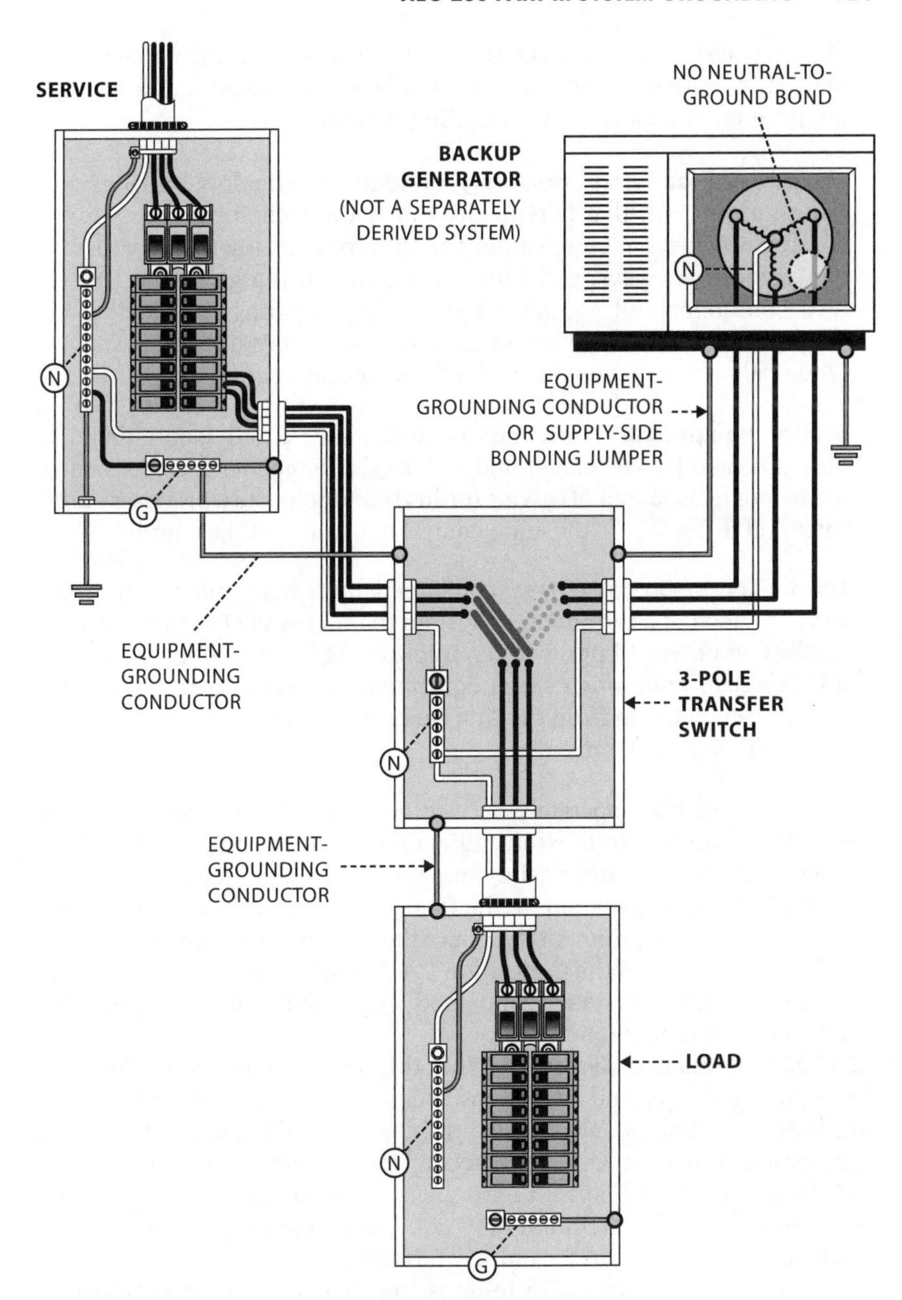

Figure 5.7 Proper wiring for a 3-pole transfer switch.

the first service disconnect(s). The Code uses several names for neutral-to-ground connections: system bonding jumper, main bonding jumper, and supply-side bonding jumper.

Equipment-grounding (bonding) conductor The low-impedance fault-current path that is typically run with or encloses the circuit conductors, used to connect the non-current-carrying metal parts of equipment, raceways, and other enclosures to the grounded (neutral) conductor and equipment-grounding (bonding) conductor at service equipment or at the source of a separately derived system. Often called the *green-wire* or *third-wire* ground conductor.

Service equipment The necessary equipment, usually consisting of circuit breaker(s), switch(es), and/or fuse(s) and their accessories, connected to the load end of service (utility) conductors to a premises and intended to constitute the main control and cutoff of the supply.

Load The electrical energy consumed by a component, circuit, device, piece of equipment, or system that is connected to a source of electric power to perform its function. This term often refers to all electrical components and equipment that consume electrical energy downstream from the first service disconnect. It can also be called an *electric load.*

In both cases (the separately derived generator and the not separately derived generator), you must install equipment-grounding conductors (a properly sized green ground wire) from the service equipment to the transfer switch, from the generator to the transfer switch, and from the transfer switch to the load equipment (typically the primary electrical panel.

The 2014 edition of the Code added verbiage to confirm that multiple separately derived systems connected in parallel must also meet the rules found in Art. 250.30.

250.30(A). Grounded Systems. What this section tells us is when you have separately derived AC system (such as a transformer or a generator, but also solar panels, wind turbines, etc.), the neutral-to-ground connection must be installed according to code, specifically Art. 250.28(A) to (D). A neutral-to-ground connection must occur at each separately derived system but may not occur in not separately derived systems. See Art. 205.30 for more information.

This part of the Code also tells us that if you have an application wherein the utility supply is 480-V power, and you have a step-down delta-to-wye transformer to generate 208/120-V power, you will commonly need to have a neutral-to-ground connection at the low side of the transformer, as the electrical power it generates will be a separately derived system.

Separately derived system(s) A premises wiring system whose power is derived from a source of electrical energy other than a (utility) service. Examples include solar panels, generators, and wind turbines that have no direct connection to another electrical source. Specifically, the neutral conductor must not be solidly bonded to the neutrals of other sources. This term is also used in regard to isolation and/or step-down transformers that generate a distinct power source separate from the utility.

Not separately derived system(s) A premises wiring system whose power is derived from a source of electrical energy other than a (utility) service. Examples include solar panels, generators, and wind turbines that have a direct connection to another electrical source. Specifically the neutral conductor is solidly bonded to the neutrals of another source.

Neutral-to-ground connection (bond) is where a grounded current-carrying conductor, the neutral, is intentionally bonded to earth/ground. In general, this should only occur at the transformer and at the first service disconnect(s). The Code uses several names for neutral-to-ground connections: system bonding jumper, main bonding jumper, and supply-side bonding jumper.

Please note that for impedance-grounded neutral systems, the grounding connections should be made in accordance with Art. 250.36 or 250.186, as applicable. Please see Arts. 250.36 and 250.186 for more information.

Impedance grounded neutral system An electrical system wherein a grounding impedance, usually a resistor or an inductor, is added between the neutral point (XO terminal) of the transformer and the grounding electrode to limit the amperage of fault currents.

And, of course, this section does not apply to ungrounded electrical systems.

250.30(A)(1). System Bonding Jumper. This section tells us that for separately derived systems, a single neutral-to-ground connection must be installed and that it must be an unspliced connection and in compliance with Art. 250.28 (A) to (D), specifically 250.28(D)(3). This means that the bonding jumper must primarily comply with Table 205.66 (see Appendix B). See Art. 250.28(D)(3) for more information.

Additionally, this section tells us that for separately derived systems, you have a choice as to where to install the System Bonding Jumper. You may have it at either the source (Art. 250.30(A)(1)(a)) or at the first service disconnect (Art. 250.30(A)(1)(b)).

If you choose to have the neutral-to-ground connection at the source, the system bonding jumper must occur within the enclosure of the separately derived system (where the power source originates). This neutral-to-ground connection must be bonded to a grounding electrode located as close as possible to the system, per Art. 250.30(C).

If you choose to have the neutral-to-ground connection at the first service disconnect (or at the overcurrent device), the system bonding jumper must occur within the enclosure of the first service disconnect. This neutral-to-ground connection must be bonded to a grounding electrode located as close as possible to the system, per Art. 250.30(C).

In both scenarios, you must have an equipment-grounding conductor bonding the chassis of the separately derived system to the first service disconnect. See Art. 250.30 (2) for more information.

Equipment-grounding (bonding) conductor The low-impedance fault-current path that is typically run with or encloses the circuit conductors, used to connect the non-current-carrying metal parts of equipment, raceways, and other enclosures to the grounded (neutral) conductor and equipment-grounding (bonding) conductor at service equipment or at the source of a separately derived system. Often called the *green-wire* or *third-wire* ground conductor.

Overcurrent device Usually a fuse or circuit breaker, a device that limits the maximum amount of current that can flow in a circuit. Sometimes called an *overcurrent protection device* or *OCPD.*

Main bonding jumper A conductor, screw, or strap that bonds the neutral (grounded) conductor to a grounding conductor of some type. The term *main bonding jumper* is confusingly used in the Code in various places to indicate both a *supply-side bonding jumper* and a *system bonding jumper* (load side). For more details, see Arts. 250.24(A)(4), 250.24(B), 250.28, and 408.3(C).

System bonding jumper A conductor, screw, or strap that bonds the equipment-grounding conductor to the neutral (grounded) conductor; sometimes called the *main bonding jumper.* The term *system bonding jumper* is used to distinguish the bond at the load side of the service (at the first service disconnect) from the *supply-side bonding jumper,* which is the bonding used on the utility or supply side of the service. For more details, see Arts. 250.24(A)(4), 250.24(B), 250.28, and 408.3(C).

Supply-side bonding jumper A conductor installed on the supply (utility) side of a service, or within a service equipment enclosure(s), or for a separately derived system, that ensures the required electrical conductivity between metal parts required to be electrically connected; sometimes called a *main bonding jumper.* A conductor, screw, or strap that bonds the neutral (grounded) conductor to the service equipment enclosures on the supply side. The term *supply-side bonding jumper* is used to distinguish the bond at the supply (utility) side of the service from the *system bonding jumper*, which is used on the load side of the service (at the first service disconnect). For more details, see Arts. 250.24(A)(4), 250.24(B), 250.28, 250.28(D)(2), and 408.3(C).

Note *The NEC occasionally uses this term to indicate an equipment-grounding conductor (see Art. 250.30).*

Neutral-to-ground connection (bond) is where a grounded current-carrying conductor, the neutral, is intentionally bonded to earth/ground. In general, this should only occur at the transformer and at the first service disconnect(s). The Code uses several names for neutral-to-ground connections: system bonding jumper, main bonding jumper, and supply-side bonding jumper.

Just to clarify, the Code uses at least three different names to describe the various neutral-to-ground connections (bonds): main bonding jumper, system bonding jumper, and supply-side bonding jumper. All three are just different names for neutral-to-ground bonds that must occur at different locations in your electrical system. Frankly, one can argue what the proper name of a given neutral-to-ground bond should be, and to many engineers and electricians the Code appears to use these proper names differently at differing points throughout the Code. However, the proper name is irrelevant. What is important is that the neutral-to-ground bonds be installed in the right locations and are sized properly to handle the fault currents. See Art. 250.28 for more information.

There are three exceptions in the Code for this section.

Exception no. 1: When you have systems with secondary ties (see NEC Art. 450.6), a single system bonding jumper is permitted to tie each power source.

Exception no. 2: You may have a system bonding jumper at both the source (transformer) and the first service disconnect, as long as it does not establish a parallel path for the neutral conductor (grounded conductor). This is sometimes called an *unintentional neutral current path*, and it is not allowed. The system bonding jumper must be sized per Table 250.102(C) (see Appendix B) and does not need to be larger than the largest phase conductor. This exception does not consider the earth to be a parallel path.

The 2014 edition of the Code reworded this exception to improve clarity.

Exception no. 3: The third and final exception is for class 1, class 2, or class 3 circuits, in which the transformer is rated for less than 1000-VA (example: 120-V, 8-A transformer), the system bonding jumper shall not be smaller than the phase wire, or at least a 14 AWG copper or 12 AWG aluminum, whichever is larger.

a. The system bonding jumper must connect the neutral (grounded conductor) to the supply-side bonding jumper and the metal enclosure, at the source.
b. The system bonding jumper must connect the neutral (grounded conductor) to the supply-side bonding jumper and the metal enclosure, at the first service disconnect.

Unintentional neutral current path Where an inappropriately applied neutral-to-ground connection has established a parallel path for neutral conductor currents to travel back to the source on the ground conductors and/or metal enclosures of an electrical system.

250.30(A)(2). Supply-side Bonding Jumper. This section is where the Code really confuses people, as it should be properly titled "Equipment-grounding Conductor." If we look back into the Code at Art. 250.28, it uses the term *supply-side bonding jumper* to indicate a neutral-to-ground connection. However, in this part of the Code it uses the term to indicate a required equipment-grounding conductor. To clarify, Art. 250.30(2) is NOT talking about a neutral-to-ground connection but is discussing a mandatory green-wire ground connection.

Equipment-bonding jumper or load-side bonding jumper A conductor, screw, or strap that bonds the equipment-grounded conductor (neutral) to the equipment enclosures that are on the load side of the first service disconnect enclosure(s), in accordance with Art. 250.102(D). The term is also used to indicate a conductor, screw, or strap that bonds the equipment-grounding conductor of a circuit to the enclosure or raceway of that circuit, in accordance with Art. 250.146.

What this section of the Code tells us is that when your separately derived system and the first service disconnect are in separate enclosures, you must have a green-wire ground run in the conduit with the circuit conductors. It also states that the green-wire ground is not required to be larger than the phase conductors and that you can use the nonflexible metal raceway (or wire- or bus-type) as the green-wire ground (the equipment-grounding conductor) under the following conditions:

a. Wire-type conductors used to bond the disconnect to the separately derived system (supply-side bonding jumper) must comply with Table 102(C) (see Appendix B).
b. Bus-type conductors used to bond the disconnect to the separately derived system (supply-side bonding jumper) must have a cross-sectional area not smaller than the equivalent wire-type conductor, in accordance with Table 102(C) (see Appendix B).

Raceway An enclosed channel of metal or nonmetallic material designed expressly for holding wires, cables, bus bars, and so on. Raceways include all types of conduit, flexible tubing, raceways, floor raceways, busways, and so on.

Exception no. 1: A supply-side bonding jumper as discussed in this section is not required if the enclosures are installed in accordance with Art. 250.30(A)(1), Exception no. 2.

The 2014 edition of the Code changed this section to reference Table 250.102(C) (see Appendix B).

250.30(A)(3). Grounded Conductor. This section of the Code is discussing the neutral conductor (grounded conductor), which is the ground fault–current path, for separately derived systems. See NEC Art. 220.61 for calculating the maximum neutral load.

Ground fault-current path An electrically conductive path from the point of a ground fault on a wiring system through normally non-current-carrying conductors, equipment, or the earth to the electrical supply source. A ground fault–current path may not use the earth/soil as a conductor or current path, as it is a violation of the NEC Arts. 250.4(A)(5) and 250.4(B)(4). This term is used for both intentional and unintentional current paths.

This section tells us that for separately derived systems that do not have the neutral-to-ground connection installed in the enclosure at the source (installed at the first service disconnect), you have some additional requirements. See Art. 205.30(A)(1), which gives you a choice as to where to install the neutral-to-ground connection: either at the source or at the first service disconnect. This section of the Code applies when you install the neutral-to-ground bond at the first service disconnect.

a. For single raceways, the neutral wire (grounded conductor) must be sized according to Table 250.102(C) (see Appendix B) and is not required to be larger than the largest phase conductor (ungrounded conductor).

b. For parallel conductors in two or more raceways, the neutral must be installed parallel with the phase conductors and must be at least 1/0 AWG copper in size. In both cases, for sets of phase conductors that are larger than 1100 kcmil copper or 1750 kcmil aluminum, the neutral conductor (grounded conductor) shall be at least 12½ percent the circular mil size of the largest phase conductor.

Raceway An enclosed channel of metal or nonmetallic material designed expressly for holding wires, cables, bus bars, and so on. Raceways include all types of conduit, flexible tubing, raceways, floor raceways, busways, and so on.

Ungrounded conductor A current-carrying conductor that has not been grounded. A phase conductor is an ungrounded conductor.

Grounded conductor (neutral) A system or circuit conductor that is intentionally grounded. This is a current-carrying conductor typically called the *neutral wire*.

This section of the Code is very similar to Arts. 250.24(C)(1) and 250.24(C)(2), in that it has the same requirements for sizing the neutral conductor (grounded conductor) and when dealing with multiple raceways, the neutral (grounded conductor) cannot be less than 1/0 AWG for each raceway and must be installed in parallel with the ungrounded conductors (the phase conductors). See NEC Art. 310.10 (H) for more information.

For delta-connected systems, the neutral wire must be equal in size to the phase conductors. Specifically, the Code states that the ampacity of the neutral wire (grounded conductor) must be equal to that of the phase conductors (ungrounded conductors).

Ampacity The maximum amount of electrical current a conductor or device can carry before sustaining immediate or progressive deterioration. Ampacity is also described as current rating or current-carrying capacity. Ampacity is the RMS electric current that a device or conductor can continuously carry while remaining within its temperature rating.

For impedance-grounded neutral systems, the neutral wire (grounded conductor) must be installed according to Art. 250.36 or 250.186, as applicable.

The 2014 edition of the Code changed this section to reference Table 250.102(C) (see Appendix B).

250.30(A)(4). Grounding Electrode. This section of the Code tells us that for separately derived systems (such as a transformer, generator, solar array, wind turbine, etc.), you must have a grounding electrode installed as close as practicable to the system. The goal is to keep the grounding electrode conductor (the wire that bonds the equipment to the electrode) as short as possible. Short conductors have lower impedance than long conductors, and when dealing with ground faults, higher impedance creates greater hazardous voltages on the entire system.

Grounding (earthing) electrode A device that establishes an electrical connection to the earth/soil.

Grounding (earthing) electrode conductor The conductor used to connect the grounding (earthing) electrode(s) to the equipment-grounding (bonding) conductor, to the neutral (grounded) conductor, or to both in accordance with Arts. 250.66 and 250.142.

This part of the Code also tells us that the electrode system for a separately derived system shall consist of any metal water pipes as defined in Art. 250.52(A)(1) and any existing structural metal grounding electrode as specified in Art. 250.52(A)(2) in the vicinity of the system.

This means that if your separately derived source (transformer, generator, wind turbine, solar array, etc.) is located inside a building or structure with water, you must have a grounding electrode conductor routed from the separately derived source to the water pipe at a point within 5 ft of the entrance to the building/structure (see Art. 250.68(C)(1)), and you must bond to any existing grounding electrodes (see Art. 250.52(A)(2)). If you do not have a water pipe or an existing grounding electrode, you must install a grounding electrode system that is in compliance with Art. 250.52(A)(2).

If your separately derived source (transformer, generator, wind turbine, solar array, etc.) is located outside the building/structure that is being supplied power, you must install a grounding electrode system that is in compliance with Art. 250.52(A)(2). Of course, if for some reason your outside separately derived source has a water pipe, you must bond to it at a point within 5 ft of the entrance to the building/structure (see Art. 250.68(C)(1)).

Structural metal grounding electrode A structural metal member that establishes an electrical connection to the earth/soil as defined in Art. 250.52(A)(2). Specifically, the metal frame of a building or structure that is properly bonded to the steel rebar in the building's or structure's foundation (concrete-encased electrode without a vapor barrier), as required in Art. 250.52(A)(3).

An additional note *if you are installing an isolation transformer, even with the express intent of reducing electrical noise for computer systems, you must still follow the electrode requirements* *found in Art. 250.30(A)(4). However, if the isolation transformer is part of a power supply for a data-processing room and is specifically listed for such a use, it is exempt from Art. 250.30(A)(4) but must be grounded in accordance with the manufacturer's instructions.*

Listed or listed item An item, product, or device that has been specifically developed for a particular function and has been independently evaluated by a product-testing organization, such as the Underwriters Laboratory (UL) or the Conformité Européenne (CE).

250.30(A)(5). Grounding Electrode Conductor, Single Separately Derived System. This fairly short section of code tells us that the conductor (the grounding electrode conductor) that connects the grounding electrode to the separately derived source (transformer, solar array, etc.) must be sized according to Table 250.66 (see Appendix B) and must be bonded at the same point where the neutral-to-ground connection is made (see Art. 250.30(A)(1)). The electrode must, of course, be made in compliance with Art. 250.60(A)(4). Figure 5.8 illustrates the proper wiring and grounding required when installing a separately derived power source, such as an emergency generator.

Grounding (earthing) electrode conductor The conductor used to connect the grounding (earthing) electrode(s) to the equipment-grounding (bonding) conductor, to the neutral (grounded) conductor, or to both in accordance with Arts. 250.66 and 250.142.

Separately derived system(s) A premises wiring system whose power is derived from a source of electrical energy other than a (utility) service. Examples include solar panels, generators, and wind turbines that have no direct connection to another electrical source. Specifically, the neutral conductor must not be solidly bonded to the neutrals of other sources. This term is also used in regard to isolation and/or step-down transformers that generate a distinct power source separate from the utility.

Neutral-to-ground connection (bond) is where a grounded current-carrying conductor, the neutral, is intentionally bonded to earth/ground. In general, this should only occur at the transformer and at the first service disconnect(s). The Code uses several names for neutral-to-ground connections: system bonding jumper, main bonding jumper, and supply-side bonding jumper.

There are three exceptions in the Code for this section.

Exception no. 1: When your neutral-to-ground connection is a wire or bus bar, you may use the grounding terminal (ground bar) to

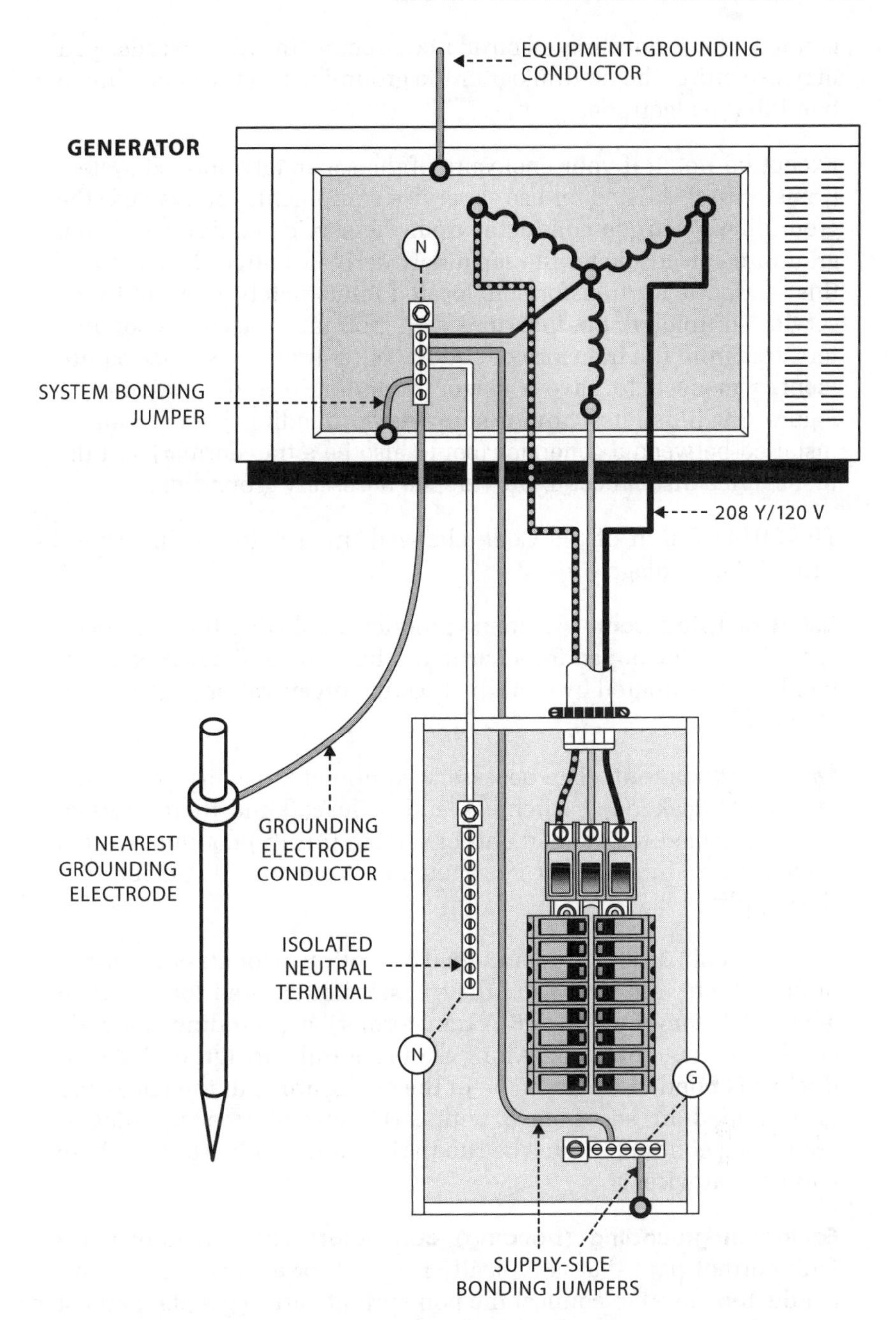

Figure 5.8 Grounding for a separately derived system with the grounding electrode conductor connection made at the separately derived system.

connect the grounding electrode conductor. In other words, you may use either the neutral bar or the ground bar in the enclosure to bond to the electrode.

Exception no. 2: If your enclosure of the separately derived system (transformer) is listed for use as service equipment, you may use the grounding electrode conductor from the service equipment as the grounding electrode for the separately derived source. The intent of this section is for transformers located immediately adjacent to the service equipment. So, instead of two grounding electrode conductors, one from the transformer and the other from the service equipment, you need to have a single grounding electrode conductor. Figure 5.9 illustrates how a supply-side bonding jumper can be installed between a generator (could also be a transformer) and the first service disconnect to provide the necessary grounding.

The 2014 edition of the Code changed this section to improve the clarity of the verbiage.

Listed or listed item An item, product, or device that has been specifically developed for a particular function and has been independently evaluated by a product-testing organization, such as the UL or CE.

Service equipment The necessary equipment, usually consisting of circuit breaker(s), switch(es), and/or fuse(s) and their accessories, connected to the load end of service (utility) conductors to a premises and intended to constitute the main control and cutoff of the supply.

Exception no. 3: The third and final exception is for class 1, class 2, or class 3 circuits for which the transformer is rated for less than 1000-VA (example: 120-V, 8-A transformer), a grounding electrode conductor is not needed, as long as the neutral wire (grounded conductor) is bonded to the frame of the enclosure, and the enclosure is grounded in accordance with Art. 250.134 (the equipment-grounding conductor must be run in the same conduit as the phase and neutral wires).

Equipment-grounding (bonding) conductor The low-impedance fault-current path that is typically run with or encloses the circuit conductors, used to connect the non-current-carrying metal parts of equipment, raceways, and other enclosures to the grounded (neutral) conductor and equipment-grounding (bonding) conductor at service equipment or at the source of a separately derived system. Often called the *green-wire* or *third-wire* ground conductor.

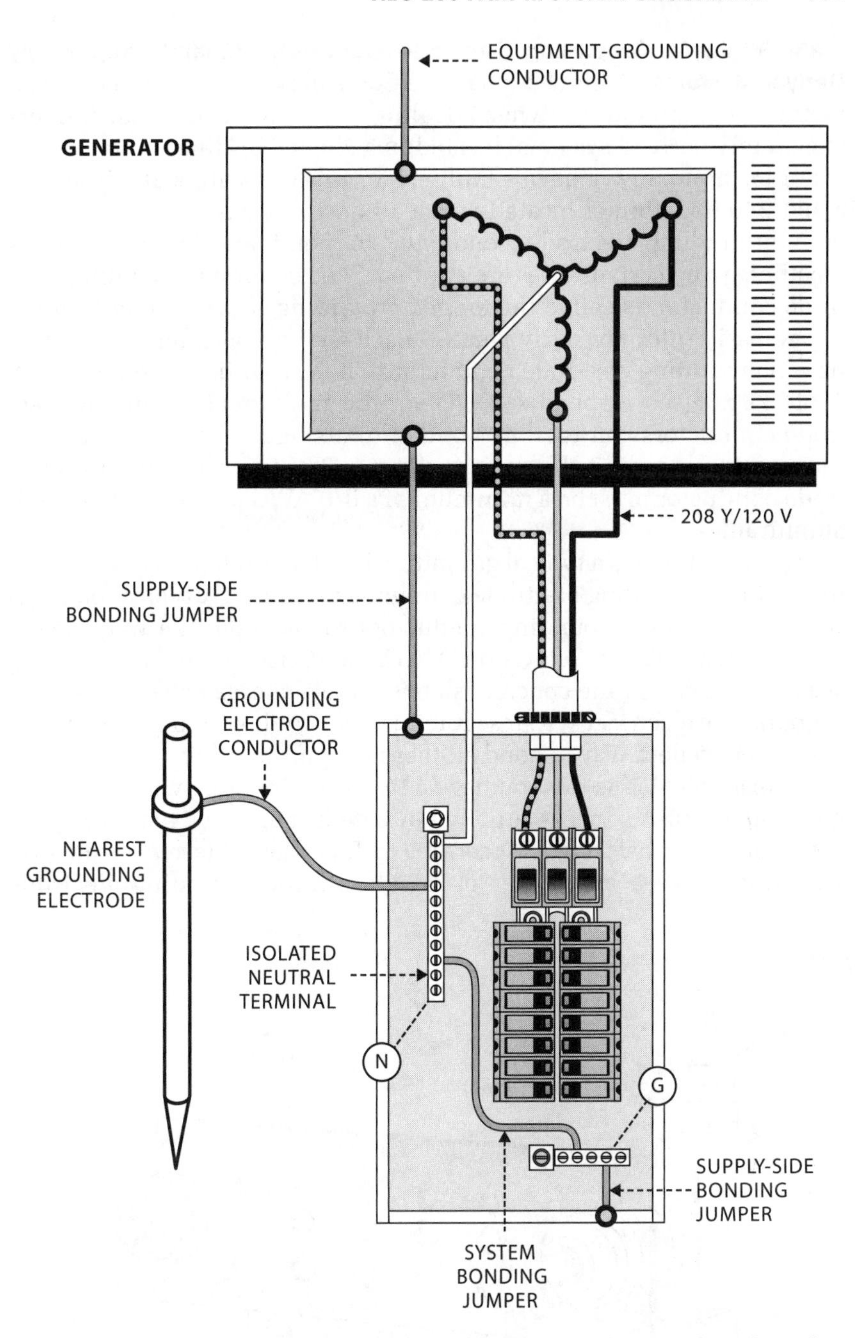

Figure 5.9 Grounding for a separately derived system with the grounding electrode conductor connection made at the first system disconnect.

250.30(A)(6). Grounding Electrode Conductor, Multiple Separately Derived Systems. This section of the Code allows us to install a logical method of grounding when dealing with multiple transformers (separately derived systems) located in a single building.

For example, if you have a multistory building, with a 480-V feed to a 480/240 transformer located on each floor, you can use a single common grounding electrode conductor to bond all the transformers together, given certain rules are applied. This common grounding electrode conductor, is called the *tapped grounding electrode conductor.*

The basic rules are fairly simple. Each individual transformer must have a grounding electrode conductor that is sized in accordance with Table 250.66(see Appendix B). Then, the individual grounding electrode conductors can then be bonded to a single common conductor, the tapped grounding electrode conductor. The tapped grounding electrode conductor must be a minimum of a 3/0 AWG copper or 250-kcmil aluminum.

Figure 5.10 shows a typical grounding bar that is often used to connect the various grounding electrode conductors together. A typical building will need to have grounding conductors routed from the water main, Telco systems, Cable Television (CATV) systems, gas mains, building steel, steel rebar in the concrete slab (concrete-encased electrodes), and of course the electrical systems grounding electrode conductor. A ground bar is an excellent way to bond all these systems together.

For example, let us say you have a two-story building with each floor having two 150-A services supplied by a dedicated 480/240 transformer, a total of four transformers. According to Table 250.66 (see Appendix B), you must have a #6 copper or #4 aluminum grounding electrode

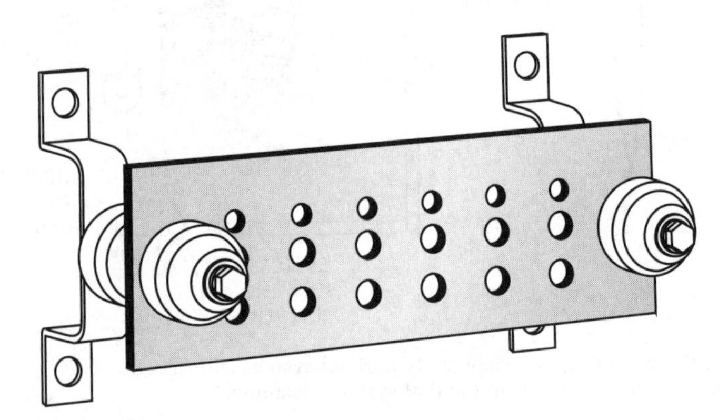

Figure 5.10 Typical copper ground bar.

conductor for each of these transformers. In the center of the building, you have a single 3/0 AWG copper or 250-kcmil aluminum conductor running from the top floor to the bottom floor, where the grounding electrode system is located. You may bond the four #6 copper or #4 aluminum conductors to the single 3/0 copper of 250 kcmil aluminum conductor. Figure 5.11 shows how a two story building with multiple electrical services and transformers can use a common grounding electrode conductor for each service to tap in to and provide the required earthing.

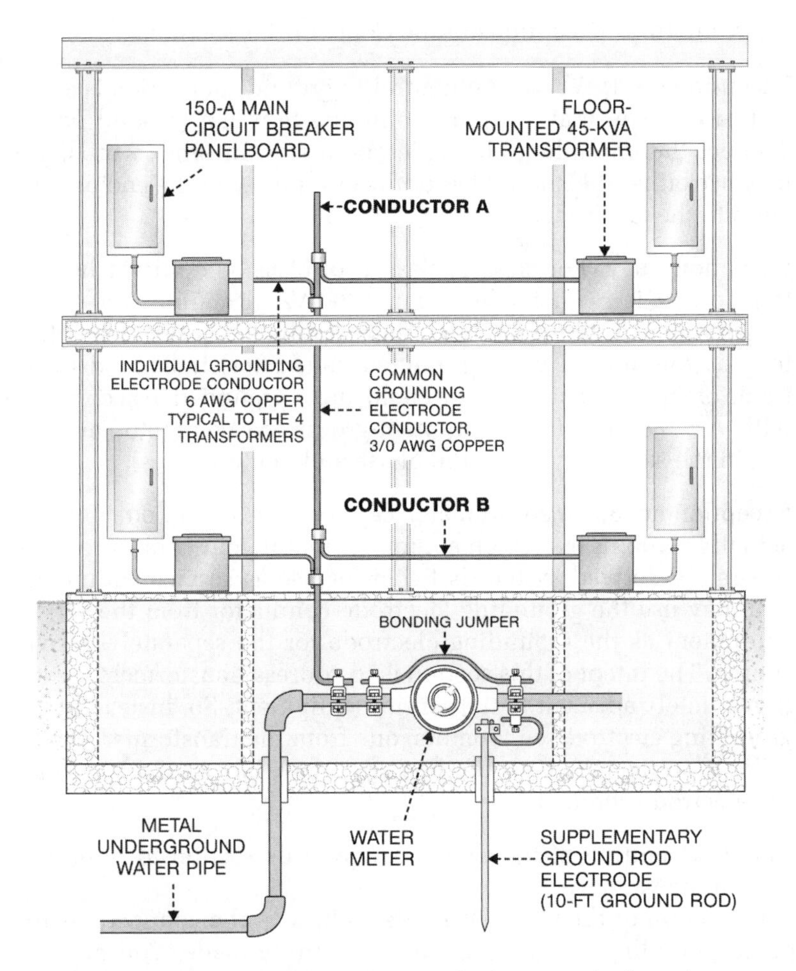

Figure 5.11 Grounding arrangement for multiple separately derived systems using a common grounding electrode conductor.

Tapped grounding electrode conductor A grounding electrode conductor sized in accordance with Art. 250.66 that is used to bond multiple transformers (separately derived systems) located in a single building to the grounding electrode system.

Grounding (earthing) electrode conductor The conductor used to connect the grounding (earthing) electrode(s) to the equipment-grounding (bonding) conductor, to the neutral (grounded) conductor, or to both in accordance with Arts. 250.66 and 250.142.

Grounding (earthing) electrode A device that establishes an electrical connection to the earth/soil.

There are three exceptions in the Code for this section:

Exception no. 1: When your neutral-to-ground connection is a wire or bus bar, you may use the grounding terminal (ground bar) to connect the grounding electrode conductor. In other words, you may use either the neutral bar or the ground bar in the enclosure to bond to the electrode.

Exception no. 2: For class 1, class 2, or class 3 circuits, when the transformer is rated for less than 1000-VA (example: 120-V, 8-A transformer), a grounding electrode conductor is not needed, as long as the neutral wire (grounded conductor) is bonded to the frame of the enclosure, and the enclosure is grounded in accordance with Art. 250.134 (the equipment-grounding conductor must be run in the same conduit as the phase and neutral wires).

Exception no. 3: An additional exception states (the Code does not actually give this exception a number) that if your enclosure of the separately derived system is listed for use as service equipment, you may use the grounding electrode conductor from the service equipment as the grounding electrode for the separately derived source. The intent of this section is to address transformers located immediately adjacent to the service equipment. So, instead of two grounding electrode conductors, one from the transformer and the other from the service equipment, you may have a single grounding electrode conductor.

The 2014 edition of the Code changed this section to improve the clarity of the verbiage.

This section of the Code also tells us how to the connections to the tapped grounding electrode conductor can be made. You have three choices: (1) You may use a connector listed for grounding and bonding of equipment. (2) You may bond to a copper bus bar that is at least ¼ in. × 2 in. (6 mm × 50 mm) in size and complies with Art. 250.64.

(3). You may use an exothermic welding process to bond the grounding electrode conductors together.

250.30(A)(7). Installation. This short section tells us that the installation of all grounding electrode conductors shall comply with Art. 250.64(A) through (E), which instructs us to protect the conductors from corrosion and physical damage and to ensure they are continuous, without a splice or joint. See Art. 250.64 for more information.

250.30(A)(8). Bonding. You are required to bond the neutral (grounded conductor) of a separately derived system to the structural steel and the metal piping of the building, in accordance with Art. 250.104(D). This is similar to the requirements found in many other areas of the Code.

250.30(B). Ungrounded Systems. Ungrounded systems (i.e., delta power) are to be grounded in a manner identical to the way we ground and bond grounded systems. Remember, the main difference between a grounded system and an ungrounded system is that there is no neutral-to-ground bond (connection) in ungrounded systems.

Ungrounded system An electrical service that does not have any of the current-carrying conductors intentionally grounded (i.e., no neutral wire). Delta-type electrical systems and some standby generators are examples of ungrounded systems.

Grounded system An electrical service that has one of the current-carrying conductors intentionally grounded (i.e., the neutral wire). Floating or delta-type electrical systems are not grounded systems, whereas wye-type systems are grounded.

Neutral-to-ground connection (bond) is where a grounded current-carrying conductor, the neutral, is intentionally bonded to earth/ground. In general, this should only occur at the transformer and at the first service disconnect(s). The Code uses several names for neutral-to-ground connections: system bonding jumper, main bonding jumper, and supply-side bonding jumper.

Ground fault-current path An electrically conductive path from the point of a ground fault on a wiring system through normally non-current-carrying conductors, equipment, or the earth to the electrical supply source. A ground fault–current path may not use the earth/soil as a conductor or current path, as it is a violation of the NEC Arts. 250.4(A)(5) and 250.4(B)(4). This term is used for both intentional and unintentional current paths.

The metal chassis of electrical equipment must have an effective ground-fault path, even for ungrounded electrical systems.

250.30(B)(1). Grounding Electrode Conductor. The grounding electrode conductor for ungrounded systems must be sized in accordance with Table 250.66 (see Appendix B) and must be connected to the metal chassis of the transformer's (separately derived systems) enclosure. If the transformer or other source of the separately derived system is located outdoors, it must comply with Art. 250.30(C).

> **Grounding (Earthing) Electrode Conductor** The conductor used to connect the grounding (earthing) electrode(s) to the equipment-grounding (bonding) conductor, to the neutral (grounded) conductor, or to both in accordance with Arts. 250.66 and 250.142.

250.30(B)(2). Grounding Electrode. This section of the Code says that for ungrounded separately derived systems, a grounding electrode is required, and it must be installed in accordance with Art. 250.30(A)(4). This means that the electrode must be installed as close as practicable to the system. The goal is to keep the grounding electrode conductor (the wire that bonds the equipment to the electrode) as short as possible. Short conductors have lower impedances than long conductors, and when dealing with ground faults, higher impedances create greater hazardous voltages on the entire system.

> **Grounding (earthing) electrode** A device that establishes an electrical connection to the earth/soil.

> **Grounding (earthing) electrode conductor** The conductor used to connect the grounding (earthing) electrode(s) to the equipment-grounding (bonding) conductor, to the neutral (grounded) conductor, or to both in accordance with Arts. 250.66 and 250.142.

This part of the Code also tells us that the electrode system for a separately derived system shall consist of any metal water pipes, as defined in Art. 250.52(A)(1), and any existing structural metal grounding electrode, as specified in Art. 250.52(A)(2), in the vicinity of the system.

This means that if your separately derived source (transformer, generator, wind turbine, solar arrays, etc.) is located inside a building or structure with water, you must have a grounding electrode conductor routed from the separately derived source to the water pipe at a point within 5 ft of the entrance to the building/structure (see Art. 250.68(C)(1)), and you must bond to any existing grounding electrodes (see Art. 250.52(A)(2)). If you do not have a water pipe or an existing grounding electrode, you must install a grounding electrode system that is in compliance with Art. 250.52(A)(2).

If your separately derived source (transformer, generator, wind turbine, solar array, etc.) is located outside the building/structure to which it is supplying power, you must install a grounding electrode system that is in compliance with Art. 250.52(A)(2). Of course, if for some reason your outside separately derived source has a water pipe, you must bond to it at a point within 5 ft of the entrance to the building/structure (see Art. 250.68(C)(1)).

Structural metal grounding electrode A structural metal member that establishes an electrical connection to the earth/soil as defined in Art. 250.52(A)(2). Specifically, the metal frame of a building or structure that is properly bonded to the steel rebar in the building's or structure's foundation (concrete-encased electrode without a vapor barrier), as required in Art. 250.52(A)(3).

An additional note is that if you are installing an isolation transformer, even with the express intent of reducing electrical noise for computer systems, you must still follow the electrode requirements found in Art. 250.30(A)(4). However, if the isolation transformer is part of a power supply for a data-processing room and is specifically listed for such a use, it is exempt from the Art. 250.30(A)(4) but must be grounded in accordance with the manufacturer's instructions.

Listed or listed item An item, product, or device that has been specifically developed for a particular function and has been independently evaluated by a product-testing organization, such as the UL or CE.

The exception for installing a grounding electrode for ungrounded separately derived systems is for vehicle-mounted and portable generators that meet the requirements found in Art. 250.34. Please see Art. 250.34 for more information.

250.30(B)(3). Bonding Path and Conductor. This section of the Code tells us that you must have an equipment-grounding conductor installed in accordance with Art. 250.30(A)(2) for all ungrounded systems (a mandatory green-wire ground connection).

Equipment-grounding (bonding) conductor The low-impedance fault-current path that is typically run with or encloses the circuit conductors, used to connect the non-current-carrying metal parts of equipment, raceways, and other enclosures to the grounded (neutral) conductor and equipment-grounding (bonding) conductor at service equipment or at the source of a separately derived system. Often called the *green-wire* or *third-wire* ground conductor.

When your separately derived ungrounded system and the first service disconnect are in separate enclosures, you must have a green-wire ground run in the conduit with the circuit conductors.

250.30(C). Outdoor Source. If you have a separately derived system (either grounded or ungrounded), you must have a grounding electrode installed as close as possible to the source and must bond to any structural steel and metal pipes, in accordance with Arts. 250.50 and 250.24(A)(2).

Of course, impedance-grounded neutral systems must comply with Art. 250.36 or 250.186 as applicable. Please see those sections for more information.

250.32. BUILDINGS OR STRUCTURES SUPPLIED BY A FEEDER(S) OR BRANCH CIRCUIT(S)

This section discusses proper grounding practices for situations wherein you have multiple buildings or structures being fed by a single utility service. You could think of a large home with three buildings: the home itself, a guesthouse in the back, and a separate garage, all being supplied with power from the home.

Figure 5.12 shows that when there are no metallic connections (no shared water pipe, gas pipe, electrical conduit, Telco ground lines, cable television (CATV) shield lines, etc.) between Building 1 and Building 2, you are not required to run a grounding conductor between the two buildings, but you must have a neutral-to-ground bond in each building. The figure also shows how to properly wire buildings that

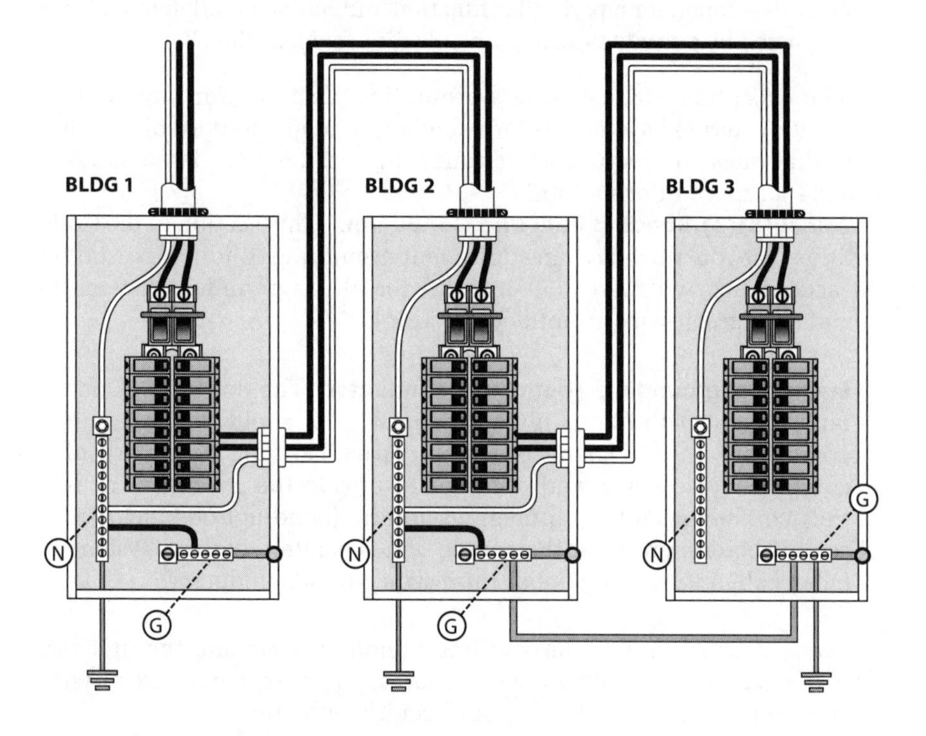

Figure 5.12 Neutral-to-ground bonding requirements for buildings 2 and 3 with power supplied by building 1.

may or may not have metallic connections. Buildings 2 and 3 shows that a grounding conductor must be routed between the two buildings, and most important it shows that Building 3 does not have a neutral-to-ground connection.

Circuit or branch circuit The electrical system downstream from the last OCPD. The outlet you plug your computer into would be part of a circuit.

Feeder circuit All circuit conductors between the first service disconnect and the final OCPD. The circuits after the final OCPD are called the *branch circuits*.

The most important thing to determine regarding the buildings or structures being supplied by feeders or branch circuits is whether or not there is a continuous metallic path between the structures. To be clear, you have two options: either you have a continuous metal path between the structures, or you do not. Whether you need an additional neutral-to-ground bond or not will depend on which scenario you have at your site.

A continuous metallic path between the buildings could be formed from metal conduit containing electrical circuits, a green-wire grounding conductor, water pipes, gas pipes, rebar in concrete, coax lines, telecommunication and data shield wires, structural building connections, fences, and any other metallic connection that could allow electrical energy to flow between the two electrical panels. You should expect to find that most, if not all, structures have a continuous metal path back to the first service disconnect. This is a very safe way to set up your structures, and you should have a green-wire grounding conductor routed in the conduit from the first service disconnect at the house to the substructure's electrical box.

When you have a structure that has a continuous metallic path back to the first service disconnect (the home), you must have a properly sized green-wire ground routed in the conduit, and you may NOT have a neutral-to-ground bond in the secondary structure's electrical panel.

Neutral-to-ground connection (bond) is where a grounded current-carrying conductor, the neutral, is intentionally bonded to earth/ground. In general, this should only occur at the transformer and at the first service disconnect(s). The Code uses several names for neutral-to-ground connections: system bonding jumper, main bonding jumper, and supply-side bonding jumper.

Truly isolated structures occur very infrequently, and it could be argued that the Code should simply ban the practice. But, if you do believe that you have an isolated structure that does not have a continuous metallic path back to the first service disconnect, you will need to install an additional neutral-to-ground bond in the electrical panel at the isolated structure.

It is very important to test the isolated structure by measuring the resistance of the structure's building steel against the building steel of the house (the main building). If the resistance is low, you probably have a continuous metallic path. It the resistance is very high, you may in fact have an isolated structure.

Isolated building or structure A building or structure that has no continuous metallic path to any other building or structure. A continuous metallic path between buildings/structures could be formed from metal conduit, a green-wire grounding conductor, metal pipes (water/gas), rebar in concrete, coaxial lines, data and communication shield wires, and so on. Isolated buildings/structures should be tested using a resistance measurement method.

Again, if you have an isolated structure, this section of the Code requires you to install an additional neutral-to-ground bond in the electrical panel of the isolated building. You may not have a green ground wire routed from the first service disconnect at the house to the electrical panel in the isolated structure. You may only have the neutral and the phase wires coming from the house to the structure, routed in PVC conduit or free air (e.g., a farm with multiple buildings being supplied power from a pole-mounted main service, as allowed in Art. 547).

Even if you do have an isolated structure, it is almost always better to run a green ground wire in the conduit leading over to the isolated structure to bond the structures together. It is difficult to predict what systems may need to be added in the future to the isolated structure. A simple phone line or data line added at a future date can be all it takes to generate a continuous metal path, and then you will wish you had routed that ground wire in the conduit.

Additionally, if your structures are within 10 ft of each other, you could have a human safety issue, as you have an increased risk of electrocution. Electrically isolated buildings that are within reaching distance (both arms stretched out) will have differences in potential that could possibly cause electrocution in people.

The bottom line is, electrically isolated buildings being fed by a common source are bad news. You do not want them. If you think you have them or are being asked to install a system for an electrically isolated structure, you will want to ensure that you have an experienced electrical engineer looking at the system for you before you install or approve such a system.

250.32(A). Grounding Electrode. This section tells us that when you have multiple buildings or structures supplied by a common electrical source (think of the house, guesthouse, and garage discussed above for

Art. 250.32), each of the structures must have a properly installed grounding electrode, whether the building is isolated or not. A properly installed electrode is one that meets the requirements found in Part III of Art. 250. Of course, the grounding electrode conductor must be installed according to Art. 250.32(B) or (C). Please see Art. 250.32(B) or (C) for additional information.

There is one exception to this rule: if the electrical panel in your separate building or structure has only one circuit breaker or fuse (OCPD), you do not have to install a grounding/earthing electrode. You can and probably should install a ground rod, but the Code allows you to skip it. However, you must have a green-wire grounding conductor routed in the conduit from the first service disconnect, to the electrical panel in the separate building.

First service disconnect The very first electrical panel that will disconnect (turn off) the power coming in from the utility company. Sometimes this term is used to describe the first electrical panel that can turn off power on the secondary or low-voltage side of a transformer.

250.32(B). Grounded Systems. This section of the Code tells us that we must have an equipment-grounding conductor (a green-wire conductor) installed for all subpanels and electrical enclosures installed in separate buildings and structures supplied with power by the system. This grounding conductor must be the proper type/size and must run from the grounding bar in the first service disconnect enclosure to every subpanel and electrical enclosure in the system.

Equipment-grounding (bonding) conductor The low-impedance fault-current path that is typically run with or encloses the circuit conductors, used to connect the non-current-carrying metal parts of equipment, raceways, and other enclosures to the grounded (neutral) conductor and equipment-grounding (bonding) conductor at service equipment or at the source of a separately derived system. Often called the *green-wire* or *third-wire* ground conductor.

250.32(B)(1). Supplied by a Feeder of Branch Circuit. This part of the Code tells us that buildings and structures supplied by an electrical system must have properly sized equipment-grounding conductors and that there are specific types of grounding conductors authorized for use. The proper wiring configuration for a single branch circuit (fuse throw box for a single load) is shown in Fig. 5.13.

This mandatory equipment-grounding conductor must be routed from the ground bar on the first service disconnect panel to the equipment panel's ground bar in the separate building or structure and to the grounding electrode. And, of course, you may *not* have a neutral-to-ground bond in any subpanel.

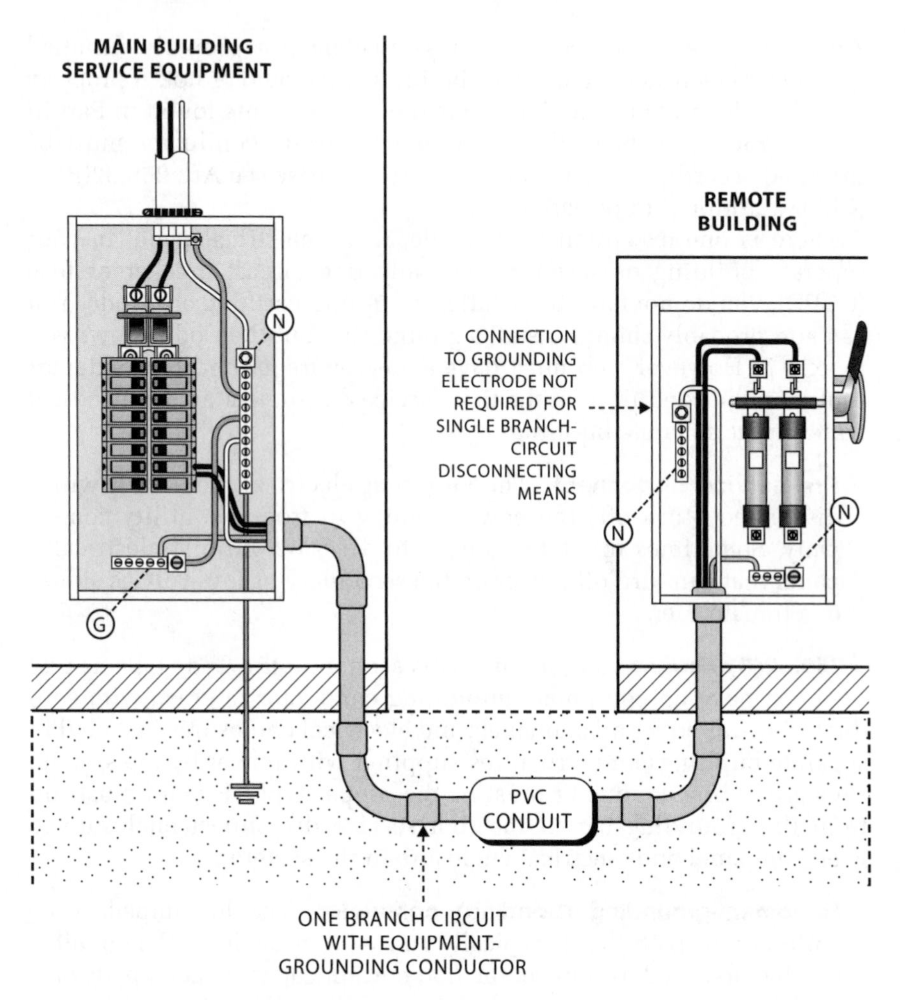

Figure 5.13 Proper remote building wiring using a 4-wire system to a single disconnect.

First service disconnect The very first electrical panel that will disconnect (turn off) the power coming in from the utility company. Sometimes this term is used to describe the first electrical panel that can turn off power on the secondary or low-voltage side of a transformer.

Equipment-grounding (bonding) conductor The low-impedance fault-current path that is typically run with or encloses the circuit conductors, used to connect the non-current-carrying metal parts of equipment, raceways, and other enclosures to the grounded (neutral) conductor and equipment-grounding (bonding) conductor at service equipment or at the source of a separately derived system. Often called the *green-wire* or *third-wire* ground conductor.

The Code also tells us that the authorized types of equipment-grounding conductor(s) can be found in Art. 250.118 (green insulated wire, rigid conduit, certain flexible conduits, etc.). Please see Art. 250.118 for more information. It also tells us that the size/gauge of the equipment-grounding conductor must be in accordance with Art. 250.122. Please see Art. 250.122 for more information. The proper wiring configuration for multiple branch circuits (circuit breaker box for a multiple loads) is shown in Fig. 5.14.

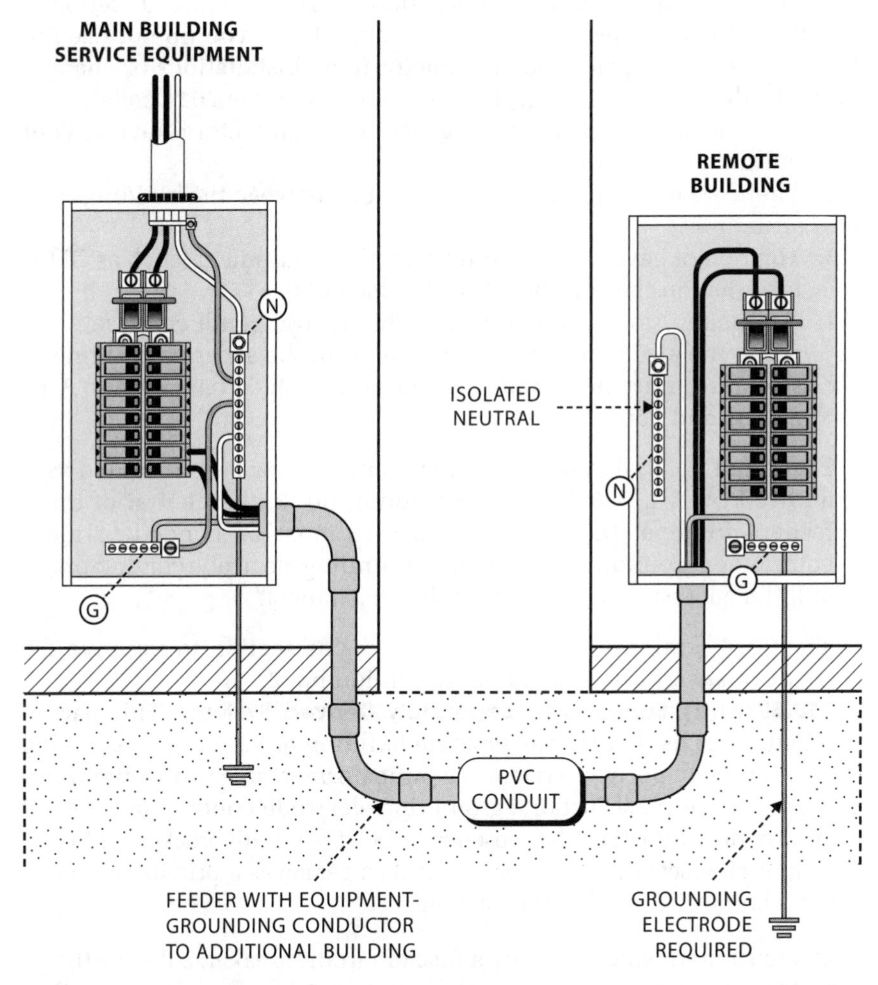

Figure 5.14 Proper remote building wiring using a 4-wire system to a circuit breaker panel.

Exception no. 1: This first exception to the rules in this section is mostly due to changes from the 2008 version of the Code. This exception has a few requirements that basically require the separate structure or building to have no metallic connections back to the primary building, as discussed in Art. 205.32. Only installations made in compliance with the Code prior to 2008 may meet these exceptions (however, an upgrade is recommended).

Feeder circuit All circuit conductors between the first service disconnect and the final OCPD. The circuits after the final OCPD are called the *branch circuits*.

If your structure or building was installed prior to implementation of the 2008 NEC and meets the following guidelines, you do not need to have the grounding electrode conductor (new installations or changes to the feeder circuit will require compliance with the 2014 code):

1. There is no existing grounding electrode conductor routed to your building or structure.
2. There are no continuous metallic paths between the buildings and structures.
3. You do not have ground-fault protection equipment (such as GFCI) installed on the supply side of the feeder(s).
4. The neutral-to-ground bond and the neutral conductor itself that are in the electrical cabinet (subpanel) in the separate building or structure must be properly sized and in accordance with Arts. 220.61 and 250.122.

Exception no. 2: If system bonding jumpers have been installed as allowed by Art. 250.30(A)(1), Exception no. 2, the neutral of the feeder circuit at the building/structure must be bonded to the equipment-grounding conductors, grounding electrode conductors, and the enclosure at the first service disconnect.

The 2014 edition of the Code changed this section to improve the clarity of the verbiage, and added Exception no. 2.

250.32(B)(2). Supplied by Separately Derived System. This part of the Code tells us that if your separate building or structure is supplied via a separately derived system (i.e., transformer, solar array, wind turbine, etc.), you must install a grounding electrode conductor in one of two ways determined by the presence of OCPDs (circuit breakers or fuses). Consider an electrical generator powering a garage as a prime example of what is being discussed in this section.

Overcurrent device Usually a fuse or circuit breaker, a device that limits the maximum amount of current that can flow in a circuit. Sometimes called an *overcurrent protection device* or *OCPD*.

If your generator has a circuit breaker installed directly on its chassis, you must run a properly sized green-wire ground conductor with the phase wires to the garage's electrical panel, in compliance with Art. 250.32(B)(1).

If your generator does not have circuit breakers installed directly on the generator, you must have a circuit breaker at the garage's electrical panel that is in compliance with NEC Arts. 700.12, 700.12(B)(5) and (6), 225.31, and 225.32. The Code states that you do not need to run a green-wire ground conductor with the phase wires from the generator to the garage's electrical panel (although you may certainly do so).

250.32(C). Ungrounded Systems. Ungrounded systems (i.e., delta power) are to have equipment-grounding conductors installed identically to the way we install them for grounded systems. Remember, the main difference between a grounded system and an ungrounded system is there is no neutral-to-ground bond (connection) in ungrounded systems.

Equipment-grounding (bonding) conductor The low-impedance fault-current path that is typically run with or encloses the circuit conductors, used to connect the non-current-carrying metal parts of equipment, raceways, and other enclosures to the grounded (neutral) conductor and equipment-grounding (bonding) conductor at service equipment or at the source of a separately derived system. Often called the *green-wire* or *third-wire* ground conductor.

Neutral-to-ground connection (bond) is where a grounded current-carrying conductor, the neutral, is intentionally bonded to earth/ground. In general, this should only occur at the transformer and at the first service disconnect(s). The Code uses several names for neutral-to-ground connections: system bonding jumper, main bonding jumper, and supply-side bonding jumper.

250.32(C)(1). Supplied by a Feeder of Branch Circuit. This part of the Code tells us that if you have an ungrounded system (delta power) and it is providing power to a separate building or structure, you must have an equipment-grounding conductor routed with the phase wires, bonding the first service disconnect in the main building to the separate building's or structure's equipment subpanel and to the grounding electrodes.

250.32(C)(2). Supplied by Separately Derived System. This part of the Code tells us that if your separate building or structure is supplied via an ungrounded (delta power) separately derived system (i.e., transformer, solar array, wind turbine, etc.), you must install a grounding electrode conductor in one of two ways determined by the presence or absence of OCPDs (circuit breakers or fuses). Consider a delta-power electrical generator powering a garage as a prime example of what is being discussed in this section.

Overcurrent device Usually a fuse or circuit breaker, a device that limits the maximum amount of current that can flow in a circuit. Sometimes called an *overcurrent protection device* or *OCPD.*

If your delta-power generator has a circuit breaker installed directly on the generator, you must run a properly sized green-wire ground conductor with the phase wires to the garage's electrical panel, in compliance with the Art. 250.32(C)(1).

If your delta-power generator does not have circuit breaker installed directly on the generator, you must have a circuit breaker at the garage's electrical panel that is in compliance with NEC Arts. 700.12(B)(5) and (6), 702.12, 225.31, and 225.32. Which means that you do not need to run a green-wire ground conductor with the phase wires from the generator to the garage's electrical panel (although you certainly may do so).

250.32(D). Disconnecting Means Located in Separate Building or Structure on the Same Premises. This section of the Code applies to both grounded (wye-type transformer) and ungrounded (delta-type transformer) systems and simply tells us that if your separate building or structure has power supplied to it, but the circuit breaker to turn the power off is located in another building, you must have an equipment-grounding conductor routed with the phase wires and bonded to the metal chassis of whatever electrical enclosures are present. You must also have a grounding electrode installed at the separate building or structure that is bonded to the equipment-grounding conductor, and you must have a connection to the building steel (rebar in concrete, steel frame, metal piping, etc.).

Equipment-grounding (bonding) conductor The low-impedance fault-current path that is typically run with or encloses the circuit conductors, used to connect the non-current-carrying metal parts of equipment, raceways, and other enclosures to the grounded (neutral) conductor and equipment-grounding (bonding) conductor at service equipment or at the source of a separately derived system. Often called the *green-wire* or *third-wire* ground conductor.

You can imagine a house with a single electrical panel located at the house that supplies power to both the house and a separate remote garage structure as an example of what the Code is discussing here. The same rules apply even when dealing with multiple buildings or structures located on the same premises. In Fig 5.15 we see the proper wiring for a separate building being fed by a branch circuit.

When you have such a scenario, the circuit breaker at the house panel must be in compliance with Art. 225.32. You may, of course, NOT have a neutral-to-ground bond in any of the subpanels. You must have a

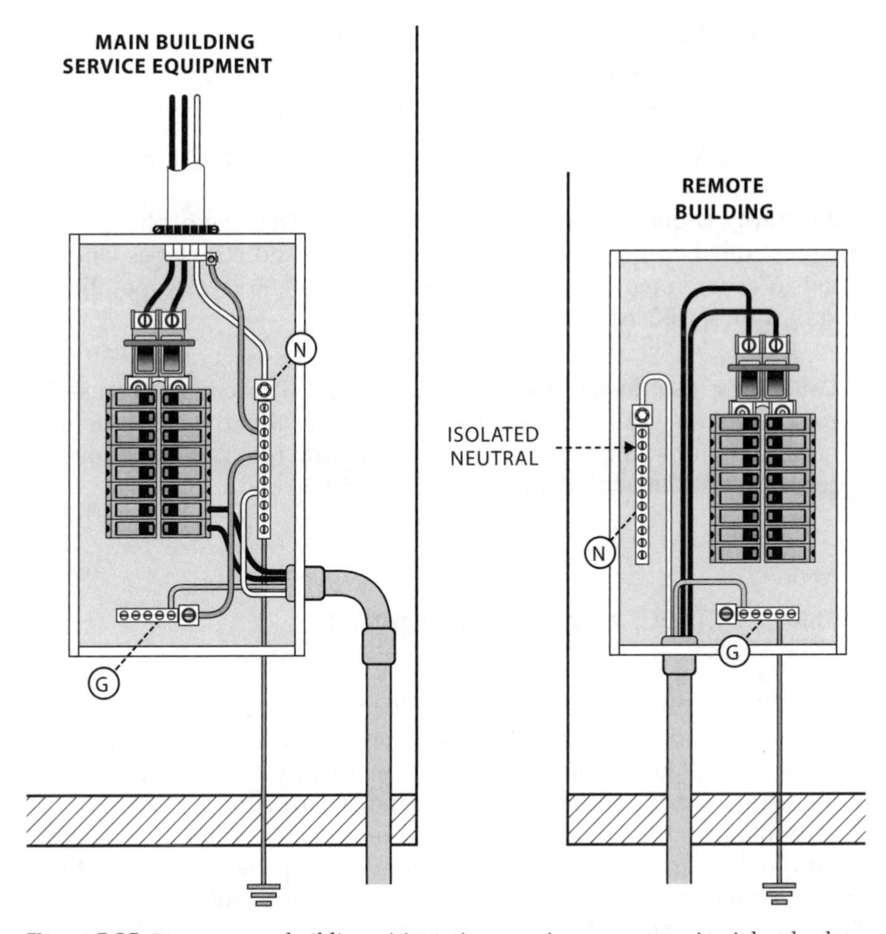

Figure 5.15 Proper remote building wiring using a 4-wire system to a circuit breaker box.

grounding electrode installed at each separate building or structure that is in compliance with Part III of the NEC Art. 250, and the connection between the mandatory equipment-grounding conductor and the grounding electrode must occur inside an approved electrical junction box. It is highly recommended that you add a local means of disconnection for each separate building/structure.

Grounding (earthing) electrode A device that establishes an electrical connection to the earth/soil.

Neutral-to-ground connection (bond) is where a grounded current-carrying conductor, the neutral, is intentionally bonded to earth/ground. In general, this should only occur at the transformer and at the first service disconnect(s). The Code uses several names for neutral-to-ground connections: system bonding jumper, main bonding jumper, and supply-side bonding jumper.

Please see Arts. NEC 700.12(B)(5) and (6), 702.12, 225.31, and 225.32 for more information.

250.32(E). Grounding Electrode Conductor. The size of the grounding electrode conductor for separate buildings and structures shall be sized in accordance with Art. 250.66, Table 250.66 (see Appendix B), and Part III of NEC Art. 250.

Grounding (earthing) electrode conductor The conductor used to connect the grounding (earthing) electrode(s) to the equipment-grounding (bonding) conductor, to the neutral (grounded) conductor, or to both in accordance with Arts. 250.66 and 250.142.

250.34. PORTABLE AND VEHICLE-MOUNTED GENERATORS

Article 250.34 discusses how to properly ground and bond portable and vehicle-mounted generators. The Code makes a distinction between permanently installed electric generators and temporary installations designed to provide power during special events and construction activities.

Portable is defined by the Code as being easily carried by personnel from one location to another. *Vehicle-mounted* generators are defined as being *mobile*, as they can be moved on wheels or rollers.

250.34(A). Grounded Conductor Bonding. This part of the Code tells us that when installing vehicle-mounted or portable generators, the neutral-to-ground bond required in Arts. 250.20 and 250.26 must occur at the frame of the generator and is mandatory for both portable and vehicle-mounted generators. See below for more specifics on portable and vehicle-mounted generators.

Neutral-to-ground connection (bond) is where a grounded current-carrying conductor, the neutral, is intentionally bonded to earth/ground. In general, this should only occur at the transformer and at the first service disconnect(s). The Code uses several names for neutral-to-ground connections: system bonding jumper, main bonding jumper, and supply-side bonding jumper.

Grounding (earthing) electrode A device that establishes an electrical connection to the earth/soil.

The 2014 edition of the Code changed the order of Sections A, B, and C to improve clarity.

250.34(B). Portable Generators. A portable generator is one that has electrical outlets installed directly on the generator and is not designed to have phase conductors powering a separate electrical panel (it must provide cord and plug service only). It must also be easily carried by personnel from one location to another. You can imagine a small gas-powered generator with several 120-V outlets mounted on the front.

The Code tells us here that the grounding prongs of the outlets on the front of the generator must be bonded to the steel frame of the generator. If this is true, you do not need to have a grounding electrode installed for this setup. However, the following must be true:

1. The generator can only supply power to equipment that is mounted directly to the generator and/or equipment that uses a cord and plug through receptacles.
2. The non-current-carrying metal parts of the equipment receiving power from the generator are connected to the frame of the generator through an equipment-grounding conductor. In other words, you must use a cord with a ground wire.

In summary, if you have a portable generator that meets code requirements, you do not need a grounding rod as required in Art. 250.52.

Please see Art. 250.30 for required grounding and bonding connections if portable or vehicle-mounted generators are installed to supply fixed wiring systems.

The 2014 edition of the Code changed the order of Sections A, B, and C to improve clarity.

250.34(C). Vehicle-mounted Generators. A vehicle-mounted generator is one that has electrical outlets installed directly on the generator (it must provide cord-and-plug service and/or power equipment directly mounted on the vehicle) and is not designed to have phase conductors powering a separate electrical panel. You can imagine a work truck with a small gas-powered generator that powers a small air compressor and has several 120-V outlets mounted on the front of the generator.

The Code tells us that all of the following must be met:

1. The frame of the generator must be bonded to the steel frame of the vehicle
2. The generator can only supply power to equipment that is mounted directly to the generator and/or equipment that uses a cord and plug through receptacles.
3. The non-current-carrying metal parts of the equipment receiving power from the generator are connected to the frame of the generator through an equipment-grounding conductor. In other words, you must use a cord with a ground wire.

In summary, if you have a vehicle-mounted generator that meets code requirements, you do not need a grounding rod as required in Art. 250.52.

Please see Art. 250.30 for required grounding and bonding connections if portable or vehicle-mounted generators are installed to supply fixed wiring systems.

The 2014 edition of the Code changed the order of Sections A, B, and C to improve clarity.

250.35. PERMANENTLY INSTALLED GENERATORS

This Section of the Code tells us that we must have an effective ground fault–current path installed with the phase conductors and that the neutral-to-ground bond must be in accordance with Art. 250.30.

Ground fault-current path An electrically conductive path from the point of a ground fault on a wiring system through normally non-current-carrying conductors, equipment, or the earth to the electrical supply source. A ground fault–current path may not use the earth/soil as a conductor or current path, as it is a violation of the NEC Arts. 250.4(A)(5) and 250.4(B)(4). This term is used for both intentional and unintentional current paths.

Equipment-grounding (bonding) conductor The low-impedance fault-current path that is typically run with or encloses the circuit conductors, used to connect the non-current-carrying metal parts of equipment, raceways, and other enclosures to the grounded (neutral) conductor and equipment-grounding (bonding) conductor at service equipment or at the source of a separately derived system. Often called the *green-wire* or *third-wire* ground conductor.

Neutral-to-ground connection (bond) is where a grounded current-carrying conductor, the neutral, is intentionally bonded to earth/ground. In general, this should only occur at the transformer and at the first service disconnect(s). The Code uses several names for neutral-to-ground connections: system bonding jumper, main bonding jumper, and supply-side bonding jumper.

In other words, you must have an equipment-grounding conductor (green-wire ground conductor) routed in the conduit with the phase wires from the generator to the switchgear.

If the generator switchgear has a neutral bar that bonds the power company's neutral and the generator's neutral to the first service disconnect's neutral, you have a not separately (non-separately) derived system and you may NOT have a neutral-to-ground bond in the generator. Please see Art. 250.30 for more information.

If the generator switchgear has a neutral system that switches between the power company's neutral and the generator's neutral to provide power to the first service disconnect's neutral, you have a separately derived system and you MUST have a neutral-to-ground bond in the generator. Please see Art. 250.30 for more information.

Note *The Code adds a new terminology here,* non-separately, *which is the same as* not separately *used earlier in the Code.*

250.35(A). Separately Derived Systems. If the generator switchgear has a neutral system that switches between the power company's neutral and the generator's neutral to provide power to the first service disconnect's neutral, you have a separately derived system and you MUST have a neutral-to-ground bond in the generator. Please see Art. 250.30 for more information.

Separately derived system(s) A premises wiring system whose power is derived from a source of electrical energy other than a (utility) service. Examples include solar panels, generators, and wind turbines that have no direct connection to another electrical source. Specifically, the neutral conductor must not be solidly bonded to the neutrals of other sources. This term is also used in regard to isolation and/or step-down transformers that generate a distinct power source separate from the utility.

An equipment-grounding conductor is required in accordance with Art. 250.102(C).

250.35(B). Non-separately Derived Systems. If the generator switchgear has a neutral bar that bonds the power company's neutral and the generator's neutral to the first service disconnect's neutral, you have a not separately (non-separately) derived system and you may NOT have a neutral-to-ground bond in the generator. Please see Art. 250.30 for more information.

Not separately derived system(s) A premises wiring system whose power is derived from a source of electrical energy other than a (utility) service. Examples include solar panels, generators, and wind turbines that have a direct connection to another electrical source. Specifically the neutral conductor is solidly bonded to the neutrals of another source.

If the permanently installed nonseparately derived generator has an integral OCPD (circuit breaker, fuses, etc.), it should be bonded in accordance to the manufacturer's instructions.

An equipment-grounding conductor is required in accordance with Art. 250.102(C).

Note *The Code adds a new terminology here,* non-separately, *which is the same as* not separately *used earlier in the Code.*

250.36. HIGH-IMPEDANCE GROUNDED NEUTRAL SYSTEMS

High-impedance grounding systems (sometimes called *resistance grounding systems*) are a type of grounding method that involves adding impedance (usually a resistor) between the neutral point of a transformer and the ground. These systems are permitted for three-phase (single-phase loads are not allowed) AC systems operating between 480 and 1000 V and must meet the following conditions:

1. Only qualified personal may service and maintain the installation. This is generally considered to mean that the electrical system is located in a locked electrical room with proper signage.
2. Ground detectors are installed and monitor the system
3. Line-to-neutral (single-phase) loads are not served by the electrical system

Keep in mind that the Code is discussing a global system. This means that your entire electrical system from the transformer on down to the last circuit would be a high-impedance grounded neutral system. You cannot have a mixed system with some circuits being grounded typically, others ungrounded, and yet more being resistance grounded. (You, of course, may always install a new separately derived transformer downstream, thereby reestablishing a new electrical system, and change from one type of electrical system to another). The decision concerning the system you will have is made at the transformer. A schematic for a typical high-impedance grounded neutral system can be seen in Fig. 5.16, where a ground impedance device, typically a resistor, is added between the neutral point (XO terminal) of the transformer and the grounding electrode conductor.

For most applications, you will be dealing with a grounded system. Very rarely will you ever deal with ungrounded or resistance-grounded systems. But to be clear, here are the three types of systems the Code discusses in Art. 250:

- Grounded systems: wye-type transformers (the most common system)
- Ungrounded systems: delta-type transformers

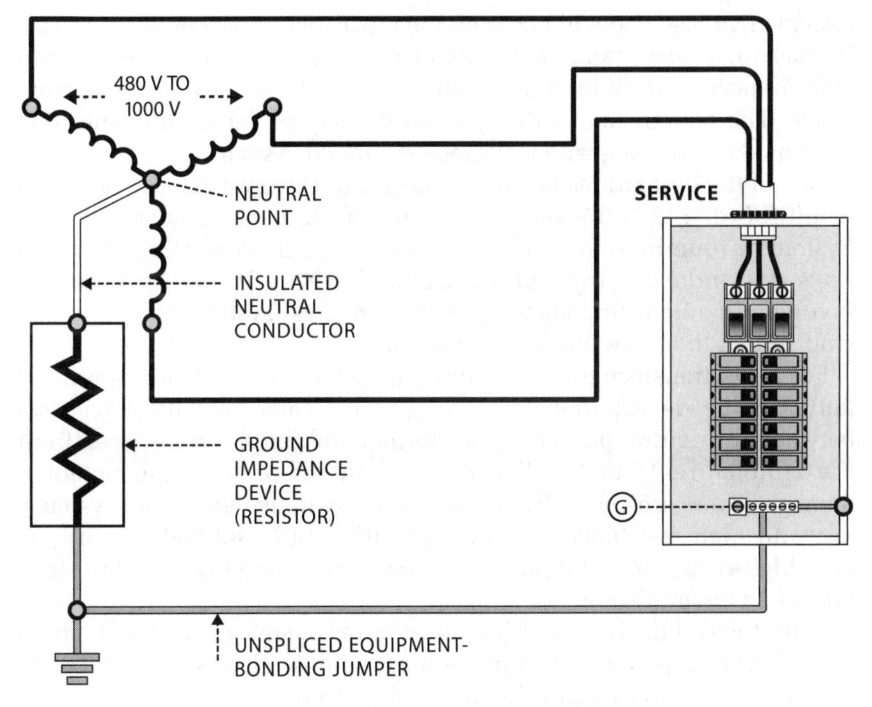

Figure 5.16 High-impedance grounded neutral electrical system.

- High-impedance grounded neutral systems or resistance-grounded systems: wye- or delta-type transformer with an added impedance (resistor) at the XO terminal (neutral point)

NEC Art. 250.36 is discussing high-impedance grounded neutral systems, a type of grounding practice that is typically only used at electrical substations and large industrial transformers that supply other smaller transformers in order to limit phase-to-ground fault currents. Resistance-grounded systems are good for loads that require a continuity of power (such as a hospital). When a ground fault occurs on one of these systems, it is more likely to cause a ground-fault alarm than it is to trip a circuit breaker. It is very unusual to find a resistance grounding system in the typical commercial/industrial setting, except in a few specialized industries, and would be illegal in a residential setting.

While it may sound like a great thing to be able to limit fault currents, one of the main reasons resistance-grounded systems are not in greater use is because the overcurrent protection and ground-fault detection equipment required to run these systems are complex, expensive, and require greater monitoring.

The biggest concern for high-impedance grounded neutral systems is in their poor ability to provide effective overcurrent protection (circuit breakers). When a fault occurs on one phase conductor of resistance-grounded system, a voltage rise will occur on the other two phase conductors. This increase in voltage can allow more electrical energy through the circuit than is desired, meaning the circuit breaker will not trip at the desired levels. The bottom line is that you need very special circuit protection systems when dealing with resistance-grounded systems.

One of the best guides for resistance grounding systems can be found in IEEE Std. 142-2007 (soon to become IEEE 3003 Standards: Power Systems Grounding) and an article by Robert B. West (*IEEE Transactions and Industry Applications* IA-15(2), 1979). If you find yourself involved in a high-impedance grounded neutral system, you would be well served to review these documents.

For large transformers that handle massive current loads, a ground fault can be devastating. By adding impedance (usually a resistor) between the neutral point of a transformer and the ground, we can limit the fault currents that will occur and therefore limit the potential damage. There are generally two types of resistance-grounded systems: low and high resistance. According to IEEE Std. 142-2007, both low- and high-resistance systems will have the following benefits, to a greater or lesser degree:

a. Reduces burning, melting, heat stress, and mechanical stress effects in switchgear, transformers, cables, motors, and so on, in electrical circuits carrying the fault currents
b. Reduces but does not eliminate, electric-shock hazards such as step-and-touch voltages that could harm personnel and equipment
c. Reduces but does not eliminate, the arc flash/blast hazard to personnel and equipment from phase-to-ground faults (but does not reduce phase-to-phase faults)
d. Reduce the momentary line-voltage dip caused by the clearing of a ground fault
e. Can help to control transient overvoltages and the shutdown of the circuit

Please note that resistance grounding systems (both high and low) will probably reduce arc flash hazards for phase-to-phase faults but does not eliminate them. However, phase-to-ground faults in both systems can be limited to predictable and manageable levels.

Ground fault An unintentional connection between an energized conductor and earth or metallic parts of enclosures, raceways, or equipment.

This section of the Code governs three-phase, AC, high-impedance grounded systems for between 480 and 1000 V. It requires that only qualified and trained personnel may service and operate the equipment. It also requires that ground detectors are installed on the circuits in the system. Most important, it strictly forbids any line-to-neutral loads (such as single-phase, 277-V lighting circuits).

A final word on high-impedance grounded neutral systems: a standard grounded wye-type transformer would be a *zero-impedance grounded neutral system.* A delta-type transformer would the *highest possible impedance grounded neutral system.* A high-impedance or low-impedance grounded neutral system should be basically treated just like a delta-type system. In general, high-impedance grounded neutral systems are in place of delta systems.

250.36(A). Grounding Impedance Locations. This section tells us where in the circuit the impedance must be added. Generally, the manufacturer of the transformer will have either already integrated an impedance into the transformer or will have instructions for adding a separate resistor system. You should follow those instructions.

But in short, for wye-type transformers, the impedance, usually a resistor, must be added between the neutral point of the transformer (XO terminal) and the grounding electrode conductor.

> **Grounding (earthing) electrode conductor** The conductor used to connect the grounding (earthing) electrode(s) to the equipment-grounding (bonding) conductor, to the neutral (grounded) conductor, or to both in accordance with Arts. 250.66 and 250.142.

For delta-type transformers, the impedance, usually a resistor, must be added between the neutral point derived from the transformer (the center point of one phase) and the grounding electrode conductor.

250.36(B). Grounded System Conductor. This section of the Code requires that you fully insulate the conductor and connections involved in bonding the transformer to the impedance (resistor), as outlined in Art. 250.36(A). You must also size the conductors according to Table 250.66 (minimum size of 8 AWG copper or 6 AWG aluminum; see Appendix B).

250.36(C). System Grounding Connection. This section tells us that the connection to the grounding electrode (earthing electrode), must occur through the impedance (resistor). In other words, you cannot jump over the impedance and bond the XO terminal to earth, or you would no longer have a resistance-grounded system, you would have a grounded system.

> ***Informational note*** *The impedance is typically set to limit the ground-fault current to a value slightly higher than the capacitive charging current of the electrical system, which will also limit transient overvoltages to safe levels. One of the best guides for resistance grounding systems can be found in IEEE Std. 142-2007 (soon to become IEEE 3003 Standards: Power Systems Grounding) and an article by Robert B. West (*IEEE Transactions and Industry Applications *IA-15(2), 1979). If you find yourself involved in a high-impedance grounded neutral system, you would be well served to review these documents.*

250.36(D). Neutral Point to Grounding Impedance Conductor Routing. This part of the Code simple says that the location of the impedance, usually a resistor, may be in a location outside the transformer's or generator's enclosure. So you could have the transformer in one enclosure, with a separate resistor system in a completely different enclosure at some distance from the transformer.

You may also route the conductor from the neutral point on the transformer (XO terminal) to the impedance, in separate raceways/conduit from the phase wires. And, of course, this conductor does not need to go to the first service disconnect device.

First service disconnect The very first electrical panel that will disconnect (turn off) the power coming in from the utility company. Sometimes this term is used to describe the first electrical panel that can turn off power on the secondary or low-voltage side of a transformer.

250.36(E). Equipment-Bonding Jumpers. This part of the Code tells us that if you do have the impedance in a separate enclosure from the transformer, as in Art. 250.36(D), you must have a properly sized green-wire grounding conductor (called the equipment-bonding jumper or conductor) bonding the chassis of the two enclosures together. In other words, you must have a ground wire bonding the transformer enclosure to the impedance enclosure.

Equipment-grounding (bonding) conductor The low-impedance fault-current path that is typically run with or encloses the circuit conductors, used to connect the non-current-carrying metal parts of equipment, raceways, and other enclosures to the grounded (neutral) conductor and equipment-grounding (bonding) conductor at service equipment or at the source of a separately derived system. Often called the *green-wire* or *third-wire* ground conductor.

The proper size of this conductor must be in accordance with Table 250.66 (see Appendix B). It is also discussed below in Art. 250.36(G).

250.36(F). Grounding Electrode Conductor Locations. This section tells us that for both a service and a separately derived system, the service equipment enclosure (transformer) and the impedance enclosure (if separate) must have a properly sized grounding electrode conductor bonding the chassis of the enclosures to a properly installed grounding electrode, as described in Part III of NEC Art. 250.

Grounding (earthing) electrode conductor The conductor used to connect the grounding (earthing) electrode(s) to the equipment-grounding (bonding) conductor, to the neutral (grounded) conductor, or to both in accordance with Arts. 250.66 and 250.142.

In the case of high-impedance grounded neutral systems, you should expect to have a human safety grounding grid based on IEEE Std. 80-2000 (or most current version). This grounding grid would meet and exceed the requirements found in Part III of Art. 250.

The 2014 edition of the Code clarified that this section applies to both services and separately derived systems.

250.36(G). Equipment-Bonding Jumper Size. This section tells us that the equipment-bonding jumper mentioned in Art. 250.36(E) must be:

1. Properly sized in accordance with Table 250.66 (see Appendix B).
2. The same size as the neutral conductor, as outlined in Art. 250.36(B), if the grounding electrode conductor is connected at the first service disconnect or the overcurrent device.

Chapter Six

NEC 250 PART III: GROUNDING ELECTRODE SYSTEM AND GROUNDING ELECTRODE CONDUCTOR

Part III of Art. 250 is the section that is most commonly associated with grounding, and that is the earthing side of grounding. *Earthing* is the intentional connection of electrical components to the earth/soil.

Remember that in the United States, and several other parts of the world, we use the single word "grounding" to indicate both the above-grade copper/aluminum conductor system and the below-grade copper/steel earthing electrode system. In Europe and many other countries, these terms are separated from each other; the above-grade copper/aluminum conductor system (green-wire ground) is called *grounding*, and the below-grade copper/steel electrode system is called earthing.

Electrical systems are earthed for many reasons: to limit the voltages caused by lightning or by accidental contact with supply conductors, to stabilize line voltages under normal operation relative to the ground, to provide a reference point in an electrical circuit from which other voltages are measured, and to limit the buildup of static electricity (which can be especially important when dealing electrostatic-sensitive devices (computers, televisions, stereos, etc.) and flammable products.

A grounding/earthing electrode can be any electrically conductive object that is in contact with the earth. This includes ground rods, ground plates, water pipes, gas pipes, other metal pipes, steel rebar in concrete (such as a building foundation), copper wires in contact with the earth, coaxial cables and telephone grounds, and more.

The primary point of the National Electrical Code (NEC) Art. 250, Part III, is to ensure that your electrical system is grounded/earthed using multiple electrodes and that these electrodes are bonded together to form a single continuous electrode network.

Please note that aluminum is not allowed for use below grade. Only copper and steel items are allowed for direct burial. When steel is used, it must have some form of listed corrosion protection, often concrete or a copper/zinc coating.

250.50. GROUNDING ELECTRODE SYSTEM

This section of the Code tells us that for each building and/or structure, all of the electrodes listed in Art. 250.52(A)(1) through (A)(7) must be bonded together to form a single electrode system. If the building and/or structure do not have any of these electrodes, you must install one or more of the electrodes listed in Art. 250.52(A)(4) through (A)(8).

The following electrodes from Art. 250.52(A) must be bonded together for each building/structure:

1. Metal underground water pipe
2. Metal frame of the building or structure
3. Concrete-encased electrodes (including rebar in concrete foundations)
4. Ground ring
5. Rod and pipe electrodes
6. Other listed electrodes
7. Plate electrodes

If you do not have the above items, you must install one or more of the following from Art. 250.52(A):

1. n/a
2. n/a
3. n/a
4. Ground ring
5. Rod and pipe electrodes
6. Other listed electrodes
7. Plate electrodes
8. Other local metal underground systems or structures (underground tanks, well casings, etc.)

The vital concept in this section of the Code is that all of the available electrodes at a building or structure, must be bonded together to form a single electrode system. Figure 6.1 shows how to properly connect the required grounding electrodes for a building or structure using a copper water pipe as the bonding medium.

There is one exception to this rule: you do not need to break up concrete to expose the steel rebar in the concrete foundations of existing buildings and/or structures. The intent of this code is for older construction, not for new construction. Newly constructed buildings/structures must have the rebar in the concrete bonded to the grounding system. This may necessitate some preplanning to ensure that pieces of rebar are exposed for bonding near the building or the structure's first service disconnect.

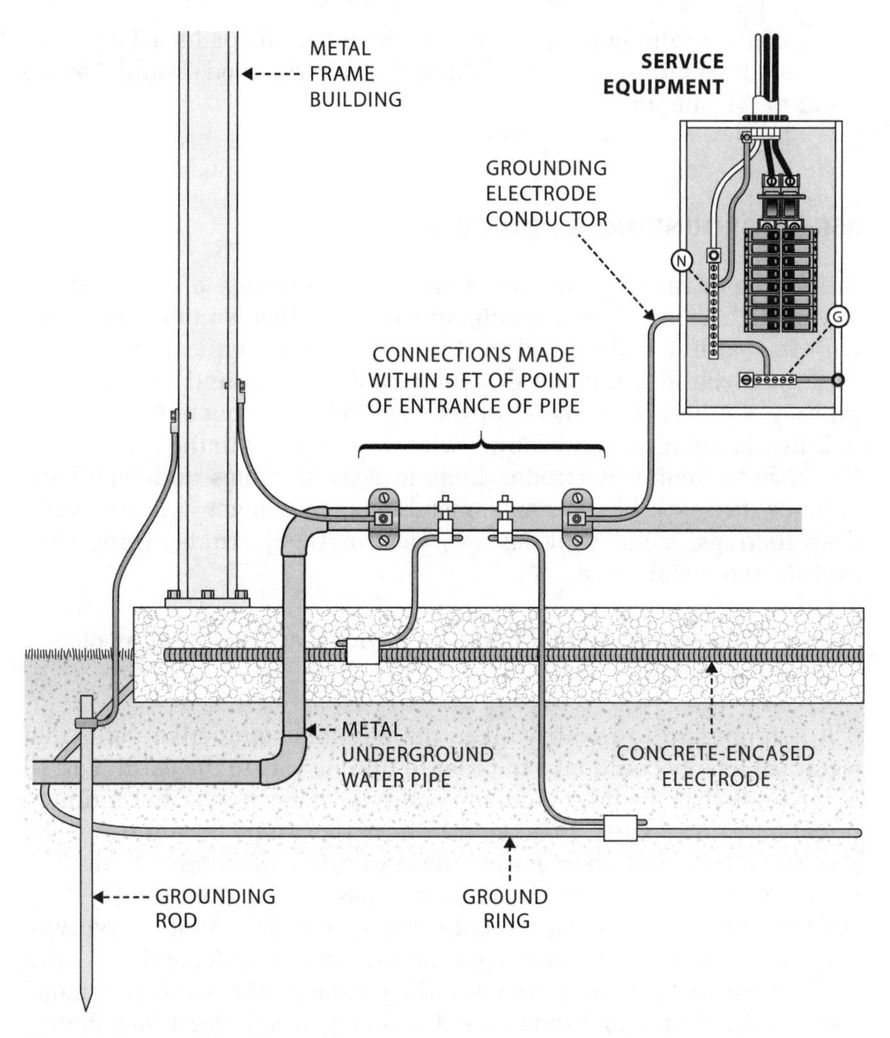

Figure 6.1 Typical building grounding system using copper water pipe for bonding.

First service disconnect The very first electrical panel that will disconnect (turn off) the power coming in from the utility company. Sometimes this term is used to describe the first electrical panel that can turn off power on the secondary or low-voltage side of a transformer.

It is not uncommon to see a piece of rebar bent upward and sticking out of the concrete near or even inside an electrical panel, as visual evidence of the bond. Sometimes you will see a copper ground wire routed through a PVC protective sleeve and down into the concrete. Lately, the trend has been to install a bronze pad that is flush-mounted to the concrete and bonded directly to the rebar; the pad has bolt holes so that jumper wires can be easily installed and replaced should damage occur to the jumper.

250.52. GROUNDING ELECTRODES

Grounding or earthing electrodes can be intentionally or unintentionally made from nearly any conductive material that is in direct contact with the earth. As discussed in the section above, typical grounding/earthing electrodes include driven ground rods, ground plates, water pipes, gas pipes, other metal pipes, steel rebar in concrete (such as a building foundation), and copper wires in contact with the earth. Some less than common electrodes include coaxial cables and telephone grounds, fire sprinkler systems, metal tanks in contact with the earth, deep footings, metal well casings, electrolytic electrodes (pipe electrodes), and metal fences.

> **Grounding (earthing) electrode** A device that establishes an electrical connection to the earth/soil.

It is important to identify all of the possible conductive paths that electrical energy could take in order to find its way to the earth. Part III of the Code lists the most common and typical electrodes, but unintentional paths may exist. Does your building for some reason use metal pipes instead of PVC for watering the grass? If so, the footings would be an electrode, and you must bond these pipes to your grounding system. Does your building and/or structure supply power to a tall tower with deep footings? If so, the footings would be an electrode, and you must bond these footings to your grounding system. What about a metal water tank? What about the foundation of a garage/barn? What about a below-grade swimming pool? You bet! The metal structures for all of these items must be bonded to your grounding system.

A thorough understanding of your building/structures complex electrode system will ensure the safety of the occupants.

Arguably, the most commonly confused electrode is the steel rebar found in the concrete foundation of a building. Not only is the steel conductive, but the concrete itself will also readily conduct electrical energy. You must bond the steel rebar in the concrete foundation to your grounding system. Vapor barriers between the concrete and the earth and plastic-coated rebar will negate the use of your foundation as an electrode; however you will still be required to bond the steel rebar to the grounding system of your building and/or structure. This will be discussed below in greater detail in Art. 250.52(A)(3).

250.52(A). Electrodes Permitted for Grounding. This section of the Code lists the electrodes that are permitted for grounding. In other words, the electrodes in Art. 250.52(A) are electrodes that you may (and must) intentionally tie to earth, and use as an electrode for your electrical system.

A permitted electrode is an intentional connection to the earth of a copper and/or protected steel system of sufficient size/length that is capable of handling likely objectionable currents from lightning strikes, electrical faults, and the like, safely into the earth.

There are electrodes that are not permitted, such as gas pipes and anything aluminum. These are discussed in greater detail in Art. 250.52(B). Remember that gas pipes must still be bonded to your grounding system; you just may not use them as an electrode. You must bond gas pipes in such a way as to ensure that there are no differences in potential and that they will not be an electrical path to carry objectionable currents to the earth.

Permitted electrode An intentional connection to the earth of a copper and/or protected steel system of sufficient size/length that is capable of handling likely objectionable currents from lightning strikes, electrical faults, and the like, safely into the earth. These electrodes can be metal structures in direct contact with the earth that may be used as an electrode to carry objectionable currents to the earth.

Not-permitted electrode system A metal structure in direct contact with the earth that may not be used as an electrode to carry objectionable currents to the earth. Gas pipes and anything aluminum are not-permitted electrodes.

Electrode A conductor used to establish electrical contact with the earth.

Keep in mind that your building and/or structure must have a minimum number of permitted (qualified) electrodes. Article 250.52(A) tells us what electrodes are permitted and lists the rules for qualifying each electrode.

250.52(A)(1). Metal Underground Water Pipe. Metal underground water pipes may be used as a permitted electrode if there are at least 10 ft (3 m) of metal pipe in direct contact with the earth and it is electrically continuous to the point where the grounding electrode conductor is bonded.

Grounding (earthing) electrode conductor The conductor used to connect the grounding (earthing) electrode(s) to the equipment-grounding (bonding) conductor, to the neutral (grounded) conductor, or to both in accordance with Arts. 250.66 and 250.142.

Bonding (bond, bonded) The permanent joining of metallic parts together to form an electrically conductive path. This path must have the capacity to safely conduct any fault current likely to be imposed on it.

Bonding jumper A reliable conductor sized per Art. 250 to ensure electrical conductivity between metal parts of the electrical installation.

Any insulating joints in the pipe must be made electrically continuous by bonding around the joint (bonding jumpers). Also, you must bond your metal water pipe to the building/structure's grounding system, whether there is 10 ft of buried pipe or not. If there is not enough pipe in direct contact with the earth, you must still bond it to your grounding system; you just may not count it as a permitted electrode.

Metal underground water pipe must be bonded to your building/structure's grounding system according to Art. 250.50.

250.52(A)(2). Metal Frame of the Building or Structure. This section of the Code tells us the requirements for using the metal frame of a building or structure (building steel) as an electrode. For starters, this section is a bit out of order, because you must have a qualifying concrete-encased electrode (steel rebar in your foundation) from Art. 250.52(A)(3) below before you can qualify the metal frame as an electrode.

But, that said, if your building/structure has a metal frame with bolted/riveted or welded connections forming a continuous metal object, and for some reason it has a structural steel member that is in direct contact with the earth for at least 10 ft, with or without a concrete encasement, you may use the building steel as an electrode.

However, if your building/structure has a metal frame with bolted/riveted or welded connections forming a continuous metal object, but has structural steel columns that are connected to the concrete foundation, you must have the following:

a. The concrete foundation must meet the requirements found in Art. 250.52(A)(3).
b. The hold-down bolts for the structural steel column must be connected to the steel rebar in the concrete foundation by welding, exothermic welding, steel-wire ties, and/or other approved means.

The Code only requires that you have the hold-down bolts for a single column bonded to the steel rebar in the concrete foundation for any given building. While the Code only requires this at one place, you would be well advised to ensure that the hold-down bolts at every column are bonded to the steel rebar. This is especially important when your building/structure is located in areas prone to lightning strikes.

Remember, just because the metal frame of your building is a permitted electrode (qualified), does not mean that it can be used to replace your green-wire ground conductor requirements found in Part VI of NEC Art. 250. However, you may use the metal frame as a grounding electrode conductor in accordance with Art. 250.68(C)(2).

Grounding (earthing) electrode conductor The conductor used to connect the grounding (earthing) electrode(s) to the equipment-grounding (bonding) conductor, to the neutral (grounded) conductor, or to both in accordance with Arts. 250.66 and 250.142.

Grounding (earthing) electrode A device that establishes an electrical connection to the earth/soil.

Structural metal grounding electrode A structural metal member that establishes an electrical connection to the earth/soil as defined in Art. 250.52(A)(2). Specifically, the metal frame of a building or structure that is properly bonded to the steel rebar in the building's or structure's foundation (concrete-encased electrode without a vapor barrier), as required in Art. 250.52(A)(3).

If the metal frame of your building/structure does not qualify, you must still bond it to your grounding system; you just may not count it as a permitted electrode.

The metal frame of the building or structure must be bonded to your building/structure's grounding system according to Art. 250.50.

Remember, the NEC does not consider structural steel to be a grounding electrode unless it is connected to one (such as a qualifying foundation with steel rebar in concrete). But let us be honest here, building steel is simply not a good electrode and it should be removed from the list found in Arts. 250.52 and 250.53. It is at best a poor-quality grounding electrode conductor, and should be bonded to the grounding system for safety but should not be used as a grounding electrode or as a grounding electrode conductor.

250.52(A)(3). Concrete-encased Electrode. There are two types of grounding electrodes discussed in this section of the Code. The first is another form of ground rod; where you trench 20 ft of earth, install 20 ft of 4 AWG or larger copper conductor, and encase it in concrete. The other, and the most common form, is the steel rebar located in the foundation of your building or structure.

We will start with the easiest one, which is the independent concrete-encased copper conductor that may be used as an electrode. Often, in very rocky soil, it is difficult to drive a ground rod into the earth. When this occurs, it is easier to trench 20 ft of earth and install a 4 AWG or larger copper conductor to act as a permitted electrode. All that is required is that you trench 20 ft of earth to a depth of at least 30 in. (2.5 ft) below grade, and encase a 4 AWG or larger bare copper conductor in at least 2 in. of concrete and, when it is dry, cover it back up. The Code allows you to use this just as you would a driven ground rod.

You may also simply install 20 ft of #4 AWG or larger bare copper conductor directly into the concrete foundation of your building. This is useful as it creates a grounding electrode and provides a mechanism for bonding the steel rebar to the ground system. If you do not use this copper conductor to bond the steel rebar to the building's grounding system, you will still be required to use some other mechanisms to bond the steel rebar to the building's electrical grounding system; this is discussed in more detail later in this section. Figure 6.2 shows the required connection to the steel rebar in a building or structure concrete foundation. A non-metallic protective sleeve is required to provide strain-relief and protect the conductor from shearing.

Please note however, that concrete-encased copper conductor electrodes are NOT considered to be an ideal choice as electrodes. The water that is inherent in concrete can flash into steam under high-current

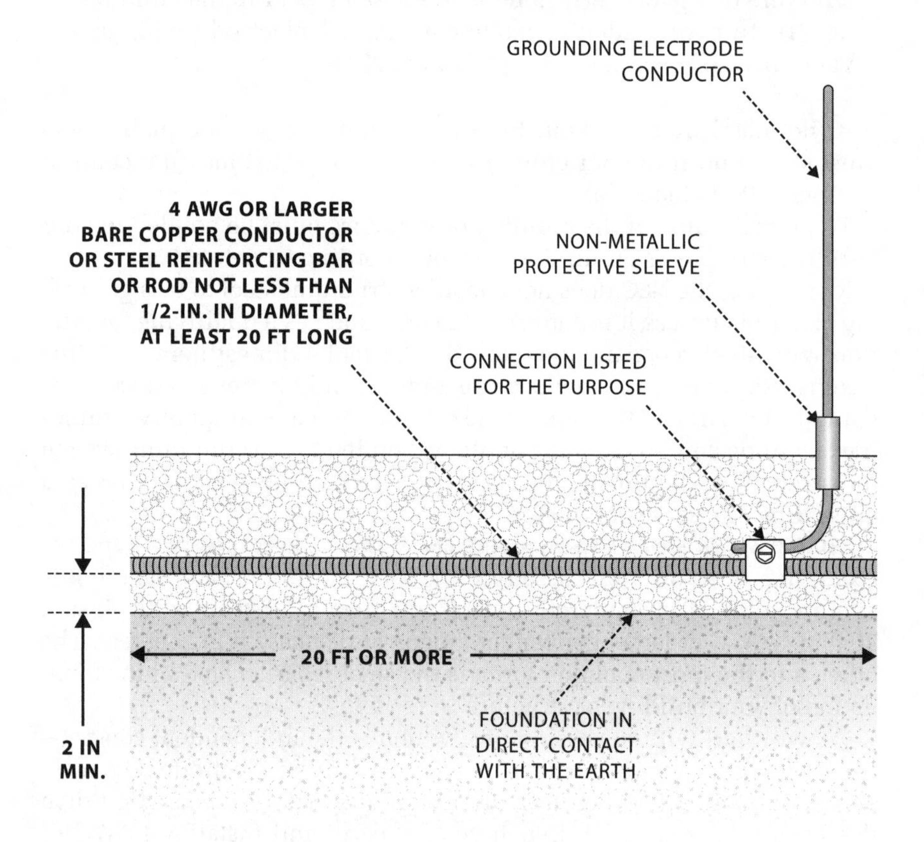

Figure 6.2 Concrete-encased electrode with stress-relief for conductor.

discharges such as lightning strikes, cracking the concrete and destroying the electrode. You would be better served to install 20 ft of bare 2 AWG or larger copper conductor in the trench without the concrete; see Art. 250.52(A)(4).

Concrete-encased electrode An electrode encased in concrete that is located within a concrete footing or foundation; the electrode, which may consist of reinforcing bars of steel and/or copper conductors, of at least 20 ft in length and encased in 2 in. of concrete, must be in electrical contact with the earth. Often called a Ufer ground.

Concrete-encased copper conductor electrode A permitted electrode composed of 20 ft of 4 AWG bare copper conductor trenched into the earth to a depth of at least 30 in. (2.5 ft) and encased in concrete, that is in direct contact with the earth. This is not considered to be a good electrode, due to the cracking of concrete under high-current fault conditions.

Ufer ground A concrete-encased electrode system integral to a building's structural system, such as the steel rebar in a concrete foundation. Copper conductors added to increase the conductivity of the steel rebar would also be a Ufer ground. Ufer grounds are named after Herbert G. Ufer, who developed the system in World War II to ground bomb storage vaults. See Concrete-encased electrode.

The other electrode discussed in this section, and clearly of primary interest, is the steel rebar located in a building/structure's foundation. This tends to be a very misunderstood section of the Code, as bonding the steel rebar in the foundations of buildings is commonly ignored. This is typically because the concrete slabs are poured long before the electrical contractors show up to install the service, and poor coordination prevents ready access to the steel rebar.

The Code does allow for an exception in Art. 250.50 for existing building/structures to skip bonding of the steel rebar in the foundation, if it requires the disturbance of the concrete. However, this was intended for older existing structures. For new construction, you must bond the steel rebar to the grounding system, even where poor coordination resulted in the failure to allow proper steel rebar bonding. Sorry, but if you have new construction, and they forgot to allow for proper rebar bonding, you must break up the concrete and get to the rebar. Even if you do have an older building, you would be well-advised to break up the concrete and get that rebar bonded.

To qualify your building/structure's foundation as a permitted electrode, you must have the following:

a. You must bond to the steel rebar in your building's foundation with a qualified copper grounding electrode conductor. Aluminum is not allowed in this circumstance.

b. The copper grounding electrode conductor must pass through a nonmetallic protective sleeve or other approved stress-relief device. Copper conductors may not exit/enter concrete slabs without a system to protect them from shear forces. Typically, a short piece of PVC pipe is used to transition the copper wire in/out of the concrete; the pipe is then filled with silicon or other flexible nonconductive material to relieve forces that may shear the copper off.
c. The concrete foundation must be in direct contact with the earth. Vapor barriers, films, moisture protection, plastic sheeting, and so on may not be used. You may not have any medium between the concrete and the earth that could prevent (insulate) electrical energy from flowing to earth.
d. You must have steel rebar in the concrete foundation that is electrically conductive. Galvanized, stainless-steel clad, and other similar types of rebar are considered to be electrically conductive. Concrete foundations that use fiberglass for structural support instead of steel rebar, do not qualify as a permitted electrode. Steel rebar that is coated with plastic, epoxy, or other nonconductive coatings does not qualify as a permitted electrode. You may not have any medium around the steel rebar that could prevent (insulate) electrical energy from flowing to earth.
e. The steel rebar must be at least ½ in. (13 mm) in diameter and must be at least 20 ft in continuous length or made continuous through the use of welding, steel-wire ties, or another approved method.
f. There must be at least 2 in. (50 mm) of concrete encasing the steel rebar.
g. There must be at least 20 ft of concrete in direct contact with the earth.

If the foundation of your building/structure does not qualify as a concrete-encased electrode, you must still bond any steel rebar (such as plastic-coated rebar or a foundation with a vapor barrier) to your grounding system; you just may not count it as a permitted electrode.

Grounding (earthing) electrode conductor The conductor used to connect the grounding (earthing) electrode(s) to the equipment-grounding (bonding) conductor, to the neutral (grounded) conductor, or to both in accordance with Arts. 250.66 and 250.142.

Permitted electrode An intentional connection to the earth of a copper and/or protected steel system of sufficient size/length that is capable of handling likely objectionable currents from lightning strikes, electrical faults, and the like, safely into the earth. These electrodes can be metal structures in direct contact with the earth that may be used as an electrode to carry objectionable currents to the earth.

Not-permitted electrode system A metal structure in direct contact with the earth that may not be used as an electrode to carry objectionable currents to the earth. Gas pipes and anything aluminum are not-permitted electrodes.

Electrode A conductor used to establish electrical contact with the earth.

Note *There are a number of products designed to reduce the resistance of concrete and theoretically enhance the ability of the concrete to conduct electricity. These products should be avoided, as they tend to utilize carbon-based materials, which will accelerate the rate of corrosion for copper.*

Concrete-encased electrodes, including the steel rebar in your foundation, must be bonded to your building/structure's grounding system according to Art. 250.50.

250.52(A)(4). Ground Ring. The ground ring is arguably the single best grounding/earthing electrode system you can install. It solves a variety of electrical engineering problems and when properly installed, is the backbone of any good grounding design. Not only does a ground ring provide excellent protection from lightning strikes and accidental electrical faults, but it will also reduce hazardous differences in potential across your building/structure, help to protect personnel from hazardous step-and-touch voltage, protect data lines from transient currents, and reduce the magnetic fields that can form in your steel structure. In short, a ground ring is your best choice.

To be a permitted electrode, a ground ring must encircle the building or structure, be buried to a depth of at least 30 in. (2.5 ft), be at least 20 ft in length, and be composed of 2 AWG or larger bare copper conductor. See Art. 250.53(F) for the depth requirement.

It is typical to incorporate ground rods with your ground ring at some given interval (20 ft or greater); however, the Code does not require the incorporation of ground rods into the ground ring.

You should also bond your ground ring into the steel rebar in the foundation of your building or to the steel columns of your building, at the corners, and at regular intervals around the perimeter of your building. While the Code only requires a single connection, you would be well advised to bond the ground ring to your building steel at regular intervals (100-ft intervals are common) as it travels around the perimeter.

Ground ring An earthing electrode, composed of 2 AWG or larger bare copper wire at least 20 ft in length and buried at least 2.5 ft below grade and in direct contact with the earth, that circumnavigates a structure or building and is reconnected to itself, typically with multiple connections to the building's or structure's steel rebar and/or the building's steel system, and used in conjunction with standard ground rods. It is considered to be the most effective electrode system in use today.

While the Code does allow it, the ground ring should not be encased in concrete, as the water that is inherent in concrete can flash into steam under high-current discharges such as lightning strikes, cracking the concrete and destroying the electrode. See 250.52(A)(3) for more information.

If you have a ground ring, it must be bonded to your building/structure's grounding system according to Art. 250.50.

250.52(A)(5). Rod and Pipe Electrodes. This section of the Code tells us the requirements for using rods and/or pipes as an electrode. Standard driven grounding rods are governed under this section of the Code, as are all pipe electrodes, including electrolytic electrode systems.

In order to qualify as a permitted electrode, all ground rod and pipe electrodes must be at least 8 ft in length (or 2.44 m). Keep in mind that Art. 250.53(A)(1) will require that you install ground rods (and pipe electrodes) below permanent moisture level and that you have at least 8 ft of rod in contact with the earth. This is why 10-ft ground rods are recommended. When a 10-ft ground rod is installed, you can have 6 in. exposed above grade, 18 in. in the permanent moisture level (commonly thought of as being 18 in. deep), and the mandatory 8 ft of rod in contact with the earth. Please see Art. 250.53(A)(1) for more information.

Permitted electrode An intentional connection to the earth of a copper and/or protected steel system of sufficient size/length that is capable of handling likely objectionable currents from lightning strikes, electrical faults, and the like, safely into the earth. These electrodes can be metal structures in direct contact with the earth that may be used as an electrode to carry objectionable currents to the earth.

Permanent moisture level Undefined within the NEC but commonly linked to either the frost line or the permanent wilting point of the site. Permanent moisture level is also often thought of as being either 18 in. (1.5 ft) or 30 in. (2.5 ft) in depth, due to various interpretations of several codes, regulations, and industry standards.

Permanent wilting point The minimal point of soil moisture a plant requires not to wilt; typically water content of soil at any point lower than −1500 J/kg of suction pressure.

Frost line The depth to which the groundwater in soil is expected to freeze; also known as *frost depth* or *freezing depth*. Typically, frost lines are legally required by local building codes.

Listed or listed item An item, product, or device that has been specifically developed for a particular function and has been independently evaluated by a product-testing organization, such as the Underwriters Laboratory (UL) or CE.

Electrolytic electrode A pipe electrode, commonly made from 2-in. copper piping, that incorporates a desiccant within the hollow portion of the pipe to extract moisture from the air and deliver the water to the base of the electrode, thereby maintaining high levels of moisture to the surrounding backfill material without the need for human assistance. See Art. 250.52(A)(6).

Ground rod An earthing electrode composed of stainless steel, copper-coated steel, or zinc-coated steel of at least 5/8 in. (15.87 mm) in diameter with at least 8 ft (2.44 m) of length in direct contact with the earth below the permanent moisture level. This is the most common grounding electrode in use today.

Ground rods (rod-type electrodes) must have a diameter of at least 5/8 in. (15.87 mm), and must be made of either zinc or copper-coated steel or solid stainless steel. Of course, there are a few other listed electrode systems that fall into this category that may be used.

Pipe-type electrodes must have a diameter of at least trade size ¾ in. (metric designator 21) and must be coated with an electrically conductive and corrosion-resistant metal such as zinc (galvanization), copper, gold, magnesium, cadmium, platinum, or similar material that is both conductive and intrinsically resistant to corrosion. Copper pipes are considered to be corrosion resistant.

All ground rods and pipe electrodes (electrolytic electrodes) must be bonded to your building/structure's grounding system according to Art. 250.50.

250.52(A)(6). Other Listed Electrodes. There are a number of grounding products available for use. As long as they are "listed" items, they are acceptable according to the Code. Some people feel that electrolytic electrodes fall into this category. Whether found in Art. 250.52(A)(6) or 250.52(A)(5), clearly listed electrolytic electrodes are valid for use by the Code.

Listed or listed item An item, product, or device that has been specifically developed for a particular function and has been independently evaluated by a product-testing organization, such as the UL or CE.

Electrolytic electrode A pipe electrode, commonly made from 2-in. copper piping, that incorporates a desiccant within the hollow portion of the pipe to extract moisture from the air and deliver the water to the base of the electrode, thereby maintaining high levels of moisture to the surrounding backfill material without the need for human assistance. See Art. 250.52(A)(6).

Any other listed electrode present at your site must be bonded to your building/structure's grounding system according to Art. 250.50.

250.52(A)(7). Plate Electrodes. Plate electrodes are one of the least common forms of grounding electrodes due to their very small sphere of influence. About the only place plate electrodes are still in use today is at the bottom of wooden distribution poles. The power company will nail a metal plate to the bottom of the pole and use it as the grounding electrode for any items (such as transformers) that may be hanging on the pole.

In general, you should avoid using plate electrodes, as they are the least effective of all the electrodes. Plate electrodes will often have a sphere of influence of less than 24 in., whereas a typical ground rod will have a sphere of influence of 96 in. Please see the earthing section in Chap. 2 for more information about the sphere of influence.

Sphere of influence The hypothetical volume of soil that will experience the majority of the voltage rise of the ground electrode when that electrode discharges current into the soil. The sphere of influence is equal to the diagonal length of the electrode or electrode system.

Listed or listed item An item, product, or device that has been specifically developed for a particular function and has been independently evaluated by a product-testing organization, such as the UL or CE.

In order to qualify as a permitted electrode, plate electrodes must have at least 2 ft^2 (0.186 m^2) of surface exposed to the soil. Plate electrodes made from solid stainless steel (or other conductive and listed nonferrous metal) must be at least 0.06-in. (1.5-mm) thick. Plate electrodes made from iron or steel must be at least ¼-in. (6.44-mm) thick. Iron and/or steel plates may be bare or conductively coated.

All plate electrodes at your site must be bonded to your building/structure's grounding system according to Art. 250.50, and they must have a supplemental electrode according to Art. 250.53(A)(2).

250.52(A)(8). Other Local Metal Underground Systems or Structures. This section of the Code tells us the requirements for using other underground metal structures, such as well casings and tanks, as an electrode.

In order to qualify as a permitted electrode, other underground metal structures such as piping systems, underground tanks, and underground metal well casings, should meet the general requirements of other electrodes in Part III of Art. 250.

For example, other metal pipe systems and metal well casings, should at a minimum meet the requirements for metal underground water pipe, as required in Art. 250.52(A)(1). Underground metal tanks should have an equivalent metal structure of such shape and size as to at least meet the requirements of the metal frame of a building or structure, as required in Art. 250.52(A)(2).

Permitted electrode An intentional connection to the earth of a copper and/or protected steel system of sufficient size/length that is capable of handling likely objectionable currents from lightning strikes, electrical faults, and the like, safely into the earth. These electrodes can be metal structures in direct contact with the earth that may be used as an electrode to carry objectionable currents to the earth.

However, the Code itself has very few requirements in this section. It would be up to your local building inspectors to approve the use of other metal underground systems or structures, if you should choose to use them as a permitted electrode. That said, even if you choose not to use other local metal underground structures as an electrode, you must still bond them to your grounding system, if these structures are located on your premises.

250.52(B). Not Permitted for Use as Grounding Electrodes. This section of the Code tells us that gas pipes and anything aluminum may not be used as a grounding electrode. Keep in mind that you must bond the gas pipes for your building or structure to the grounding system; you just may not use the gas pipes as an electrode for removing objectionable currents. Article 250.104(B) requires you to bond the building/structure's gas pipes to the grounding system. NFPA 54, the *National Fuel Gas Code*, mandates that an electrical isolation system be installed between the building/structure's gas piping system and the underground portion of the gas pipes. Your local gas company should take care of this for you and should be consulted.

You may not bury aluminum, as it will readily corrode when installed in direct contact with the earth (below grade).

250.53. GROUNDING ELECTRODE SYSTEM INSTALLATION

This section of the Code governs how electrodes are to be installed. You will need to reference Art. 250.52 for additional information.

Article 250.53 is one of the areas of the Code that has undergone some recent changes that resulted in a new understanding of how electrodes are to be installed. The new understanding is that in addition to bonding to building steel, cold-water pipe, and so on, you must install two ground rods at least 6 ft apart at every first service disconnect. Whether or not the Code actually says this or not can be debated. What is clear is that cities and building inspectors nationwide interpret the

Code exactly this way, and you will have little choice in the matter. Ultimately, the practice of installing ground rods at your first service disconnect is a good one.

Note 1 *Agricultural sites often consist of multiple buildings being fed by a single pole-mounted transformer. The* **NEC** *requires that these types of sites have a single disconnecting means at the pole, so the power can be shut off quickly in case of an emergency. This shutoff system is called a site isolation device.* **NEC** *Art. 547.9(3) requires that the grounding electrode system for each building/structure located at an agricultural site be bonded to the site isolation device. Please see Art. 547.9 in the* **NEC** *for more information.*

Note 2 *Agricultural sites often require equipotential planes for the protection of sensitive livestock from electrical shock hazards. It is well known that cows can be killed or injured simply by the electrical fields that can form in the steel rebar in concrete foundations. Copper equipotential planes will reduce the magnetic fields that can form in steel.* **NEC** *Art. 547.10 sets forth the requirements for installing copper equipotential planes within the concrete foundations of buildings and mandates that these equipotential planes be bonded back to the common electrode system for the premises.*

Note 3 *Similar equipotential bonding requirements exist for swimming pools, as mentioned in Note 2. Please see* **NEC** *Art. 680.26 for more information.*

250.53(A). Rod, Pipe, and Plate Electrodes. This section tells us the requirements for installing ground rods, pipe electrodes, and plate electrodes. You may not have nonconductive coatings such as paint or enamel on your grounding electrodes. As mentioned in Art. 250.52(A)(7), plate electrodes should be avoided, as they are the least effective type of electrode.

Ground rod An earthing electrode composed of stainless steel, copper-coated steel, or zinc-coated steel of at least 5/8 in. (15.87 mm) in diameter with at least 8 ft (2.44 m) of length in direct contact with the earth below the permanent moisture level. This is the most common grounding electrode in use today.

250.53(A)(1). Below Permanent Moisture Level. This section of the Code is very confusing, as it simply states that you should install rod, pipe, and plate electrodes below the permanent moisture level, if practicable. Unfortunately, the Code does not define what "permanent moisture level" means. This has left open many interpretations as to the purpose of this provision. Some believe the Code wants grounding electrodes installed below the frost line. Some think that it is below the agricultural wilting point for vegetation. Most people believe that below permanent moisture level means either 18 in. or 30 in. in depth.

What we can say for certain is that the Code intends for electrodes to be placed at some distance below the surface of the earth. This requirement is most typically interpreted as 18 in. (1.5 ft) below grade. This is why inspectors will often require 10-ft ground rods at your site. If you leave 6 in. above grade to connect the grounding electrode conductor to the rod via a clamp, that leaves you with 9.5 ft below grade. If you subtract the permanent moisture level of 1.5 ft, it leaves you 8 ft of effective ground rod in earth, thus qualifying your electrode under Art. 250.52(A)(5).

Permanent moisture level Undefined within the NEC but commonly linked to either the frost line or the permanent wilting point of the site. Permanent moisture level is also often thought of as being either 18 in. (1.5 ft) or 30 in. (2.5 ft) in depth, due to various interpretations of several codes, regulations, and industry standards.

Permanent wilting point The minimal point of soil moisture a plant requires not to wilt; typically water content of soil at any point lower than −1500 J/kg of suction pressure.

Frost line The depth to which the groundwater in soil is expected to freeze; also known as *frost depth* or *freezing depth*. Typically, frost lines are legally required by local building codes.

Grounding (earthing) electrode A device that establishes an electrical connection to the earth/soil.

Nonconductive coatings such as paint or enamel are prohibited from use on your grounding electrodes.

250.53(A)(2). Supplemental Electrode Required. This section of the Code tells us that if you have a single ground rod, pipe, or plate electrode, you must also have a qualified electrode from Art. 250.52(A)(2) through (A)(8) bonded to the rod, pipe, or plate electrode, or bonded to the grounding electrode conductor, the neutral conductor at the service entrance, a nonflexible grounded service raceway, or any grounded service enclosure.

These electrodes in Art. 250.52(A)(2) through (A)(8) are: the metal frame of the building/structure (Art. 250.52(A)(2)), the rebar in the concrete foundation (Art. 250.52(A)(3)), a ground ring (Art. 250.52(A)(4)), a ground rod (Art. 250.52(A)(5)), other listed products (Art. 250.52(A)(6)), plate electrodes (Art. 250.52(A)(7)), and/or other structures (Art. 250.52(A)(8)).

Listed or listed item An item, product, or device that has been specifically developed for a particular function and has been independently evaluated by a product-testing organization, such as the UL or CE.

Grounding (earthing) electrode conductor The conductor used to connect the grounding (earthing) electrode(s) to the equipment-grounding (bonding) conductor, to the neutral (grounded) conductor, or to both in accordance with Arts. 250.66 and 250.142.

First service disconnect The very first electrical panel that will disconnect (turn off) the power coming in from the utility company. Sometimes this term is used to describe the first electrical panel that can turn off power on the secondary or low-voltage side of a transformer.

Service The conductors and equipment for delivering electric energy from the serving utility to the wiring system of the premises.

Service equipment The necessary equipment, usually consisting of circuit breaker(s), switch(es), and/or fuse(s) and their accessories, connected to the load end of service (utility) conductors to a premises and intended to constitute the main control and cutoff of the supply.

Ground rod An earthing electrode composed of stainless steel, copper-coated steel, or zinc-coated steel of at least 5/8 in. (15.87 mm) in diameter with at least 8 ft (2.44 m) of length in direct contact with the earth below the permanent moisture level. This is the most common grounding electrode in use today.

In other words, you may not just install a ground rod or ground plate at your electrical service (first service disconnect). You must also bond your grounding electrode system to building steel (and water pipe), the rebar in your concrete foundation, and any other electrode system (ground ring, etc.) that is in use.

In Fig. 6.3A we see the proper installation for a pair of 8- or 10-ft ground rods that are completely buried below-grade using burial-rated connections. In Fig. 6.3B we see the proper installation for a pair of 10-ft ground rods that are only partially buried with the connections to the grounding electrode conductor above-grade.

There is an exception listed in the Code at Art. 250.53(A)(2) that is the source of much confusion. This exception tells us that you only need a single ground rod if the resistance to earth is less than 25 Ω. The only logical conclusion is that the intent of Art. 250.53(A)(2) was to mandate a second ground rod when single ground rods measure over 25 Ω. The confusion is caused because the text of the Code at Art. 250.53(A)(2) does not actually say anything about installing a second ground rod. There is merely an exception that only makes sense if you interpret the Code to mean that a second ground rod is required.

But do not let this puzzle you. Your local building inspector understands what the Code writers meant and will require you to install a second electrode, unless you can prove that the resistance of your single ground rod is less than 25 Ω. What this has effectively done is to mandate

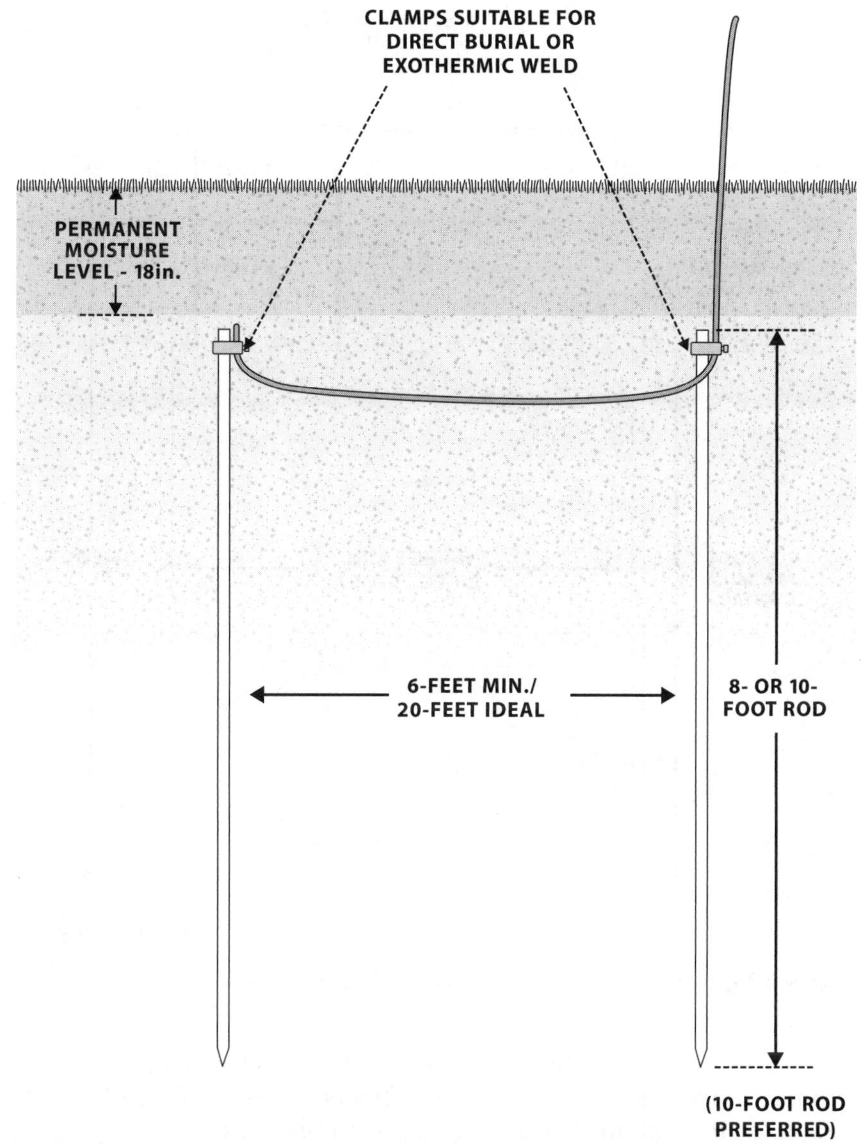

Figure 6.3A Installation method for complete burial of ground rods.

a second ground rod at each and every first service disconnect, mostly because the inspectors do not have the required equipment (or time) to measure the ground rod and confirm that it is under 25 Ω. It is much easier for the inspector to simply have you install the second ground rod.

The Code itself does not actually state how you are to test a ground rod for a resistance to earth (or resistance to ground) of 25 Ω or less. The only mention is in one of the exhibits in the *NEC Handbook* that talks about

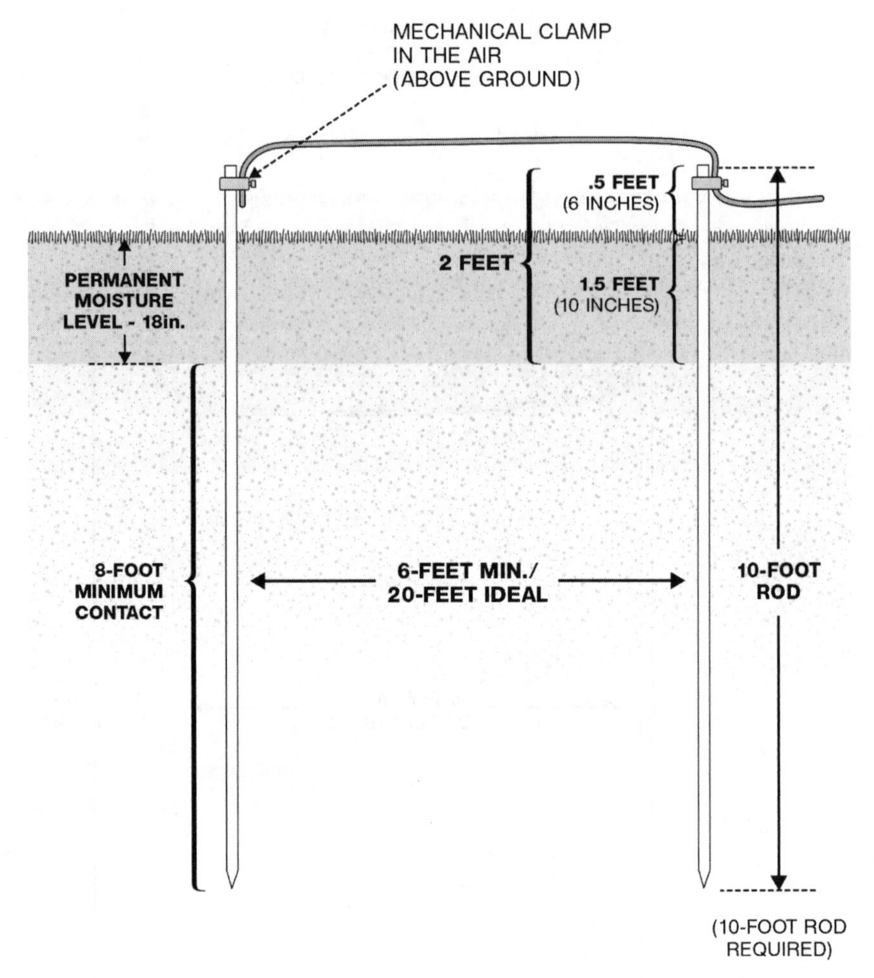

Figure 6.3B Installation method for partial burial of ground rods.

two tests; the three-point fall-of-potential method and the clamp-on ground resistance meter (or induced-frequency test). These tests are discussed in detail in the earthing section in Chap. 2 of this book.

250.53(A)(3). Supplemental Electrode. This section of the Code simply states that your ground rods (or pipe/plate electrodes) should be spaced 20 ft apart. But if room does not exist, you may bring them in as close as 6 ft from each other.

The actual rule listed in the Code at Art. 250.53(A)(3) is that your electrodes should be spaced at twice the distance of the longest electrode. For instance, if you have 8-ft ground rods (which you should *never* use), you should install them 16 ft apart. If you have 10-ft ground rods (which you should *always* use), you should install them 20 ft apart.

Ground rod An earthing electrode composed of stainless steel, copper-coated steel, or zinc-coated steel of at least 5/8 in. (15.87 mm) in diameter with at least 8 ft (2.44 m) of length in direct contact with the earth below the permanent moisture level. This is the most common grounding electrode in use today.

Again, you should not use plate electrodes in normal circumstances, even though they are allowed.

250.53(B). Electrode Spacing. This section of the Code says virtually the exact same thing as Art. 250.53(A)(3): ground rods, ground plates, and other listed electrode products must be spaced at least 6 ft apart from each other, although a distance of twice the length of the longest electrode is preferred.

There are two items the Code adds at this point. The first is that it tells us that ground rods used for lightning-protection systems (aerials, lightning rods, strike-termination devices, etc.) must also follow the 6-ft rule. Remember, the lightning-protection grounding system must be bonded to building's grounding electrode system; see Art. 250.106.

Listed or listed item An item, product, or device that has been specifically developed for a particular function and has been independently evaluated by a product-testing organization, such as the UL or CE.

Grounding (earthing) electrode system Two or more grounding electrodes that are bonded together to form a single system. For a typical building/structure, a grounding electrode system should consist of: two ground rods, water pipe, steel frame of the building, the steel rebar in the concrete foundation, and any ground rings.

Strike-termination devices A component of a lightning-protection system designed to intercept lightning strikes that is in turn connected to a conductive path to ground. Strike-termination devices include lightning rods, finials, air terminals, masts, catenary systems, and other permanent metal structures that qualify.

The other item that is added by the Code in this section is the simple statement that two or more grounding electrodes that are bonded together shall be considered a single grounding electrode system.

250.53(C). Bonding Jumpers. When you install two or more ground rods together, the wire that connects the ground rods together is called a bonding jumper. This bonding jumper must be properly sized, made from the correct type of material, and protected from damage. Specifically, the bonding jumper must meet the requirements found in Art. 250.64(A), (B), and (E), must be sized according to Art. 250.66 and Table 250.66 (see Appendix B), and must be connected in accordance with Art. 250.70.

Bonding jumper A reliable conductor sized per Art. 250 to ensure electrical conductivity between metal parts of the electrical installation.

Grounding (earthing) electrode conductor The conductor used to connect the grounding (earthing) electrode(s) to the equipment-grounding (bonding) conductor, to the neutral (grounded) conductor, or to both in accordance with Arts. 250.66 and 250.142.

Listed or listed item An item, product, or device that has been specifically developed for a particular function and has been independently evaluated by a product-testing organization, such as the UL or CE.

In other words, do not use aluminum wire. If you do, it must be in conduit to protect it from corrosion. The bonding jumper is usually below grade, so aluminum is out in any case, as it is strictly banned from below-grade earthing. Also, the bonding jumper should be the same size and gauge as your grounding electrode conductor; see Art. 250.66. And, of course, all the welds, clamps, connectors, and so on used for the bonding jumper must be listed for such use; see Art. 250.70.

250.53(D). Metal Underground Water Pipe. You must bond to the building/structure's metal underground water pipe, as it is an electrode. It must have a continuous metal path, and you must supplement it by adding an additional ground rod. And, of course, water pipes must have at least 10 ft of pipe in direct contact with the earth; see Art. 250.52(A)(1).

250.53(D)(1). Continuity. This section tells us that water meters, regulators, filtering devices, and other such items installed in the water piping path must be bypassed using a bonding jumper, as discussed in Art. 250.53(C).

Water pipe must be bonded to your building/structure's grounding electrode system within 5 ft of where the piping enters the building. If you have filters, meters, regulators, and other items between the connection to the water pipe and where it enters the earth, you must install a properly sized bonding jumper around the meters.

250.53(D)(2). Supplemental Electrode Required. This section states that not only must you bond to your incoming water pipe, but you must also have an additional grounding electrode (a second electrode). The acceptable grounding electrodes that may be used are found in Art. 250.52(A)(2) through (A)(8): the metal frame of the building/structure (Art. 250.52(A)(2)), the rebar in the concrete foundation (Art. 250.52(A)(3)), a ground ring (Art. 250.52(A)(4)), a ground rod (Art. 250.52(A)(5)), other listed products (Art. 250.52(A)(6)), plate electrodes (Art. 250.52(A)(7)), and/or other structures (Art. 250.52(A)(8)).

Grounding (earthing) electrode conductor The conductor used to connect the grounding (earthing) electrode(s) to the equipment-grounding (bonding) conductor, to the neutral (grounded) conductor, or to both in accordance with Arts. 250.66 and 250.142.

Listed or listed item An item, product, or device that has been specifically developed for a particular function and has been independently evaluated by a product-testing organization, such as the UL or CE.

Ground fault An unintentional connection between an energized conductor and earth or metallic parts of enclosures, raceways, or equipment.

Ground fault-current path An electrically conductive path from the point of a ground fault on a wiring system through normally non-current-carrying conductors, equipment, or the earth to the electrical supply source. A ground fault–current path may not use the earth/soil as a conductor or current path, as it is a violation of the NEC Arts. 250.4(A)(5) and 250.4(B)(4). This term is used for both intentional and unintentional current paths.

Ground rod An earthing electrode composed of stainless steel, copper-coated steel, or zinc-coated steel of at least 5/8 in. (15.87 mm) in diameter with at least 8 ft (2.44 m) of length in direct contact with the earth below the permanent moisture level. This is the most common grounding electrode in use today.

This second electrode must be bonded to the grounding electrode conductor, the neutral conductor at the service entrance, nonflexible grounded service raceway, any grounded service enclosure, or the ground bar found inside the first service disconnect cabinet, as in Art. 250.32(B).

In other words, you must have at least one properly installed ground rod to go along with your metal water pipe. Most important, the supplemental grounding electrode, typically a ground rod, must be installed as if it were the sole grounding electrode for the entire electrical system.

The reason the Code has this requirement is that, traditionally, the only grounding electrode for an electrical system, would be the water pipe. Water companies have for many years now been replacing the metal pipes with plastic ones, thus leaving many buildings without a proper ground. The requirement for an additional ground rod ensures electrical systems will have a proper ground-fault path.

250.53(E). Supplemental Electrode Bonding Connection. This section of the Code tells us that if the supplemental grounding electrode required in Art. 250.53(D)(2) is a ground rod, pipe, or plate electrode, the grounding electrode conductor (or bonding jumper) does not need to be larger than 6 AWG copper or 4 AWG aluminum.

> **Grounding (earthing) electrode conductor** The conductor used to connect the grounding (earthing) electrode(s) to the equipment-grounding (bonding) conductor, to the neutral (grounded) conductor, or to both in accordance with Arts. 250.66 and 250.142.

You are best off simply sizing the grounding electrode conductor in accordance with Table 250.66 (see Appendix B), even if you are allowed to go smaller due to this rule. The notes in the *NEC Handbook* also make the same recommendation; you should follow the sizing rules found in Art. 250.66, even if this rule is applicable to you.

250.53(F). Ground Ring. This part of the Code lists additional requirements to those found in Art. 250.52(A)(4), in that when installing a ground ring, it must be installed at least 2.5 ft or 30 in. (750 mm) below grade level.

You will often see requirements for ground rings to be installed only 1–1.5 ft below grade. This is often due to an electrical engineering requirement for reducing hazardous touch voltages that could compromise human safety. If this is the case for your site, you will not be able to qualify the ground ring as an electrode under the NEC.

> **Ground ring** An earthing electrode, composed of #2 AWG or larger bare copper wire at least 20 ft in length and buried at least 2.5 ft below grade and in direct contact with the earth, that circumnavigates a structure or building and is reconnected to itself, typically with multiple connections to the building's or structure's steel rebar and/or the building's steel system, and used in conjunction with standard ground rods. It is considered to be the most effective electrode system in use today.

If you do run into this very common requirement, just remember that installing a ground ring at only 1.5 ft below grade simply disqualifies it for use as an electrode under the NEC. You need to have another qualified grounding electrode, such as two 10-ft ground rods spaced at least 6 ft apart, to meet the requirements.

250.53(G). Rod and Pipe Electrodes. This part of the Code adds additional requirements to those found in Art. 250.52(A)(5), in that you must have at least 8 ft of the electrode in direct contact with the earth and below permanent moisture level; see Art. 250.53(A)(1). However, it gives some flexibility when driving rods into rocky soil. The Code gives you three options in order of preference (see Fig. 6.4):

- Method 1: install the rod straight into the earth.
- Method 2: if rock is encountered when driving in the rod (first method), you may install the rod at a 45° angle; preferably pointed away from your building/structure.
- Method 3: If the rod will still not go 8 ft at a 45° angle due to buried rocks, you may dig a trench 2.5-ft deep and bury the ground rod in it. The top and bottom of the ground rod must be completely buried to a depth of at least 2.5 ft.

In Fig. 6.4 we see the proper installation of an 8- or 10-ft ground rod in all three of the acceptable ways listed in the Code.

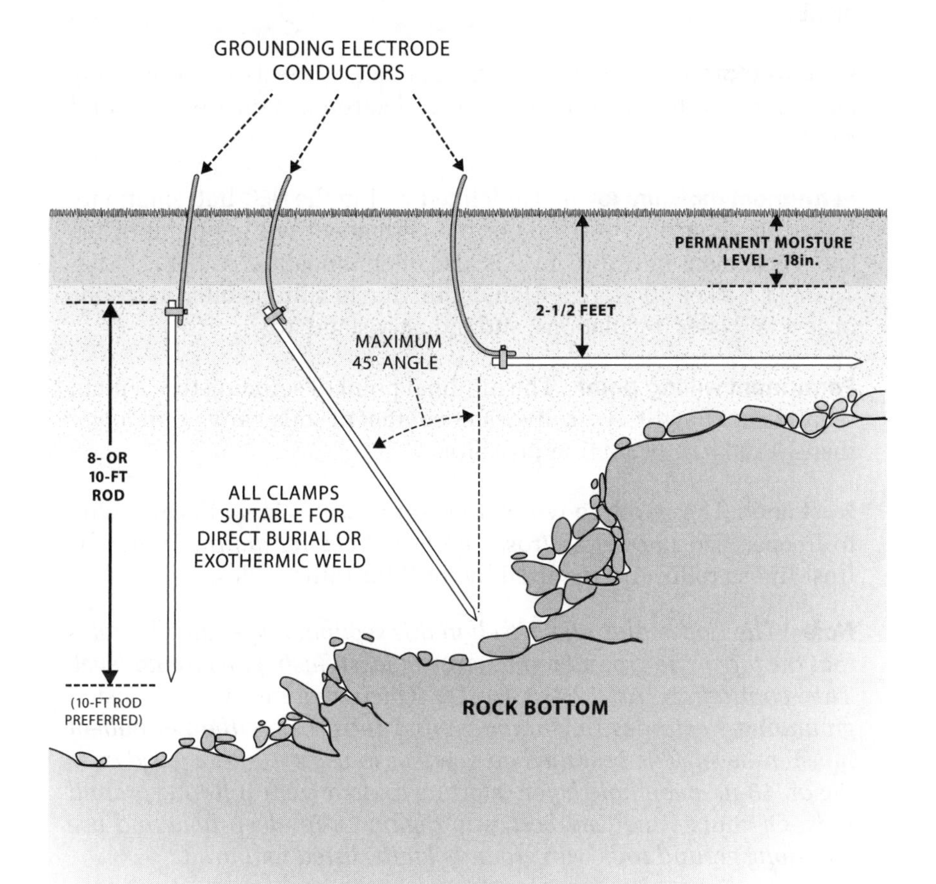

Figure 6.4 Three allowable ground rod installation methods.

Remember that Art. 250.70 requires you to use ground clamps that are listed and rated for direct burial. Also, Art. 250.10 requires you to protect ground clamps from physical damage.

Ground rod An earthing electrode composed of stainless steel, copper-coated steel, or zinc-coated steel of at least 5/8 in. (15.87 mm) in diameter with at least 8 ft (2.44 m) of length in direct contact with the earth below the permanent moisture level. This is the most common grounding electrode in use today.

Listed or listed item An item, product, or device that has been specifically developed for a particular function and has been independently evaluated by a product-testing organization, such as the UL or CE.

Ground clamp A mechanical device that connects the grounding electrode conductor to a grounding electrode, typically a ground rod.

Permanent moisture level Undefined within the NEC but commonly linked to either the frost line or the permanent wilting point of the site. Permanent moisture level is also often thought of as being either 18 in. (1.5 ft) or 30 in. (2.5 ft) in depth, due to various interpretations of several codes, regulations, and industry standards.

Permanent wilting point The minimal point of soil moisture a plant requires not to wilt; typically water content of soil at any point lower than –1500 J/kg of suction pressure.

Frost line The depth to which the groundwater in soil is expected to freeze; also known as *frost depth* or *freezing depth*. Typically, frost lines are legally required by local building codes.

Note *The Code contradicts itself in this section. It specifically states that the top of the ground rod must be at least flush with ground level. This contradicts Art. 250.53(A)(1), which requires you to install grounding electrodes (including ground rods) below the permanent moisture level. The best method is either to use a 10-ft-long rod or to dig an 18-in.-deep hole before starting to drive your 8-ft-long ground rods. Of course, the very best is to dig an 18-in.-deep hole and use 10-ft-long ground rods with ground clamps listed for burial.*

250.53(H). Plate Electrode. This part of the Code adds additional requirements to those found in Art. 250.52(A)(7), in that you must bury ground plates at least 2.5 ft below grade. Again, you should not use plate electrodes, as they are the worst grounding electrodes.

250.54. AUXILIARY GROUNDING ELECTRODES

What this section of the Code tells us is that you are allowed to install additional grounding electrodes (typically ground rods) at various points throughout your electrical system, to wherever you have an equipment-grounding conductor, without having to meet all the Code requirements found in Arts. 250.50 and 250.53(C), or the 25-Ω resistance requirement found in Art. 250.53(A)(2).

Equipment-grounding (bonding) conductor The low-impedance fault-current path that is typically run with or encloses the circuit conductors, used to connect the non-current-carrying metal parts of equipment, raceways, and other enclosures to the grounded (neutral) conductor and equipment-grounding (bonding) conductor at service equipment or at the source of a separately derived system. Often called the *green-wire* or *third-wire* ground conductor.

First service disconnect The very first electrical panel that will disconnect (turn off) the power coming in from the utility company. Sometimes this term is used to describe the first electrical panel that can turn off power on the secondary or low-voltage side of a transformer.

Grounding (earthing) electrode system Two or more grounding electrodes that are bonded together to form a single system. For a typical building/structure, a grounding electrode system should consist of: two ground rods, water pipe, steel frame of the building, the steel rebar in the concrete foundation, and any ground rings.

Ground fault-current path An electrically conductive path from the point of a ground fault on a wiring system through normally non-current-carrying conductors, equipment, or the earth to the electrical supply source. A ground fault–current path may not use the earth/soil as a conductor or current path, as it is a violation of the NEC Arts. 250.4(A)(5) and 250.4(B)(4). This term is used for both intentional and unintentional current paths.

In other words, your first service disconnect must have a grounding electrode system that meets all of the requirements found in the NEC. Any subpanels in your building's electrical system may have additional grounding electrodes installed, as best determined, that do not need to be bonded to water pipe or building steel or to measure below 25 Ω, or to meet other electrical engineering specifications.

It is quite common and always a good idea (some believe it should be mandatory) to add a ground rod at each electrical subpanel in your system. This part of the Code allows you to add these additional ground rods.

Remember, you may not use the earth/soil as a conductor for ground-fault currents; this is strictly forbidden under Art. 250.4(A)(5) and (B)(4). So you must have a green-wire ground running from your first service disconnect to your subpanels and other equipment enclosures.

250.58. COMMON GROUNDING ELECTRODE

The Code requires that all of the grounding electrodes throughout your entire electrical system must be bonded together to form a single common grounding electrode. This is true for separate services, feeders, and/or branch services.

Grounding (earthing) electrode A device that establishes an electrical connection to the earth/soil.

Grounding (earthing) electrode system Two or more grounding electrodes that are bonded together to form a single system. For a typical building/structure, a grounding electrode system should consist of: two ground rods, water pipe, steel frame of the building, the steel rebar in the concrete foundation, and any ground rings.

Service The conductors and equipment for delivering electric energy from the serving utility to the wiring system of the premises.

Feeder circuit All circuit conductors between the first service disconnect and the final OCPD. The circuits after the final OCPD are called the *branch circuits.*

Circuit or branch circuit The electrical system downstream from the last OCPD. The outlet you plug your computer into would be part of a circuit.

Ground fault-current path An electrically conductive path from the point of a ground fault on a wiring system through normally non-current-carrying conductors, equipment, or the earth to the electrical supply source. A ground fault–current path may not use the earth/soil as a conductor or current path, as it is a violation of the NEC Arts. 250.4(A)(5) and 250.4(B)(4). This term is used for both intentional and unintentional current paths.

Two or more grounding electrodes that are bonded together are considered to be a single grounding electrode system under the NEC. This ensures that you do not use the earth as a ground fault–current path, which is forbidden under Art. 250.4(A)(5) and (B)(4) of the Code.

250.60. USE OF STRIKE-TERMINATION DEVICES

A strike-termination device is commonly known as an air terminal, lightning aerial, lightning rod, or sometimes called a lightning conductor. The Code tells us that the ground rods used for lightning-protection systems may not be used as qualified electrodes. In other words, you must have a dedicated grounding electrode system for your first service disconnect.

Strike-termination devices A component of a lightning-protection system designed to intercept lightning strikes that is in turn connected to a conductive path to ground. Strike-termination devices include lightning rods, finials, air terminals, masts, catenary systems, and other permanent metal structures that qualify.

First service disconnect The very first electrical panel that will disconnect (turn off) the power coming in from the utility company. Sometimes this term is used to describe the first electrical panel that can turn off power on the secondary or low-voltage side of a transformer.

Grounding (earthing) electrode system Two or more grounding electrodes that are bonded together to form a single system. For a typical building/structure, a grounding electrode system should consist of: two ground rods, water pipe, steel frame of the building, the steel rebar in the concrete foundation, and any ground rings.

The Code additionally informs us in this section that under Art. 250.106 you are required to bond your building/structure's lightning-protection system to the common electrical service electrode system.

You are also required to bond your telecommunications grounding system under Art. 800.100(D), your radio and television grounding system under Art. 810.21(J), and your coaxial shield grounds (from cable television service) under Art. 820.100(D) to the common electrical service electrode system.

All of your grounding systems (electrical, lightning, radio, television, cable, telecommunications, etc.) in your building/structure must be bonded together to limit the potential differences between the systems.

250.62. Grounding Electrode Conductor Material. The wire used to connect your electrodes to your electrical system is called the *grounding electrode conductor.* The Code tells us that the grounding electrode conductor must be made of copper, aluminum, or copper-clad aluminum. The material must be resistant to any likely corrosive agent it could come in contact with and/or must be protected against the corrosive agent. The grounding electrode conductor may be solid, stranded, insulated, or bare.

> **Grounding (earthing) electrode conductor** The conductor used to connect the grounding (earthing) electrode(s) to the equipment-grounding (bonding) conductor, to the neutral (grounded) conductor, or to both in accordance with Arts. 250.66 and 250.142.

In other words, if you know that the grounding electrode conductor will be in a corrosive environment, you should pick either copper or aluminum based on its ability to withstand the corrosive agent and/or protect the conductor from the corrosive agent.

Remember, aluminum may not be used in direct contact with the earth (below grade), as it simply corrodes too quickly; see Arts. 250.52(B) and 250.64(A). If the top of your ground rod is going to be buried below grade using a listed direct burial clamp, your grounding electrode conductor must be copper. In fact, you may not have aluminum or copper-clad aluminum conductors within 18 in. of the earth.

250.64. GROUNDING ELECTRODE CONDUCTOR INSTALLATION

This section of the Code discusses the installation requirements for the grounding electrode conductor. Primarily, you may not bury aluminum or copper-clad aluminum, you must secure and protect the conductor from damage, and it should be one continuous piece of wire free from splices. This section also discusses how to install the grounding electrode conductor for multiple disconnects, how to tap off the conductor for multiple grounding electrodes and enclosures, and more.

250.64(A). Aluminum or Copper-clad Aluminum Conductors. This section of the Code tells us that aluminum and copper-clad aluminum conductors may not be used in direct contact with the earth (below grade) or be in direct contact with masonry, which will corrode materials too quickly; see Art. 250.52(B). Additionally, if your grounding electrode conductor is located outside, you may not have aluminum or copper-clad aluminum conductors come within 18 in. of the earth.

> **Grounding (earthing) electrode conductor** The conductor used to connect the grounding (earthing) electrode(s) to the equipment-grounding (bonding) conductor, to the neutral (grounded) conductor, or to both in accordance with Arts. 250.66 and 250.142.

If the top of your ground rod is going to be buried below grade using a listed direct burial clamp, your grounding electrode conductor must be copper.

250.64(B). Securing and Protection against Physical Damage. The Code tells us that you must securely fasten the grounding electrode conductor to the surface on which it is carried, and protect it from damage under the following conditions:

	8 or 6 AWG conductor	4 AWG conductor or larger
Free from exposure to physical damage	Does not need to be routed in conduit and is permitted to be routed along the service building/structure	Must be routed in conduit (see note below)
Exposed to physical damage	Must be in one of the following conduits: RMC, IMC, PVC, RTRC, EMT, or cable armor	Must be in one of the following conduits: RMC, IMC, PVC, RTRC, EMT, or cable armor

Note *This section of the Code is poorly written and does not explicitly state that 4 AWG or larger grounding electrode conductors must be in conduit; it merely implies that they should be. However, almost everyone understands it to mean that 4 AWG and larger conductor must be routed in securely fastened conduit.*

The Code also tells us in this section that you may install grounding electrode conductors on or through the framing members of buildings/structures. This includes holes in studs, joists, and/or rafters, whether exposed or concealed.

Table 250.66 (see Appendix B) requires a minimum size of an 8 AWG copper or 6 AWG aluminum conductors for all grounding electrode conductors and has a maximum size of a 3/0 AWG copper or 250-kcmil aluminum, although larger sizes are allowed.

Also, please see Art. 250.64(E), which has some very specific bonding requirements regarding metal raceways and conduit when grounding electrode conductors are involved.

250.64(C). Continuous. Your grounding electrode conductor should be one continuous length without splices. However, if it is necessary to splice the conductor due to construction or a remodeling project, the splice must be done as follows:

- For grounding electrode conductors made from wire, only listed irreversible compression fittings or exothermic welding is allowed.
- For grounding electrode conductors made from bus bars, the various bus bar sections may be connected together via the listed method for that bus bar system.
- For grounding electrode conductors made from structural metal frames, splices must be bolted, riveted, or welded.
- For grounding electrode conductors made from metal water piping, splices must be threaded, welded, brazed, soldered (silver solder is typically required due to the higher melting point), or connected with a bolted flange.

Remember, for new construction, you must have a single continuous grounding electrode conductor when using wire. You may only splice the wire, if necessary, due to remodeling/construction of the electrical system.

Also, it is unusual (and not recommended) to use water pipe as a grounding electrode conductor; typically we only bond to water pipe within 5 ft of its entrance to the building. If you are using water pipe as a grounding electrode conductor, Art. 250.68(C) requires that only qualified persons may have access to such a system.

There are a few exceptions to this rule. Article 250.30(A)(5) and (6) allows some exceptions for separately derived systems and multiple separately derived systems. Article 250.30(B)(1) allows some exceptions for ungrounded separately derived systems. Please refer to these sections for more information.

250.64(D). Service with Multiple Disconnecting Means Enclosures. This section of the Code tells how to properly install the grounding electrode conductor when you have multiple electrical cabinets (first service disconnects) being fed from a common electrical service.

The basic premise is that the grounding electrode conductor must be sized to match the service entrance conductor that feeds the multiple disconnects. You may then use the grounding electrode conductor as a bus system and tap into it for each of the individual disconnects. The individual grounding conductors from each of the disconnects may be sized accordingly and tapped into the main grounding electrode conductor.

The example listed in the *NEC Handbook* is for a main electrical utility feed (service entrance conductor) that is 350 AWG (310 A) acting as a bus system. Two disconnects are tapped off the 350 AWG, based on their respective sizes. Disconnect 1 is a 100-A service and uses a 2 AWG (115 A) feeder conductor tapped off the 350 AWG. Disconnect 2 is a 200-A service with a 3/0 AWG (200 A) feeder conductor tapped off the 350 AWG. These two disconnects must have the ground bar inside the cabinet bonded to the building/structure's grounding electrode system via a grounding electrode conductor. Instead of running two separate wires, we can use a single common wire to connect both disconnects to the grounding electrode system.

If we look at Table 250.66 (see Appendix B), we see that Disconnect 1 requires an 8 AWG ground conductor, and Disconnect 2 requires a 4 AWG ground conductor. These two ground conductors must be tapped into a common (main) grounding electrode conductor that is at least 2 AWG, because the main feed is 350 AWG.

Service entrance conductor The service conductors that bring power to the meter and enter the building, up to the first service disconnect.

Grounding (earthing) electrode conductor The conductor used to connect the grounding (earthing) electrode(s) to the equipment-grounding (bonding) conductor, to the neutral (grounded) conductor, or to both in accordance with Arts. 250.66 and 250.142.

First service disconnect The very first electrical panel that will disconnect (turn off) the power coming in from the utility company. Sometimes this term is used to describe the first electrical panel that can turn off power on the secondary or low-voltage side of a transformer.

The following sections detail the requirements for using a common grounding electrode conductor when installing multiple disconnects. Your system must, of course, meet the requirements found in Art. 250.64(D)(1) to (3).

250.64(D)(1). Common Grounding Electrode Conductor and Taps. This section of the Code tells us that when you have multiple electrical disconnects being fed from a common utility service, you may use a single unspliced common grounding electrode conductor to bond the various disconnects to the grounding electrode system. The common grounding electrode conductor must be sized according to the incoming utility service entrance conductor, per Table 250.66 (see Appendix B). The individual disconnects must have grounding conductors bonded to the interior ground bar in the cabinet, and the conductor must be sized according to the size of the individual disconnects, per Table 250.66 (see Appendix B).

Figure 6.5 is a schematic showing the proper sizing of the three required grounding electrode conductors for the multiple electrical disconnects shown. In the figure, we can see that the first electrical disconnect has a #8 AWG grounding electrode conductor which is sized in accordance with the incoming #3 AWG phase conductor. The second electrical disconnect has a #4AWG grounding electrode conductor which is sized to in accordance with the incoming 3/0 AWG phase conductor. The grounding electrode conductor from both electrical disconnects are then bonded to a #2 AWG common grounding electrode conductor that is sized in accordance with the main 350 kcmil utility feed.

The bonds made to the common grounding electrode conductor must be made using one of the following methods, so the common grounding electrode conductor remains without a splice or a joint:

1. Exothermic welding.
2. A connector(s) listed for grounding and bonding equipment.
3. A copper or aluminum grounding bar (bus bar); see below for details.

If you are using a grounding bar (bus bar) to connect your grounding electrode conductors, the ground bar (bus bar) must be at least ¼ in. × 2 in. (6 mm × 50 mm) and of sufficient length to accommodate the number of terminations necessary for the installation. It must be securely fastened in an accessible area using listed connections or exothermic welding.

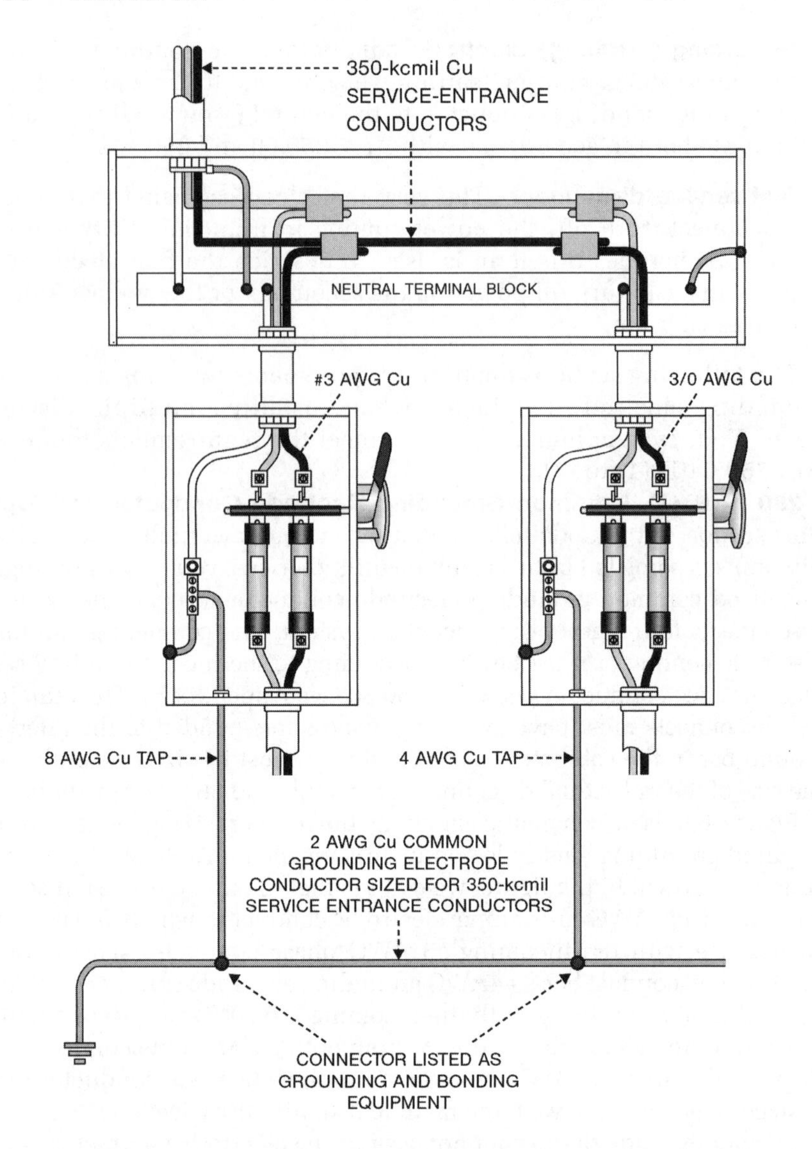

Figure 6.5 Proper grounding electrode conductor sizing.

More information can be found in Art. 250.30(A)(6). If you have an aluminum system, it must comply with Art. 250.64(A), which prohibits allowing aluminum within 18 in. of the earth.

Service entrance conductor The service conductors that bring power to the meter and enter the building, up to the first service disconnect.

Grounding (earthing) electrode conductor The conductor used to connect the grounding (earthing) electrode(s) to the equipment-grounding (bonding) conductor, to the neutral (grounded) conductor, or to both in accordance with Arts. 250.66 and 250.142.

First service disconnect The very first electrical panel that will disconnect (turn off) the power coming in from the utility company. Sometimes this term is used to describe the first electrical panel that can turn off power on the secondary or low-voltage side of a transformer.

Please see Art. 250.64(D) for additional information.

The 2014 Code changed this section slightly to include the provision that the ground bar must be of sufficient length to accommodate the number of terminations necessary for the installation.

250.64(D)(2). Individual Grounding Electrode Conductors. This section of the Code tells us that when you have multiple first service disconnects being fed from a common utility feed (service entrance conductor), the neutral wire (grounded conductor) must be bonded to ground in each enclosure. Part II of NEC Art. 250 covers this requirement for a neutral-to-ground bond as well.

Service entrance conductor The service conductors that bring power to the meter and enter the building, up to the first service disconnect.

Grounding (earthing) electrode conductor The conductor used to connect the grounding (earthing) electrode(s) to the equipment-grounding (bonding) conductor, to the neutral (grounded) conductor, or to both in accordance with Arts. 250.66 and 250.142.

First service disconnect The very first electrical panel that will disconnect (turn off) the power coming in from the utility company. Sometimes this term is used to describe the first electrical panel that can turn off power on the secondary or low-voltage side of a transformer.

Grounded conductor (neutral) A system or circuit conductor that is intentionally grounded. This is a current-carrying conductor typically called the *neutral wire.*

Neutral-to-ground connection (bond) is where a grounded current-carrying conductor, the neutral, is intentionally bonded to earth/ground. In general, this should only occur at the transformer and at the first service disconnect(s). The Code uses several names for neutral-to-ground connections: system bonding jumper, main bonding jumper, and supply-side bonding jumper.

In other words, each first service disconnect must have a neutral-to-ground bond that is sized in accordance to Table 250.66 (see Appendix B).

250.64(D)(3). Common Location. The Code tells us in this section that in addition to the neutral-to-ground bond that must be installed in each first service disconnect(s), you must also have a neutral-to-ground bond on the service side of the disconnect. Typically this is done at the XO of the transformer, but it could be done at other places. See Part II of NEC Art. 250 for more information.

250.64(E). Raceways and Enclosures for Grounding Electrode Conductors. This section of the Code gives us some very important rules regarding the routing of grounding electrode conductors through metal raceway and/or conduit. Improper installation can result in the formation of an inductive choke. If your service were to have an electrical fault, there would be a massive amount of current on the grounding electrode conductor for a short duration. If the metal conduit is not electrically in parallel with grounding electrode conductor, the magnetic fields that form on the two objects will occur at different times/rates, resulting in a massive increase in impedance (resistance will stay low, but the impedance will be high). The result could be a catastrophic failure of the grounding electrode conductor when it is needed most.

Inductive choke An electrical phenomenon that occurs when alternating current (AC) is passed through an insulated nonferrous (copper or aluminum) conductor in proximity to a ferrite material (steel/iron), causing two different electrical fields to form and resulting in a dramatic increase in impedance. The two materials will form differing magnetic fields that will resist changes to the AC passing through them (which reverses direction rapidly), resulting in an impedance increase with frequency. This typically occurs when a single conductor is routed through a metallic raceway without bonding the two together; high-frequency/high-current events, such as lightning or electrical faults, will cause the raceway to act as an inductor, thus increasing the impedance of the current path by as much as 97 percent, resulting in catastrophic failure of the conductor.

Grounding (earthing) electrode conductor The conductor used to connect the grounding (earthing) electrode(s) to the equipment-grounding (bonding) conductor, to the neutral (grounded) conductor, or to both in accordance with Arts. 250.66 and 250.142.

First service disconnect The very first electrical panel that will disconnect (turn off) the power coming in from the utility company. Sometimes this term is used to describe the first electrical panel that can turn off power on the secondary or low-voltage side of a transformer.

Raceway An enclosed channel of metal or nonmetallic material designed expressly for holding wires, cables, bus bars, and so on. Raceways include all types of conduit, flexible tubing, raceways, floor raceways, busways, and so on.

Bonding jumper A reliable conductor sized per Art. 250 to ensure electrical conductivity between metal parts of the electrical installation.

Grounding (earthing) electrode A device that establishes an electrical connection to the earth/soil.

The 2014 NEC split this section of the Code into four subsections for clarity.

250.64(E)(1). General. When your grounding electrode conductor is enclosed in metal (ferrous) raceway or conduit, it must be electrically continuous or made to be electrically continuous with additional bonding. This electrically continuous state must occur from the enclosure of the first service disconnect all the way through to the grounding electrode. You must also bond the grounding electrode clamp back to the raceway using a conductor that is the same size or larger than the grounding electrode conductor. This is the "loop" you often see at the top of a ground rod, where the grounding electrode conductor passes through the clamp and is bent back and around to the conduit bushing. Figure 6.6 illustrates the proper installation of a grounding electrode conductor encased in metal conduit. Note the two mandatory bonding jumpers connecting the metal conduit to the grounding electrode conductor on both ends.

PVC conduit (nonferrous) is exempt from this rule.

The 2014 NEC split this section of the Code into a subsection for clarity.

250.64(E)(2). Methods. Bonding methods must be in compliance with Art. 250.92(B) and ensured by one of the methods in Art. 250.92(B)(2) through (B)(4).

The Code in Art. 250.92(B) tells us that locknuts, washers, and bushings are not acceptable for use as bonding. You must have bonding jumpers around these items to ensure electrical continuity from the disconnect to the grounding electrode. Article 250.92(B)(2) and (3) tells us that threaded couplings, hubs, or connectors for enclosures or raceways are acceptable if made tight. Article 250.92(B)(4) tells us that listed devices such as bonding-type locknuts, bushings, or bushings with bonding jumpers are also acceptable

Listed or listed item An item, product, or device that has been specifically developed for a particular function and has been independently evaluated by a product-testing organization, such as the UL or CE.

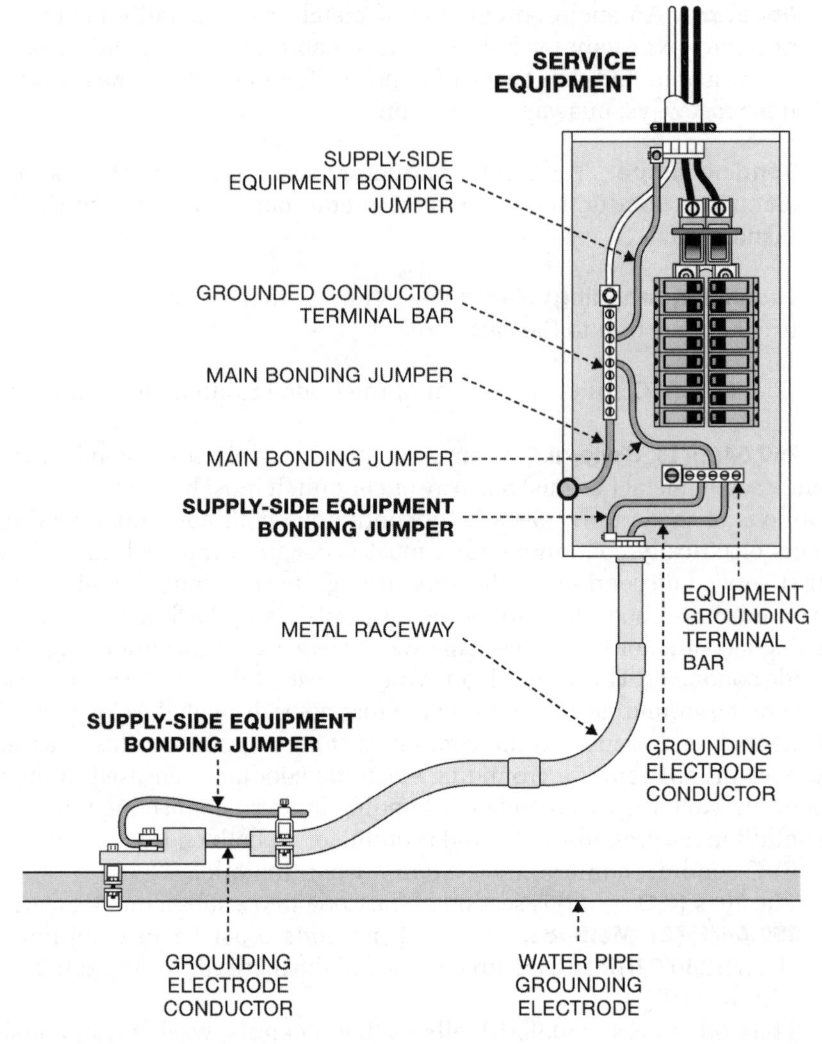

Figure 6.6 Grounding electrode conductor in metal conduit bonded on both ends.

The 2014 **NEC** split this section of the Code into a subsection for clarity.

250.64(E)(3). Size. The bonding jumper for ensuring that the raceway or enclosure protecting a grounding electrode conductor or cable armor must be the same size or larger than the encased grounding electrode conductor.

The 2014 **NEC** split this section of the Code into a subsection for clarity.

250.64(E)(4). Wiring Methods. If you use a raceway to protect the grounding electrode conductor, the installation of the raceway must be done in accordance with the appropriate raceway article in the Code.

The 2014 NEC split this section of the Code into a subsection for clarity.

250.64(F). Installation to Electrodes(s). This section tells us how the installation of the grounding electrode conductor must be made to the grounding electrodes. It gives you three options:

1. You may install the grounding electrode conductor to any convenient electrode available in your building/structure's grounding electrode system. The Code is saying that once you have your building/structure's electrode system established (building steel, rebar in your concrete foundation, cold-water pipe, ground rods, etc. all bonded together to form one system), you may connect the grounding electrode conductor to any convenient electrode in that system. See Art. 250.53(C) for more information.
2. Your grounding electrode conductor may be routed to your grounding electrodes individually. This means you may have a grounding electrode conductor routed from your first service disconnect to your ground rod system, then a separate connection to the water pipe, a third to building steel, a fourth to the rebar in your concrete foundation, and so on. Typically, this is done using a bus bar (see option 3).
3. You may use an aluminum or copper ground bar that is at least ¼ in. × 2 in. (6 mm × 50 mm) in size, securely fastened, and accessible as a bonding jumper for connecting the grounding electrode conductor to the various electrodes. You must use listed connections to the ground bar or exothermic welding. If you use an aluminum ground bar, it must comply with Art. 250.64(A), which says that aluminum may not come within 18 in. of the earth.

Grounding (earthing) electrode conductor The conductor used to connect the grounding (earthing) electrode(s) to the equipment-grounding (bonding) conductor, to the neutral (grounded) conductor, or to both in accordance with Arts. 250.66 and 250.142.

First service disconnect The very first electrical panel that will disconnect (turn off) the power coming in from the utility company. Sometimes this term is used to describe the first electrical panel that can turn off power on the secondary or low-voltage side of a transformer.

Raceway An enclosed channel of metal or nonmetallic material designed expressly for holding wires, cables, bus bars, and so on. Raceways include all types of conduit, flexible tubing, raceways, floor raceways, busways, and so on.

Bonding jumper A reliable conductor sized per Art. 250 to ensure electrical conductivity between metal parts of the electrical installation.

Grounding (earthing) electrode A device that establishes an electrical connection to the earth/soil.

Listed or listed item An item, product, or device that has been specifically developed for a particular function and has been independently evaluated by a product-testing organization, such as the UL or CE.

You need to bond all of the various grounding electrodes for your building/structure together to form a single grounding electrode system. This grounding electrode system must be tied back to your first service disconnect via a grounding electrode conductor. The grounding electrode conductor may be bonded to any convenient point on the grounding electrode system.

250.66. SIZE OF ALTERNATING-CURRENT GROUNDING ELECTRODE CONDUCTOR

The grounding electrode conductor must be properly sized so that it can handle the likely fault currents. The sizing is based on the incoming service feed of the phase wires. The following charts show you the minimum size conductor allowed. You may of course use a larger size. Many engineers specify the grounding electrode conductor be a 4/0 AWG copper conductor, as this exceeds all requirements in the NEC.

Table 250.66 Copper and Aluminum Conductors

Size of largest phase conductor for service in AWG or kcmil (MCM)		Required size of grounding electrode conductor in AWG or kcmil (MCM)	
Copper	Aluminum or copper-clad aluminum	Copper	Aluminum or copper-clad aluminum
2 or smaller	1/0 or smaller	8	6
1 or 1/0	2/0 or 3/0	6	4
2/0 or 3/0	4/0 or 250	4	2
Over 3/0 to 350	Over 250 to 500	2	1/0
Over 350 to 600	Over 500 to 900	1/0	3/0
Over 600 to 1100	Over 900 to 1750	2/0	4/0
Over 1100	Over 1750	3/0	250

Figure 6.7 shows all of the required grounding connections for a typical building from the electrical service to each of the grounding electrodes: building steel, steel rebar in concrete (concrete encased electrode), water pipe, ground ring, and ground rods.

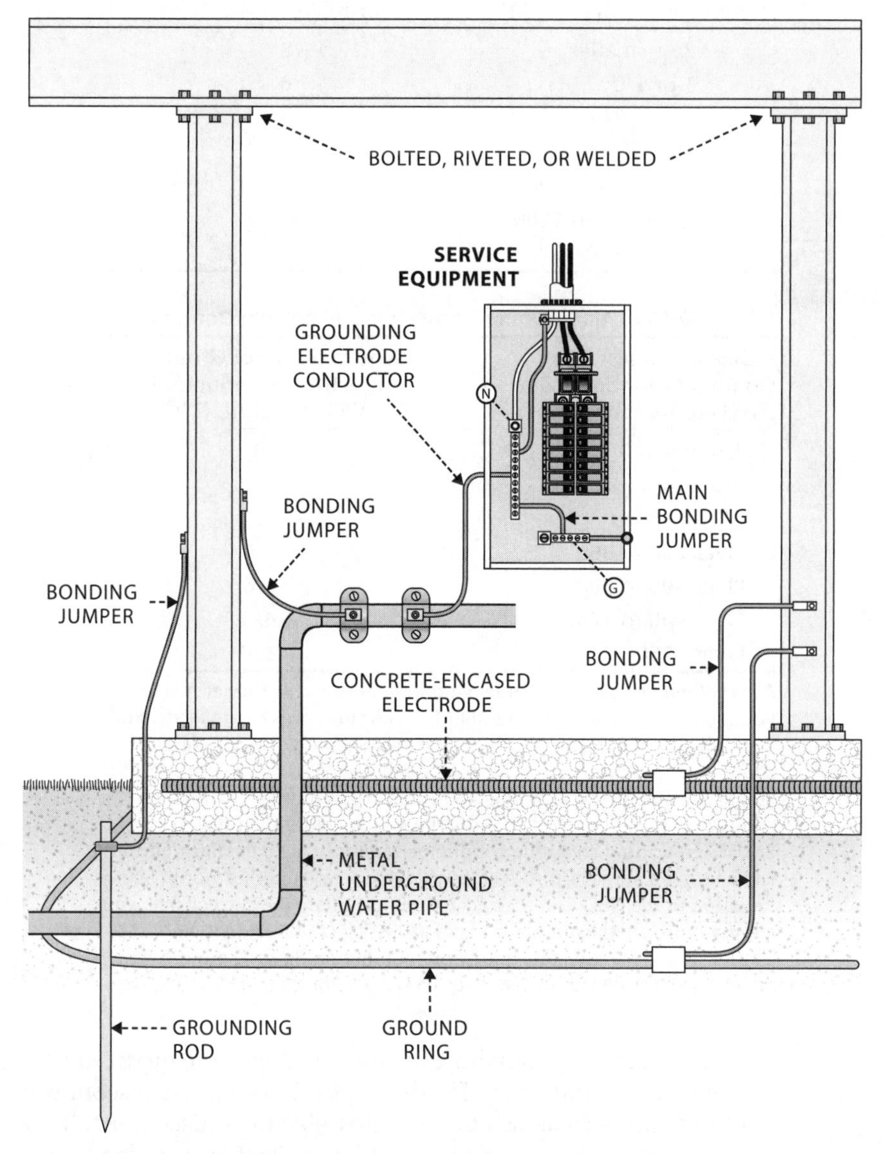

Figure 6.7 Jumper naming conventions in a typical building.

Here is an alternate breakdown for you, separating out the two different materials:

Table 250.66A Copper Conductors

Size of Largest Phase Conductor for Service in AWG or KCMIL (MCM)	Required Size of Grounding Electrode Conductor in AWG or KCMIL (MCM)
2 or smaller	8
1 or 1/0	6
2/0 or 3/0	4
Over 3/0 to 350	2
Over 350 to 600	1/0
Over 600 to 1100	2/0
Over 1100	3/0

Table 250.66B Aluminum or Copper-clad Aluminum Conductors

Size of Largest Phase Conductor for Service in AWG or KCMIL (MCM)	Required Size of Grounding Electrode Conductor in AWG or KCMIL (MCM)
1/0 or smaller	6
2/0 or 3/0	4
4/0 or 250	2
Over 250 to 500	1/0
Over 500 to 900	3/0
Over 900 to 1750	4/0
Over 1750	250

Note: When you have multiple conductors providing the service feed, you must take the combined total of the circular mils of the feeds. For example, if your electrical cabinet has two sets of 4/0 AWG copper conductors providing a service feed, each 4/0 AWG conductor is rated at 230 A, so your actual feed is 460 A. Your grounding electrode conductor must be rated to handle the 460-A feed. If you install a 2 AWG grounding electrode conductor, you only have the fault-current capacity for a single 4/0 AWG (230 A) conductor, and this would be a violation of the **NEC** Code. To accurately calculate the correct size, you simply multiply the circular mils of 4/0 AWG (211,592) by two: $211{,}592 \times 2 = 423{,}184$, or a 450-kcmil conductor. So your grounding electrode conductor must be a 1/0 AWG, and not a 2 AWG.

Figure 6.8 is a schematic showing the proper sizing of the three required grounding electrode conductors for the multiple electrical disconnects shown. In the figure we can see that the first electrical disconnect, has a #8 AWG grounding electrode conductor which is sized in accordance with the incoming #3 AWG phase conductor. The second electrical disconnect, has a #4 AWG grounding electrode conductor which is sized to in accordance with the incoming 3/0 AWG phase conductor. The grounding

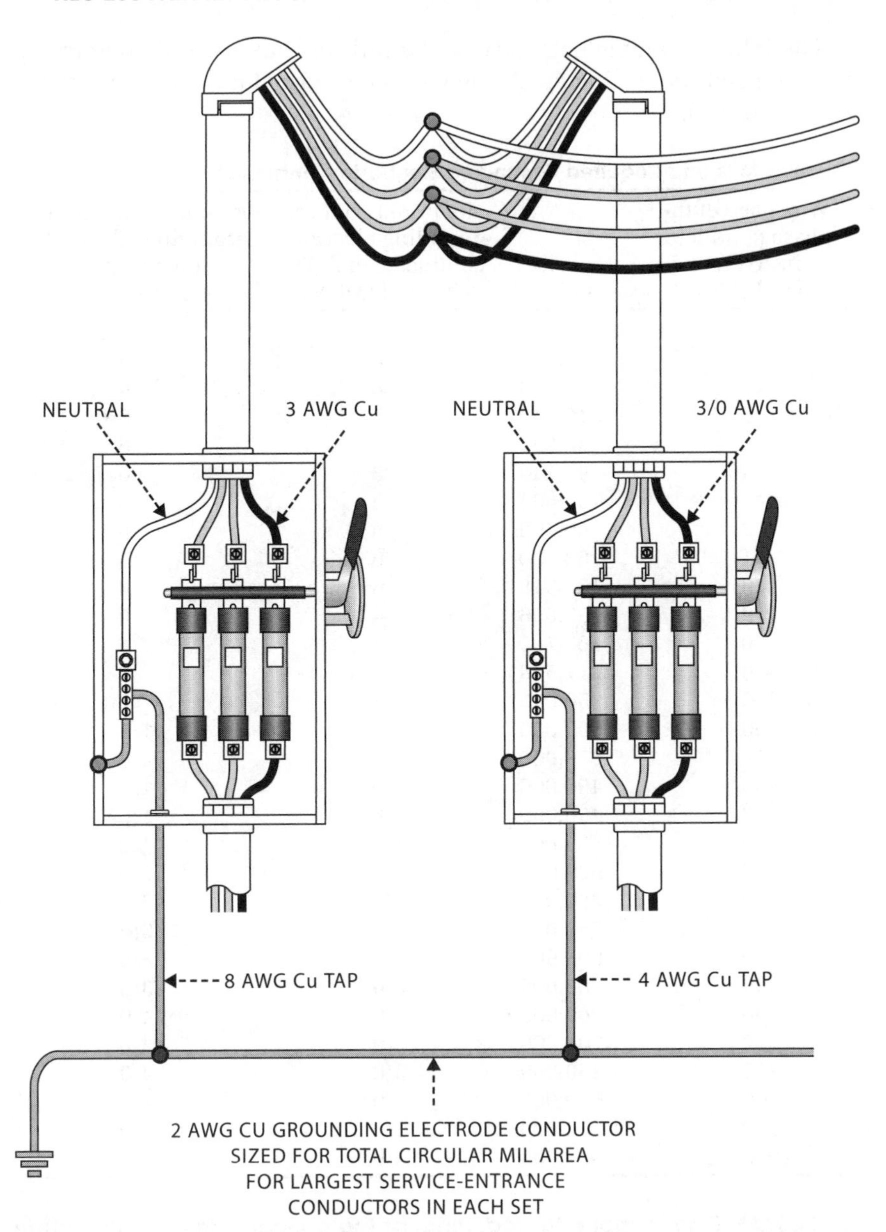

Figure 6.8 Multiple disconnects feeding a common grounding electrode conductor.

electrode conductor from both electrical disconnects are then bonded to a #2 AWG common grounding electrode conductor that is sized in accordance to the combined circular mils of the utility feeds.

If your cabinet is not being fed by an actual utility company power feed, but from a subpanel, separately derived source, and so on, you must still follow the above rules for the grounding electrode conductor.

The following table lists the circular mils of common conductors and the required grounding electrode conductor based on material (copper or aluminum).

Circular Mils and Required Grounding Electrode Conductor

Wire gage of the phase conductor in AWG or kcmil (MCM)	Circular mils	Required copper grounding electrode conductor in AWG or kcmil (MCM)	Required aluminum grounding electrode conductor in AWG or kcmil (MCM)
15	3,257	8	6
12	6,530	8	6
10	10,383	8	6
8	16,509	8	6
6	26,251	8	6
4	41,740	8	6
3	52,633	8	6
2	66,369	8	6
1	83,680	6	6
1/0	105,518	6	6
2/0	133,056	4	4
3/0	167,780	4	4
4/0	211,566	2	2
250	250,000	2	2
300	300,000	2	1/0
350	350,000	2	1/0
400	400,000	1/0	1/0
450	450,000	1/0	1/0
500	500,000	1/0	1/0
600	600,000	1/0	3/0
700	700,000	2/0	3/0
750	750,000	2/0	3/0
800	800,000	2/0	3/0
900	900,000	2/0	3/0
1000	1,000,000	2/0	4/0
1100	1,100,000	2/0	4/0
1250	1,250,000	3/0	4/0
1500	1,500,000	3/0	4/0
1750	1,750,000	3/0	4/0
2000	2,000,000	3/0	250

250.66(A). Connections to Rod, Pipe, or Plate Electrodes. This section states that when you are connecting the grounding electrode conductor to a single or multiple rod, pipe, or plate electrodes(s) in any combination as permitted in Art. 250.52(A)(5) or (A)(7), then the sole conductor leading to an individual electrode need not be larger than 6 AWG copper or 4 AWG aluminum. In Fig 6.9A we see all of the mandatory grounding connections for a building with the required sizing of each of the grounding electrode conductors. Note that instead of using the water pipe as a connection medium for all of the various bonding

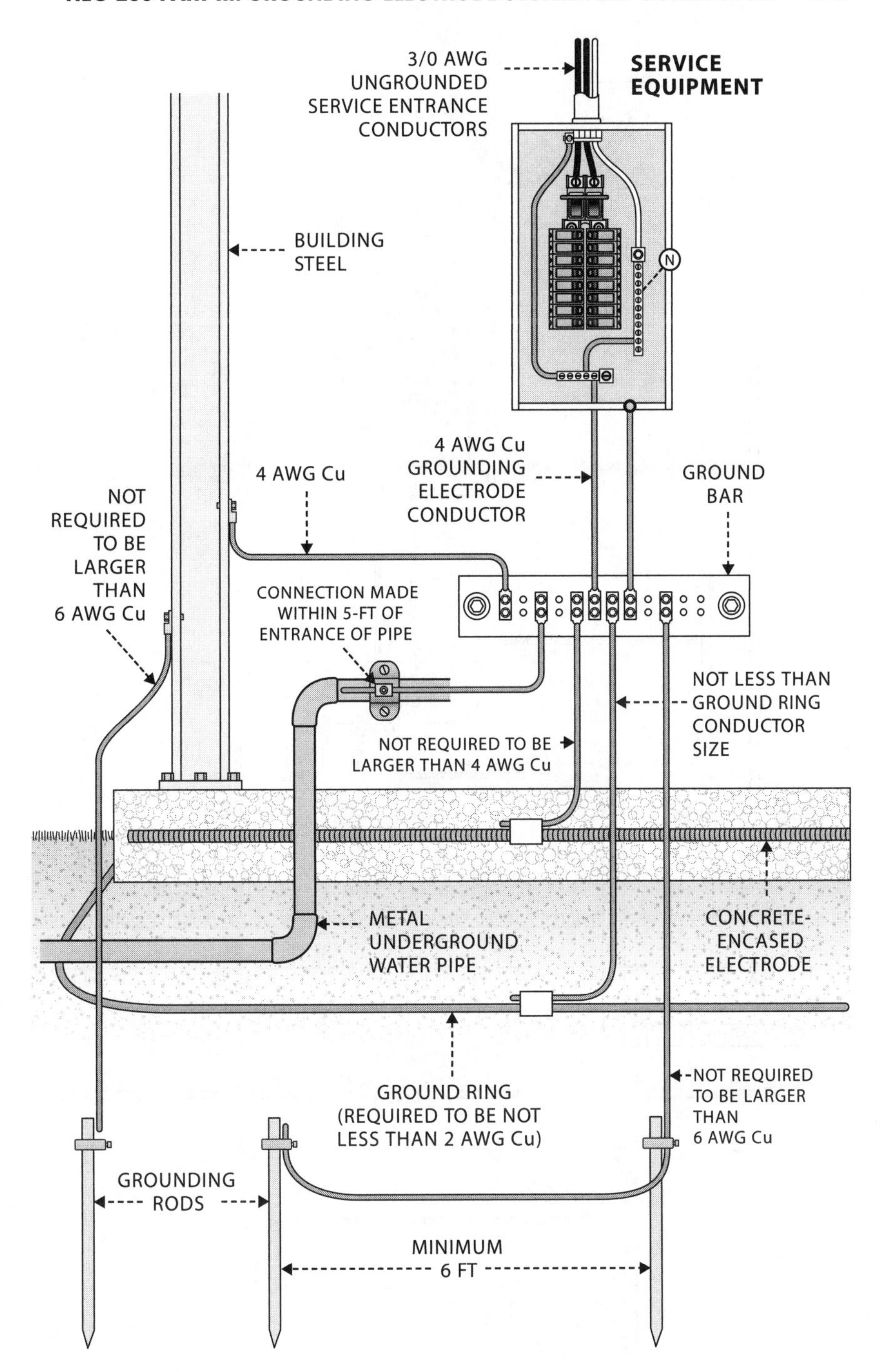

Figure 6.9A Proper conductor sizing for the associated electrode.

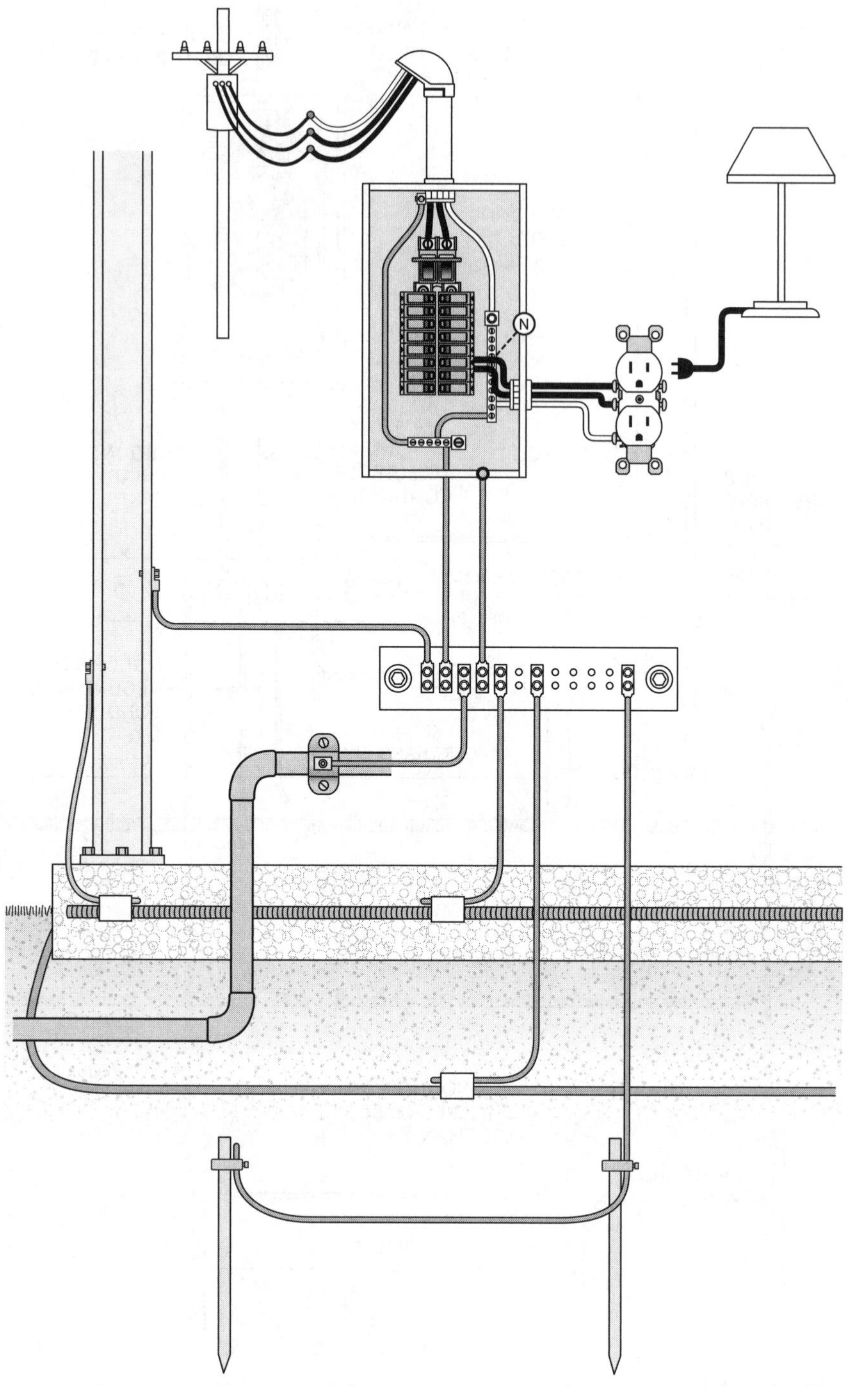

Figure 6.9B Standard grounding and bonding connections using a ground bar.

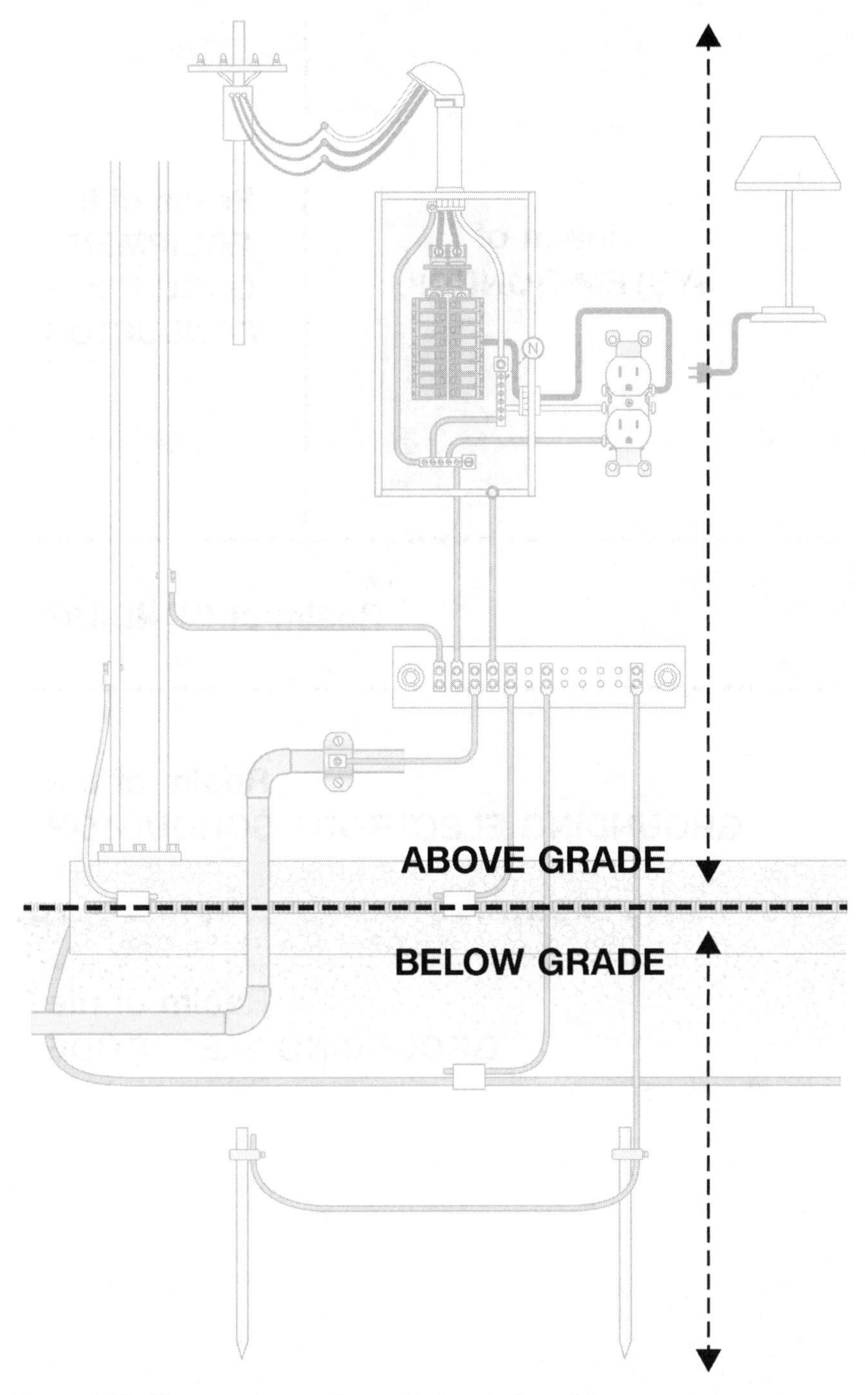

Figure 6.9C Above-grade grounding and below-grade earthing.

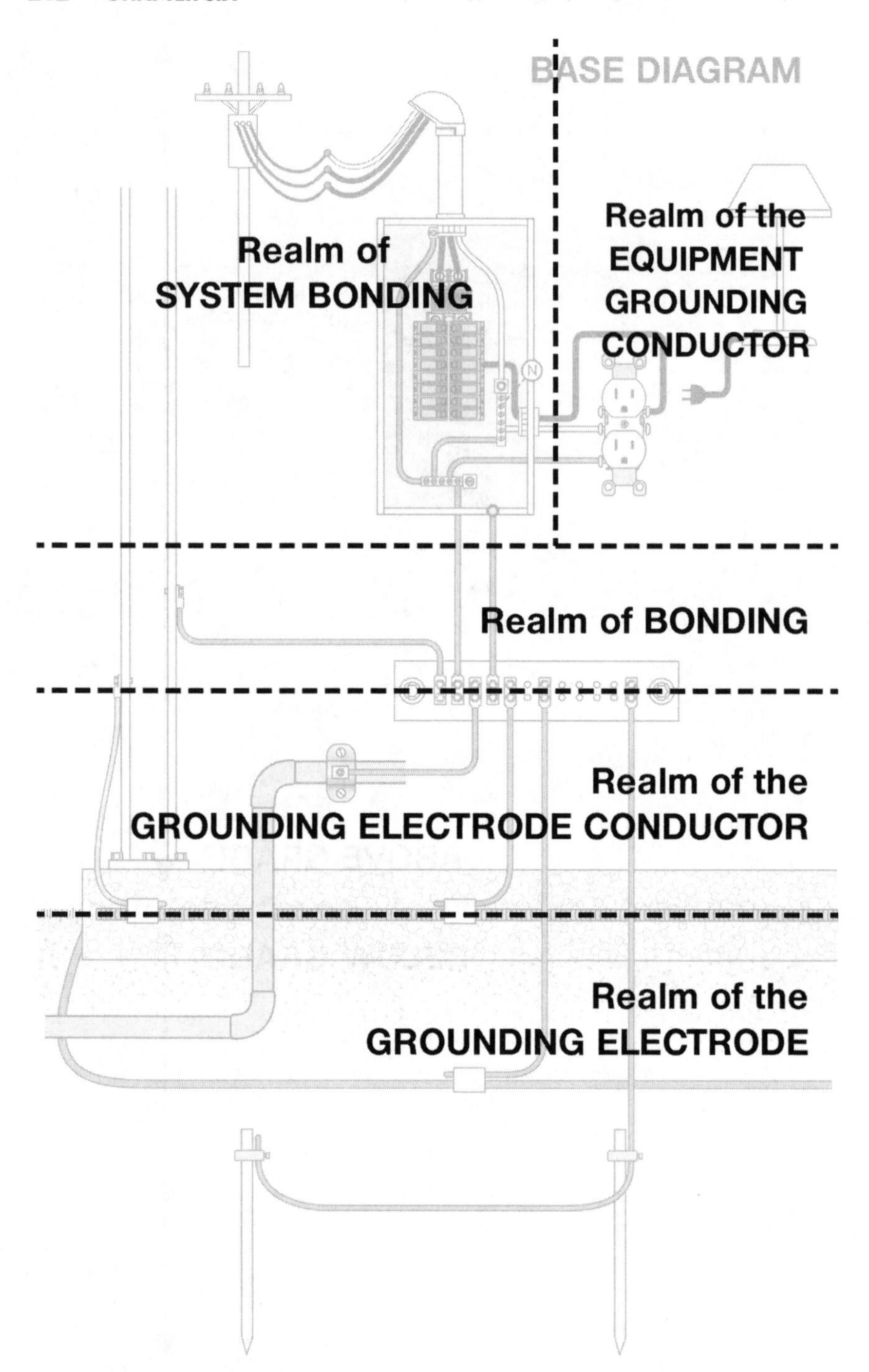

Figure 6.9D Grounding terminology for a typical building.

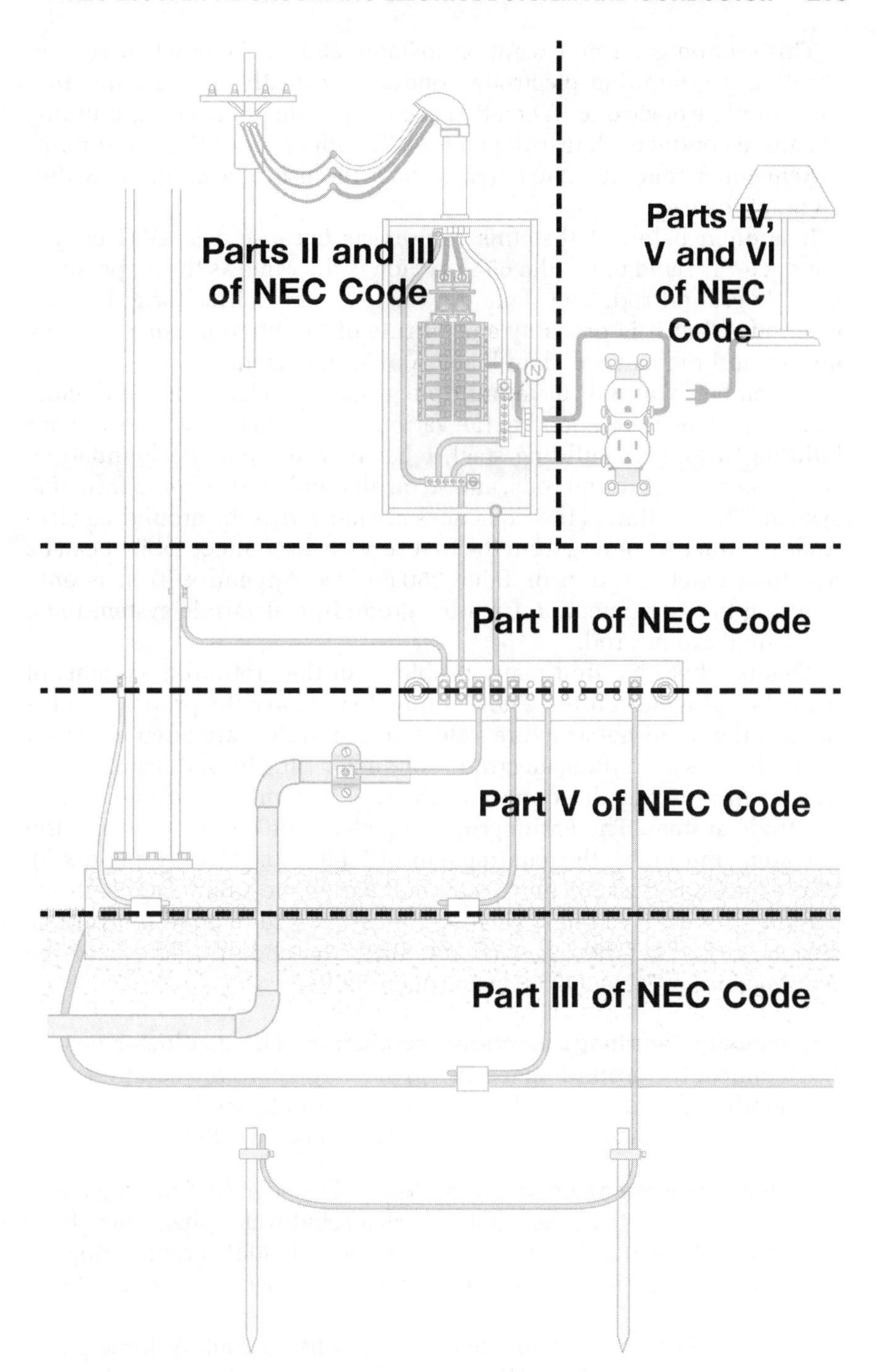

Figure 6.9E General NEC Art. 250 parts in relation to the installation.

This section gives an exception to Table 250.66 above when you are bonding a grounding electrode conductor directly to a ground rod, pipe, or plate electrode. When this occurs, you do not need a grounding electrode conductor larger than a 6 AWG copper or 4 AWG aluminum.

Remember that Art. 250.64(A) forbids the use of aluminum within 18 in. of the earth.

It is often believed that this rule exists because a 6 AWG copper conductor is rated to handle 65 A, which is the same as the ampacity of a single ground rod. The Code does not list ampacities for grounding electrodes, but it is probably a good rule of thumb to assume that any one ground rod can only handle 65 A of fault current.

Remember, you will have multiple grounding electrode conductors in your system as you bond to the various grounding electrodes in your building/structure: building steel, rebar in your concrete foundation, cold-water pipe, ground rods, and so on all bonded together to form one system. This section of the Code does not mean that the grounding electrode conductor coming from the first service disconnect cabinet can be less than what is found in Table 250.66 (see Appendix B). It is only referring to the connection from the grounding electrode system to an individual ground rod.

This is where we find many problems in the grounding systems of many building/structures. They simply do not have the proper connections to the building/structure's electrodes and they are often not sized properly. The grounding electrode conductor runs from the ground bar in your first service disconnect all the way to your grounding/earthing electrode system. The entire grounding electrode conductor, from top to bottom, must meet the requirements of Table 250.66 (see Appendix B). While the Code does not make it clear, if part of the grounding electrode conductor is split in one area and reduced in size to bond the various electrode conductors, the total circular mils should add up to the requirement in Table 250.66 (see Appendix B).

Grounding (earthing) electrode conductor The conductor used to connect the grounding (earthing) electrode(s) to the equipment-grounding (bonding) conductor, to the neutral (grounded) conductor, or to both in accordance with Arts. 250.66 and 250.142.

Grounding electrode conductor system The network of above-grade conductors used to connect building steel, cold-water pipe, concrete-encased electrodes (steel rebar in the foundation), ground rings, ground rods, and so on to the first service disconnect, per Art. 250.66.

There has been more than one occasion, when a 2000-A, three-phase service requiring a 3/0 AWG grounding electrode conductor per Table 250.66 (see Appendix B) had a single 6 AWG ground wire run to a single ground rod because of the exception found in Art. 250.66(A). This is not the intent of this exception in the Code, and it should be considered a violation.

Imagine you have an electrical service with three (3×) incoming conductors of 500-kcmil phase wires (total of 1500 kcmil), which requires you to have a minimum 3/0 AWG grounding electrode conductor, as stated in Table 250.66 (see Appendix B). The circular mils of a 3/0 AWG conductor are 167,780, which is the minimum size needed to carry the likely fault currents to earth. You should have a 3/0 AWG copper conductor running from the first service disconnect to a properly sized copper ground bar per Art. 250.68(C)(1). (You can use water pipe instead of a ground bar in certain circumstances, but it is not recommended.) From that ground bar, you would need to connect to the various grounding electrodes in your building/structure, as follows:

- The grounding electrode conductor from the ground bar to a ground rod does not have to be larger than 6 AWG at 26,251 cmil.
- The grounding electrode conductor from the ground bar to the rebar in your concrete foundation does not have to be larger than 4 AWG at 41,740 cmil.
- The grounding electrode conductor from the ground bar to your building/structure's ground ring, must be at least 2 AWG, as that is the minimum size for a ground ring, at 66,369 cmil.

If we add the circular mils of those three grounding electrode conductors, we see that the total circular mils is 134,360, which does not add up to the necessary 167,780. If all you had were the above connections, your system would not have the required circular mils to carry the likely fault currents under the NEC.

Note *The Code does not make this clear, but it should!*

Fortunately, you must also bond the following:

- The grounding electrode conductor from the ground bar to the cold-water pipe must be a 3/0 AWG according to Table 250.66 (see Appendix B), at 167,780 cmil.
- The grounding electrode conductor from the ground bar to the building steel must be a 3/0 AWG according to Table 250.66 (see Appendix B), at 167,780 cmil.

As you can see, we now have 469,920 cmil of conductor bonding our electrical system to our grounding electrode system. This easily meets our minimum 167,780 circular mils of required connection. Of course, if your water pipe is plastic and your building foundation has a vapor barrier installed, you cannot count the water pipe, building steel, or the rebar in the concrete, as an electrode.

Keep in mind, you must also bond your buildings fire sprinkler system (metal piping), the lightning-protection system, television cable ground, telecommunication system ground, and the gas pipes, to your common grounding electrode system (typically at the ground bar). See Arts. 250.94, 250.104, 250.104 (B), 250.106, and 250.52(B) for more information. Also remember, that these additional grounding systems (sprinklers, lightning, CATV, etc.) do not count toward your electrical service grounding electrode system.

The 2014 NEC made a small change in this section to clarify that a combination of differing types of electrodes or multiple electrodes of the same type could be used, but that the same rules apply.

250.66(B). Connections to Concrete-encased Electrodes. When connecting the grounding electrode conductor to the steel rebar in your building's foundation (concrete-encased electrode) at a single location or at multiple locations, as allowed in Art. 250.52(A)(3), the individual (sole) grounding electrode conductor need not be larger than 4 AWG copper. Aluminum is not allowed for concrete-encased electrodes per Arts. 250.52(B) and 250.64(A).

This section is similar to Art. 250.66(A), in that it gives an exception to Table 250.66 (see Appendix B) for the grounding electrode conductor to the rebar in your building's foundation (concrete-encased electrode), in that the conductor does not have to be larger than 4 AWG copper. Please see Art. 250.66(A) above for more information.

> **Grounding (earthing) electrode conductor** The conductor used to connect the grounding (earthing) electrode(s) to the equipment-grounding (bonding) conductor, to the neutral (grounded) conductor, or to both in accordance with Arts. 250.66 and 250.142.

The 2014 NEC made a small change in this section to clarify that single or multiple connections to the steel rebar in the foundation (concrete-encased electrode) could be used, but that the same rules apply.

250.66(C). Connections to Ground Rings. This section is similar to Art. 250.66(A), in that it gives an exception to Table 250.66 (see Appendix B) for your building/structure's ground ring, if it has one. When bonding to the ground ring, the conductor does not have to be larger than the size of the ground ring. Article 250.52(A)(4) requires a minimum of 2 AWG bare copper (with at least 20 ft of length) for your ground ring, but it can be larger. Often ground rings are sized per Table 250.66 (see Appendix B). This exception states that the grounding electrode conductor does not need to be larger than the gauge of the ground ring. Aluminum is not allowed for use in ground rings per Arts. 250.52(B) and 250.64(A). Please see Art. 250.66(A) for more information.

> **Grounding (earthing) electrode conductor** The conductor used to connect the grounding (earthing) electrode(s) to the equipment-grounding (bonding) conductor, to the neutral (grounded) conductor, or to both in accordance with Arts. 250.66 and 250.142.

250.68. GROUNDING ELECTRODE CONDUCTOR AND BONDING JUMPER CONNECTION TO GROUNDING ELECTRODES

This portion of the Code tells us that the grounding electrode conductor and any bonding jumpers that were used in your grounding electrode conductor system must be accessible, provide an effective ground-fault path, and be bonded in the correct location. In other words, you need to be able to get to the grounding electrode conductors and their mechanical connections. There are a few exceptions, in that buried ground rods and connections to rebar in concrete do not need to be accessible. Nor do connections to building steel (fireproofed structural metal) when they are made with irreversible compression fittings and/or exothermic welding.

You must also make sure that connections to your cold-water pipe are only made within 5 ft of the building entrance. You do not want to have the fault-current path pass through your entire building's piping system as it tries to find a path to ground.

250.68(A). Accessibility. This section clearly states that all the mechanical connections that connect your grounding electrode conductor to your grounding electrodes must be accessible. There are two exceptions to this rule:

1. The connections made to buried ground rods, pipe electrodes, plate electrodes, and connections to rebar in concrete do not need to be accessible.
2. The connections to building steel (fireproofed structural metal), when they are made with irreversible compression fittings and/or exothermic welding, do not need to be accessible.

Remember that Art. 250.70 require you to use connections that are properly listed for burial.

In other words, the mechanical connections to ground rods do not have been accessible, if they are properly buried. And the connections to your building steel may be covered over with drywall or other aesthetic coverings if you use the irreversible compression fittings and/or exothermic welds. Often, your local electrical inspector will want to see these connections before they are covered over.

250.68(B). Effective Grounding Path. This section of the Code tells us that when using water pipe (metal piping) as part of the grounding electrode conductor per Art. 250.68(C), you must ensure there is an effective ground fault–current path around any joints, regulators, valves, and so on that may be in the fault path. In other words, when using cold-water pipe as an electrode, it is the 20 ft of the incoming pipe that is in direct contact with the earth outside [see Arts. 250.52(A)(1) and 250.53(D) for more information] that will pass the electrical energy to ground/earth. If you have a number of regulators, valves, and pipe joints between the point where the fault will enter your pipe and the earth, you need to put a jumper sized per

Table 250.66 (see Appendix B) around the regulators and valves, so the fault current does not have to pass through items of unknown conductivity. Also, any bonding jumpers you do install around joints, valves, and regulators need to be long enough to allow for the servicing and removal of those items.

Grounding (earthing) electrode conductor The conductor used to connect the grounding (earthing) electrode(s) to the equipment-grounding (bonding) conductor, to the neutral (grounded) conductor, or to both in accordance with Arts. 250.66 and 250.142.

Bonding jumper A reliable conductor sized per Art. 250 to ensure electrical conductivity between metal parts of the electrical installation.

250.68(C). Grounding Electrode Connections. Grounding electrode conductors and bonding jumpers are allowed to be connected to the locations given in the following list. You may also use these items to extend the connection to an electrode.

1. Interior water pipe may be used as a grounding electrode conductor, the water pipe connection must be within 5 ft (1.52 m) of the entrance to the building, as found in Arts. 250.52(A)(1) and 250.53(D).
2. Building structural steel may be used as a grounding electrode conductor, as long as the structural steel meets the requirements found in Art. 250.52(A)(2), which requires that the steel rebar in your concrete foundation must be a qualified electrode per Art. 250.52(A)(3) (no vapor barrier) or must be configured as follows:
   a. Connecting the structural frame directly to the steel rebar in the concrete foundation (such as physically bonding hold-down bolts to the steel rebar in the concrete foundation or welding pieces of rebar to the steel frame) or bonding the steel structural members to a ground ring in accordance with Art. 250.52(A)(4).
   b. Bonding the structural frame to one or more grounding electrodes, ground rods as specified in Art. 250.52(A)5) or (A)(7) that have a supplemental electrode as required in Art. 250.53(A)(2).
   c. You may use other approved (listed) means of establishing a connection to earth.
3. A concrete-encased electrode (steel rebar type or copper conductor type) that is in compliance with Art. 250.52(A)(3) may be extended above the concrete to create an accessible bonding location. In other words, you are allowed to extend a piece of rebar or a copper conductor up and out of the concrete for bonding purposes. Remember to install a nonmetallic stress relief at the point where the conductor leaves the concrete, as shown in Art. 250.52(A)(3).

Exception no. 1: When using water pipe as a grounding electrode conductor, if you are in an industrial or commercial building and can limit access to only qualified personnel (such as having a dedicated electrical room), you can increase that distance; however, the water pipe must be exposed and continuous. It may pass through walls as ceilings, but must limit access to only authorized personnel and be exposed the entire way to the building entrance. You must also of course make sure that the metal pipe is electrically continuous by installing bonding jumpers around any regulators, valves, and so on, as required in Art. 250.68(B).

This section of the Code allows you to use copper water pipe and/or the structural metal of your building as part of the grounding electrode conductor system. So, instead of installing a copper ground bar as in Arts. 250.30(A)(6) and 250.64(D)(1), you may use certain portions of water pipe and building structural steel to make your connections.

In other words, the Code allows you to use the water pipe and/or structural metal as both a grounding electrode and as a grounding electrode conductor:

- In the case of the water pipe, the 20 ft section of pipe outside and in direct contact with the earth is the electrode portion. The 5 ft section of the pipe inside the building is the conductor portion.
- In the case of building steel, the rebar in the concrete is the electrode portion. The structural steel column is the conductor portion. The concrete must be in direct contact with the earth (no vapor barrier).

If your building does not meet the requirements found in Art. 250.52(A)(2) (e.g., if your foundation has a vapor barrier, it would not qualify), then you may still use building steel as a grounding electrode conductor if you meet the following requirements:

- The structural steel is bonded to the steel rebar in the concrete foundation per Art. 250.52(A)(3).
- The structural steel is bonded to a ground ring per Art. 250.52(A)(4).
- The structural steel is bonded to ground rods and/or plates per Art. 250.52(A)(5) or (7) and has a supplemental electrode that complies with Art. 250.53(A)(2).
- There are other approved means to establishing a connection to earth.

Remember, the **NEC** does not consider structural steel to be a grounding electrode unless it is connected to one. But let us be honest here, building steel is simply not an electrode; it is a conductor that is hopefully bonded to an properly earthed electrode (such as steel rebar in a concrete foundation and/or driven ground rods), and it should be removed from the list found in Arts. 250.52 and 250.53. It is at best a poor-quality grounding electrode conductor and it must be bonded to the grounding system for safety. Building steel should not be used as a grounding electrode or as a grounding electrode conductor.

Remember, just because the Code allows you to use metal water pipe and structural metal as a grounding electrode conductor, does not mean that you should. You will be far better served installing a simple ground bar, per Arts. 250.30(A)(6) and 250.64(D)(1), to use as a bonding point for the various required connections.

Grounding (earthing) electrode A device that establishes an electrical connection to the earth/soil.

Grounding (earthing) electrode conductor The conductor used to connect the grounding (earthing) electrode(s) to the equipment-grounding (bonding) conductor, to the neutral (grounded) conductor, or to both in accordance with Arts. 250.66 and 250.142.

The 2014 Code changed the title of this section and added the permission to extend rebar or conductors out of and above the concrete foundation to allow an accessible location for connecting the grounding electrode conductor to the concrete-encased electrode.

250.70. METHODS OF GROUNDING AND BONDING CONDUCTOR CONNECTIONS TO ELECTRODES

This section tells how you may connect the grounding electrode conductors to the grounding electrodes.

When bonding the grounding electrode conductor to the grounding electrode, you may use the following methods:

- Exothermic welding.
- Listed lugs.
- Listed clamps.
- Listed pressure connectors.
- Other listed mechanisms of equally substantial approved means.
- All clamps, lugs, pressure connectors, and other mechanisms must be listed for use with the materials they come in contact with (both the electrode and the conductor) to prevent galvanic corrosion.
- If the clamps, lugs, pressure connectors, and so on are to be buried, they must be listed and rated for direct soil burial and/or concrete encasement.
- A pipe fitting, pipe plug, or other approved device screwed into a pipe or pipe fitting.
- A listed bolted clamp of cast bronze or brass or plain or malleable iron.
- For indoor communications purposes only, you may use a listed sheet-metal strap-type ground clamp with a rigid base that seats on the electrode. The strap must be made of a material and with the dimensions necessary to ensure that it is not likely to stretch during or after installation.

You may not use the following methods:

- Solder (see Art. 250.8(B) for more information).
- More than one conductor connected to any given clamp or fitting, unless it is rated for multiple conductors.

Basically, all you need to do is just use the appropriate listed connectors rated for bonding ground conductors to the specific electrode(s).

Grounding (earthing) electrode A device that establishes an electrical connection to the earth/soil.

Grounding (earthing) electrode conductor The conductor used to connect the grounding (earthing) electrode(s) to the equipment-grounding (bonding) conductor, to the neutral (grounded) conductor, or to both in accordance with Arts. 250.66 and 250.142.

Listed or listed item An item, product, or device that has been specifically developed for a particular function and has been independently evaluated by a product-testing organization, such as the UL or CE.

Chapter Seven

NEC 250 PART IV: ENCLOSURE, RACEWAY, AND SERVICE CABLE CONNECTIONS

This short section of Art. 250 gives us instructions on how to properly ground all metal raceways and enclosures from the utility service to the very last metal enclosure in your building or structure. While there are a few exceptions, the governing principle is simple: all metal raceways and enclosures in your building/structure must be connected to ground to ensure the electrical safety of persons.

This will require you to properly connect the neutral conductor (grounded conductor) to the incoming service raceways, and to ensure that equipment-grounding conductors (third-wire or green-wire ground) are used throughout your electrical system.

Additionally, there are some very specific rules regarding how the utility service feed is to be connected to ground in order to ensure the safety of the occupants of the building/structure, particularly when dealing with armored or sheathed cables.

250.80. SERVICE RACEWAYS AND ENCLOSURES

This section of the Code tells us that any form of metal raceway (conduit, flexible tubing, raceways, etc.) used to route service conductors and equipment should be connected to the neutral (grounded conductor) wire. Power coming from the utility service rarely (if ever) has a third-wire ground. The neutral wire is the grounded conductor in the system and must be bonded to all metal raceways so that any accidental contact between the phase wire and the metal raceway will have an effective fault-current path back to the transformer.

If you have an ungrounded electrical system, you must bond the raceway to your system's grounding electrode conductor.

Raceway An enclosed channel of metal or nonmetallic material designed expressly for holding wires, cables, bus bars, and so on. Raceways include all types of conduit, flexible tubing, raceways, floor raceways, busways, and so on.

Grounded conductor (neutral) A system or circuit conductor that is intentionally grounded. This is a current-carrying conductor typically called the *neutral wire.*

Grounding (earthing) electrode conductor The conductor used to connect the grounding (earthing) electrode(s) to the equipment-grounding (bonding) conductor, to the neutral (grounded) conductor, or to both in accordance with Arts. 250.66 and 250.142.

Service The conductors and equipment for delivering electric energy from the serving utility to the wiring system of the premises.

There is one exception to this rule, and that is for services with underground raceways that comprise a combination of both metallic and nonmetallic components. Often, underground service raceways will be made from polyvinyl chloride (PVC) conduit (nonmetallic), as it will not corrode as quickly when in contact with the soil/earth. However, PVC elbow bends are often not strong enough to withstand the rigors of pulling conductors through them, so the PVC elbows are often replaced with metal ones. In other words, your underground service raceway may be a combination of both metal and plastic conduit, with the elbow bends being metal. The exception states that you do not need to bond those metal raceway bends to ground, if they are completely buried and covered by at least 18 in. (450 mm) of earth/soil. We can see in Fig. 7.1 that metal elbows and sweeps that are buried at least 18 inches below-grade do not have to be bonded to the grounding system.

250.84. UNDERGROUND SERVICE CABLE OR RACEWAY

This section of the Code tells us how to ground and bond underground electrical service feeds that use metal sheath or armored cables.

250.84(A). Underground Service Cable. If your underground service cable is the type with a metal sheath or armor and is continuous from the utility transformer all the way to your first service disconnect, you do not need to bond the metal sheath or armor to the neutral (grounded conductor) at the building, if the metal sheath or armor is bonded to the neutral at the utility transformer.

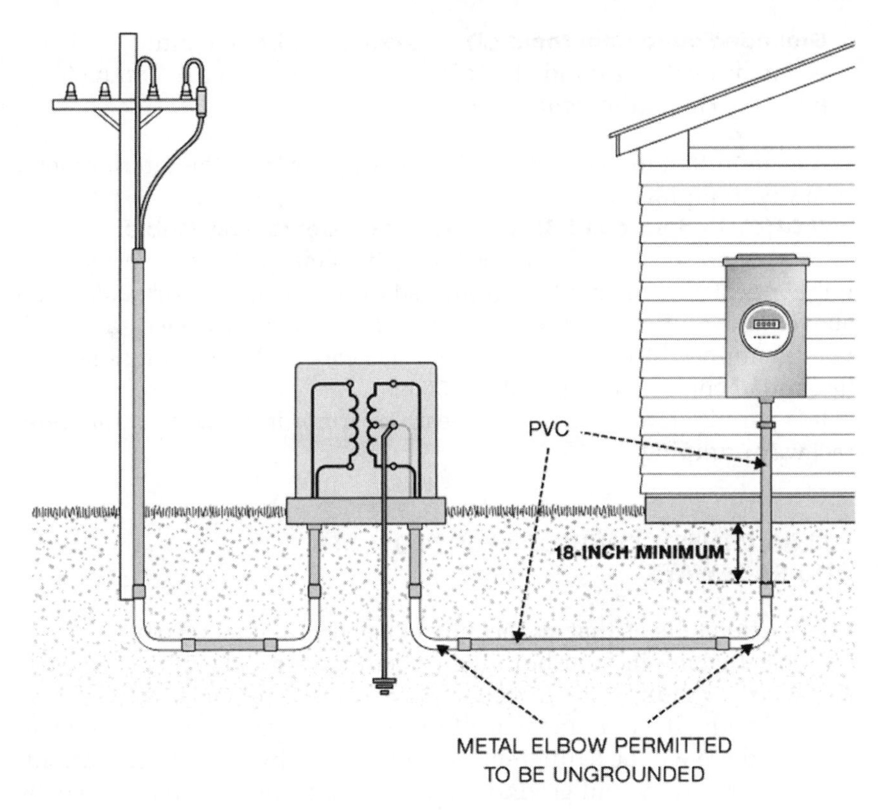

Figure 7.1 Ungrounded metal elbow 18-in. below-grade.

In other words, if your service supply conductors are the sheathed or armored type, you should bond the armor to the XO of the supply transformer but should not bond it to the neutral at the building. This is because your service conductors will have an armored cable for each phase wire and will also have a wire for the neutral (grounded conductor). If you bond the armor at both ends, you will create two neutral return paths; one through the neutral wire and the other through the armor sheathing, which is a hazardous situation.

Service The conductors and equipment for delivering electric energy from the serving utility to the wiring system of the premises.

First service disconnect The very first electrical panel that will disconnect (turn off) the power coming in from the utility company. Sometimes this term is used to describe the first electrical panel that can turn off power on the secondary or low-voltage side of a transformer.

Grounded conductor (neutral) A system or circuit conductor that is intentionally grounded. This is a current-carrying conductor typically called the *neutral wire*.

It is permitted to insulate the sheath or armor from the interior metal raceway or piping.

250.84(B). Underground Service Raceway Containing Cable. If your underground service cable is the type with a metal sheath or armor and it is connected to the neutral (grounded conductor), as in Art. 250.84(A) above, and routed in underground metal raceway (service raceway), you do not need to ground the metal raceway to the building's neutral (grounded conductor), as in Art. 250.80.

It is permitted to insulate the sheath or armor from the interior metal raceway or piping.

250.86. OTHER CONDUCTOR ENCLOSURES AND RACEWAYS

This section tells us that all the metal raceways and metal enclosures of all types that are in your building/structure need to be grounded, so if an electrical fault were to occur, there would be an effective fault-current path. If your metal enclosure or raceway is not bonded to ground, there will be a difference in potential between the non-current-carrying enclosure and ground. That difference in potential is a shock hazard to people and can result in the inability of circuit breakers to function properly.

Effective ground fault-current path An intentionally constructed, low-impedance, electrically conductive path designed and intended to carry current under ground-fault conditions from the point of a ground fault on a wiring system to the electrical supply source and that facilitates the operation of the OCPD or ground-fault detectors.

Equipment-grounding (bonding) conductor The low-impedance fault-current path that is typically run with or encloses the circuit conductors, used to connect the non-current-carrying metal parts of equipment, raceways, and other enclosures to the grounded (neutral) conductor and equipment-grounding (bonding) conductor at service equipment or at the source of a separately derived system. Often called the *green-wire* or *third-wire* ground conductor.

Raceway An enclosed channel of metal or nonmetallic material designed expressly for holding wires, cables, bus bars, and so on. Raceways include all types of conduit, flexible tubing, raceways, floor raceways, busways, and so on.

The previous two articles in Section IV, 250.80 and 250.84, addressed the service conductors and service raceways. This section covers all other metal raceways and enclosures in your building/structure's electrical system. All of the metal raceways and metal enclosures in your building/structure's electrical system must be bonded to the equipment-grounding conductor.

There are four exceptions to this rule; however, only three are numbered in the text of the Code. The primary exception (and the one that is not numbered), is from Art. 250.112(I), which states that circuits less than 50 V that are used for fire alarms, signaling, and/or remote control systems do not need to be grounded. You may, of course, ground them (and you should), but it is not required.

Another exception to this rule (called exception 1 in the Code) involves metal enclosures and raceways that have been added to existing installations in buildings/structures that have open-wire, knob-and-tube wiring, and nonmetallic sheathed cable. You do not need to bond metal raceways and enclosures to the equipment-grounding conductor if all of the following four requirements are met:

1. They do not provide an equipment ground.
2. The runs are less than 25 ft (7.5 m).
3. They are unlikely to come in contact with ground or other grounded objects.
4. They are guarded from contact by persons.

Another exception, called exception 2, tells us that short sections of metal enclosures or raceways used to provide additional support/protection to cable assemblies do not need to be tied to the equipment-grounding conductor. Cable assemblies are not defined within the NEC, but are generally considered to be purposely built multiconductor assemblies inside an armored sheathing such as Type AC, Type MC, or Type NM cables. See Arts. 200.6(E) and 200.7(C)(1) for more information on cable assemblies.

The last exception (called exception 3 in the Code), is very similar to the exception found in Art. 250.80. This exception says that an underground nonmetallic raceway may have ungrounded metal elbows, if the metal elbows are isolated from possible contact by one of the following methods:

- Completely buried in 18 in. (450 mm) of soil/earth
- Encased in at least 2 in. (50 mm) of concrete

Please see the exception in Art. 250.80 for more information.

Chapter Eight

NEC 250 PART V: BONDING

The fifth part of Art. 250 deals with bonding various metal objects together to form a single electrically continuous object. Article 250 is titled "Grounding and Bonding," and these are indeed two separate actions. There is a big difference between using a metal object as an earthing electrode and simply bonding it to eliminate hazardous differences in potential.

In general, Art. 250 has three different electrical functions that all relate to grounding and bonding. For any given metal object, Art. 250 will want the item to be one (sometimes two) of the following:

1. An earthing/grounding electrode
2. A fault-current conductor
3. A bonded normally non-current-carrying object

When a metal object is designated as an *earthing/grounding electrode*, we have placed it in direct contact with the earth/soil and expect it to be able to efficiently "leak" electrical energy out to the surrounding earth/soil.

When a metal object is designated as a *fault-current conductor*, we have purposely constructed it to be a fault current–path conductor and anticipate it being able to handle the likely electrical fault currents and to safely conduct those currents to an earthing/grounding electrode.

When a metal object is designated as *bonded*, we do not anticipate this item to ever have electrical energy on it, but we are aware that accidents happen (also static charges could build up), so we have ensured that the item has a continuous metallic path back to an earthing/grounding electrode.

For example, one of the most confusing elements of bonding involves gas pipes. Gas pipes must be bonded to ground, but they may not be used as a fault-current conductor or as an electrode. The methods on how these systems are bonded are critical for safety and must be completely understood. See Art. 250.104(B) for more information. In Fig. 8.1, we see a standard first service disconnect enclosure with two bonding jumpers, a connection bonding the ground conductor to the chassis of the enclosure, and a connection to the supply side metal conduit via a bushing connection.

In summary, Section V of Art. 250 gives us instructions on how to properly bond normally non-current-carrying metal objects so that they can safely handle accidental fault currents to ground, thereby protecting persons.

250.90. GENERAL

This simple and broad-reaching section of code tells us that you must ensure the electrical continuity and the capacity of any metal object to safely conduct likely fault currents.

In other words, all metal objects in your building or structure must have a continuous electrical path back to ground. As you will see in the sections below, this includes cold-water pipe, building structural steel, steel rebar in concrete foundations (concrete-encased electrodes), ground rings, ground rods, fire sprinkler systems, all other metal pipes, lightning protection systems, cable television (CATV) systems, alarm systems, telecommunication systems, gas pipes, and so on.

Fault current The electrical energy released by a given electrical system during an unintentional line-to-ground or line-to-line fault.

250.92. SERVICES

This section simply restates Art. 250.90 and says that all metal raceways and enclosures on the service side will be bonded to ground.

Service The conductors and equipment for delivering electric energy from the serving utility to the wiring system of the premises.

Raceway An enclosed channel of metal or nonmetallic material designed expressly for holding wires, cables, bus bars, and so on. Raceways include all types of conduit, flexible tubing, raceways, floor raceways, busways, and so on.

250.92(A). Bonding of Equipment for Services. For any system that supports enclosures or contains service conductors, you should bond together the normally non-current-carrying metal parts of all raceways, cable trays, cable bus framework, auxiliary gutters, or service cable sheath or armor. This includes the fittings, bushings, boxes, and other items that are part of the service equipment or service raceway.

Service The conductors and equipment for delivering electric energy from the serving utility to the wiring system of the premises.

Raceway An enclosed channel of metal or nonmetallic material designed expressly for holding wires, cables, bus bars, and so on. Raceways include all types of conduit, flexible tubing, raceways, floor raceways, busways, and so on.

See Arts. 250.80 and 250.86 for an exception involving buried and/or concrete-encased metal raceway elbows.

250.92(B). Method of Bonding at the Service. There are some methods of bonding metal components together that simply do not work well. This section of the Code tells us that reducing washers, oversized washers, concentric or eccentric knockouts, standard locknuts, sealing locknuts, and metal bushings are not acceptable bonding methods; you must have an additional bonding system. The National Electrical Code (NEC) states that these methods may not be relied on as the *sole* bonding connection. See Art. 250.8 for more information.

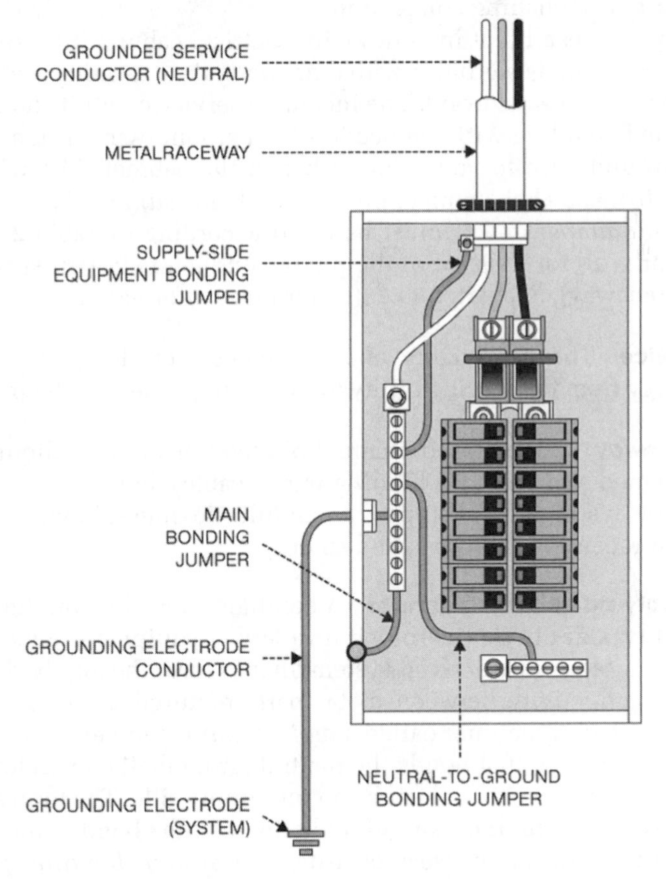

Figure 8.1 Required electrical cabinet bonding jumpers.

What the Code is talking about in this section is the metal service conduit (service raceway) coming into your first service disconnect, specifically the connection between the conduit and the enclosure. Often, the service conduit enters the enclosure through a series of bushings and washers, which simply do not provide an effective bond.

To ensure that you have an effective bond between the enclosure of your first service disconnect and the supply conduit (supply raceway), you must employ one of the following four methods:

1. Use a listed connection per Art. 250.8.
2. Use a threaded coupling or hub, and wrench it tight.
3. Use threadless connections for metal raceway and metal-clad cables, and make them tight.
4. Use other listed devices, such as bonding-type locknuts, bushings, or bushings with bonding jumpers (see item 1 and Art. 250.8).

In Fig. 8.2, we see a series of disconnect enclosures, each with two bonding jumpers, a connection bonding the ground conductor to the chassis of the enclosure, and a connection to the supply side metal conduit via a bushing connection.

You may recall that in your main electrical cabinet (first service disconnect), there is a ground wire that bonds the incoming metal service conduit (service raceway). The incoming service conduit should have a threaded coupling with connection for a ground wire. You are to bond that ground wire to your ground bar in the cabinet. The wire that is used to make that connection is called the *supply-side equipment-bonding jumper* and it must be sized according to Table 250.66 (see Appendix B) for the size of the phase conductors in that specific conduit (raceway). See Art. 250.28(D) for more information.

Service The conductors and equipment for delivering electric energy from the serving utility to the wiring system of the premises.

Raceway An enclosed channel of metal or nonmetallic material designed expressly for holding wires, cables, bus bars, and so on. Raceways include all types of conduit, flexible tubing, raceways, floor raceways, busways, and so on.

Supply-side bonding jumper A conductor installed on the supply (utility) side of a service, or within a service equipment enclosure(s), or for a separately derived system, that ensures the required electrical conductivity between metal parts required to be electrically connected; sometimes called a *main bonding jumper*. A conductor, screw, or strap that bonds the neutral (grounded) conductor to the service equipment enclosures on the supply side. The term *supply-side bonding jumper* is used to distinguish the bond at the supply (utility) side of the service from the *system bonding jumper*,

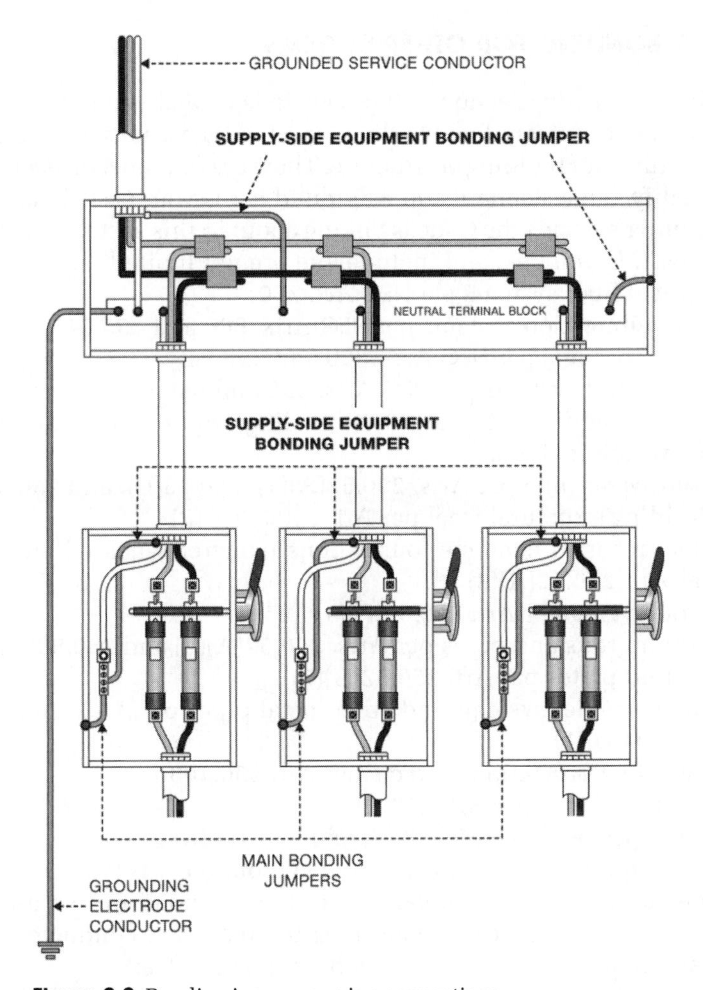

Figure 8.2 Bonding jumper naming conventions.

which is used on the load side of the service (at the first service disconnect). For more details, see Arts. 250.24(A)(4), 250.24(B), 250.28, 250.28(D)(2), and 408.3(C).

Note *The NEC occasionally uses this term to indicate an equipment-grounding conductor (see Art. 250.30).*

First service disconnect The very first electrical panel that will disconnect (turn off) the power coming in from the utility company. Sometimes this term is used to describe the first electrical panel that can turn off power on the secondary or low-voltage side of a transformer.

250.94. BONDING FOR OTHER SYSTEMS

This broad-reaching section of the Code tells us that you must provide a mechanism for intersystem bonding for all of the various grounding systems within your building or structure. There are a number of mechanisms that qualify, but a standard copper ground bar is your best choice.

The other systems the Code is talking about in this section are defined in the *NEC Handbook*, and include, but are not limited to:

- Optical fiber systems per NEC Art. 770
- Communication systems per NEC Arts. 800 and 250.94
- CATV systems per NEC Arts. 820 and 250.94
- Broadband systems per NEC Arts. 830 and 840

This is in addition to the other bonding requirements already discussed, which include:

- Cold-water pipe per Arts. 250.52(A)(1), 250.53(D), and 250.68(C)
- Building structural steel per Art. 250.52(A)(2)
- Steel rebar in concrete foundations (concrete-encased electrodes) per Art. 250.52(A)(3)
- Ground ring per Art. 250.52(A)(4)
- Ground rods and pipes per Arts. 250.52(A)(5) and 250.53(A)(2)
- Ground plates per Art. 250.52(A)(7)
- Fire sprinkler systems and other metal pipes per Arts. 250.104 and 250.104(D)(1)
- Lightning protection systems per Art. 250.106
- Alarm systems per Art. 250.94
- Gas pipes per Arts. 250.52 (B) and 250.104(B)

You can quickly see why a good-sized ground bar is typically needed in order to get all of these various systems bonded to the grounding electrode conductor. It is a good idea to label each conductor so that you can identify the system from which it originates.

This bonding mechanism (i.e., a ground bar) must be outside your electrical service equipment (it may not be inside your electrical system enclosures), located at the first service disconnect, either on the service side or on the disconnect side, and must comply with the following:

1. Accessible for making connections and for inspection.
2. Have a set of terminals capable of bonding at least three intersystem bonding conductors.
3. Not interfere with the access of electrical system enclosures (you have to be able to open and close doors).
4. Securely mounted (it can be mounted directly onto the outside of an electrical enclosure).
5. Made electrically continuous with the grounding electrode conductor, either by direct contact with a grounded enclosure, as in item 4 above, or bonded to the grounding electrode conductor via a minimum 6 AWG copper conductor.
6. Terminals must be listed for grounding and bonding of equipment.

Typically, you simply install a copper ground bar directly below your main electrical panel and bond everything together at that point, with the incoming bonds at the top of the ground bar, and the outgoing connections to the electrodes tied to the bottom of the ground bar.

First service disconnect The very first electrical panel that will disconnect (turn off) the power coming in from the utility company. Sometimes this term is used to describe the first electrical panel that can turn off power on the secondary or low-voltage side of a transformer.

Grounding (earthing) electrode conductor The conductor used to connect the grounding (earthing) electrode(s) to the equipment-grounding (bonding) conductor, to the neutral (grounded) conductor, or to both in accordance with Arts. 250.66 and 250.142.

Cable television companies are notorious for installing illegal and isolated grounding rods. The Code is very clear that CATV must enter the building/structure and install a ground rod within 20 ft of the first service disconnect and that this electrode must be connected to the building's common grounding electrode system. See Art. 820.100(A)(4), (B)(1), and (B)(2) for more information. Additionally, please see the *NEC Handbook*, Section 250.94 as they go to great length to explain the hazards of failing to bond the CATV grounding system to the building's common grounding electrode system.

250.96. BONDING OTHER ENCLOSURES

This section tells us that no matter what type of system is housed inside a metal enclosure, even if it is part of a CATV system or other system that is not part of the electrical system, you still must bond those metal enclosures and raceways to ground.

This section additionally provides instructions for the grounding of isolated grounding circuits.

250.96(A). General. This section is similar to Art. 250.94, in that it requires you to bond the metal enclosures of other systems (communication system, CATV, etc.) to ground. This includes the metal raceways, cable trays, cable armor, cable sheath, enclosures, frames, fittings, and so on that are involved in supporting other systems. You must make these metal systems electrically continuous, so they have the capacity to safely conduct fault currents. You must also remove any paint or nonconductive coatings from threads, contact points, and contact surfaces to ensure efficient contact, unless you are using a listed connection mechanism designed to make paint removal unnecessary.

This rule applies to all metal objects within your building/structure and includes, but is not limited to:

- Optical fiber systems per NEC Art. 770
- Communication systems per NEC Arts. 800 and 250.94
- CATV systems per NEC Arts. 820 and 250.94
- Broadband systems per NEC Arts. 830 and 840

And, of course, also includes:

- Cold-water pipe per Arts. 250.52(A)(1), 250.53(D), and 250.68(C)
- Building structural steel per Art. 250.52(A)(2)
- Steel rebar in concrete foundations (concrete-encased electrodes) per Art. 250.52(A)(3)
- Ground ring per Art. 250.52(A)(4)
- Ground rods and pipes per Arts. 250.52(A)(5) and 250.53(A)(2)
- Ground plates per Art. 250.52(A)(7)
- Fire sprinkler systems and other metal pipes per Arts. 250.104 and 250.104(D)(1)
- Lightning protection systems per Art. 250.106
- Alarm systems per Art. 250.94
- Gas pipes per Arts. 250.52 (B) and 250.104(B)

Ultimately, there must be a continuous conductive path from every metal object in your building or structure back to the grounding electrode conductor.

Grounding (earthing) electrode conductor The conductor used to connect the grounding (earthing) electrode(s) to the equipment-grounding (bonding) conductor, to the neutral (grounded) conductor, or to both in accordance with Arts. 250.66 and 250.142.

Listed or listed item An item, product, or device that has been specifically developed for a particular function and has been independently evaluated by a product-testing organization, such as the Underwriters Laboratory (UL) or CE.

250.96(B). Isolated Grounding Circuits. There is a lot of confusion involving isolated grounding circuits and how they are to be wired. Much of this confusion comes directly from the term *isolated*. A better term would be *dedicated grounding circuits*. But that is neither here nor there, the real process of creating isolated grounding circuits has more to do with the raceway (conduit) running to the equipment that needs protection, than it does with the grounding conductors. Figure 8.3 is a schematic showing how to properly install an isolated equipment grounding conductor to piece of equipment, such as a server rack, when being fed from a first service disconnect. Note the listed non-metallic raceway fitting that is used to connect the metal conduit to the server rack. This non-metallic fitting provides electrical isolation for the server

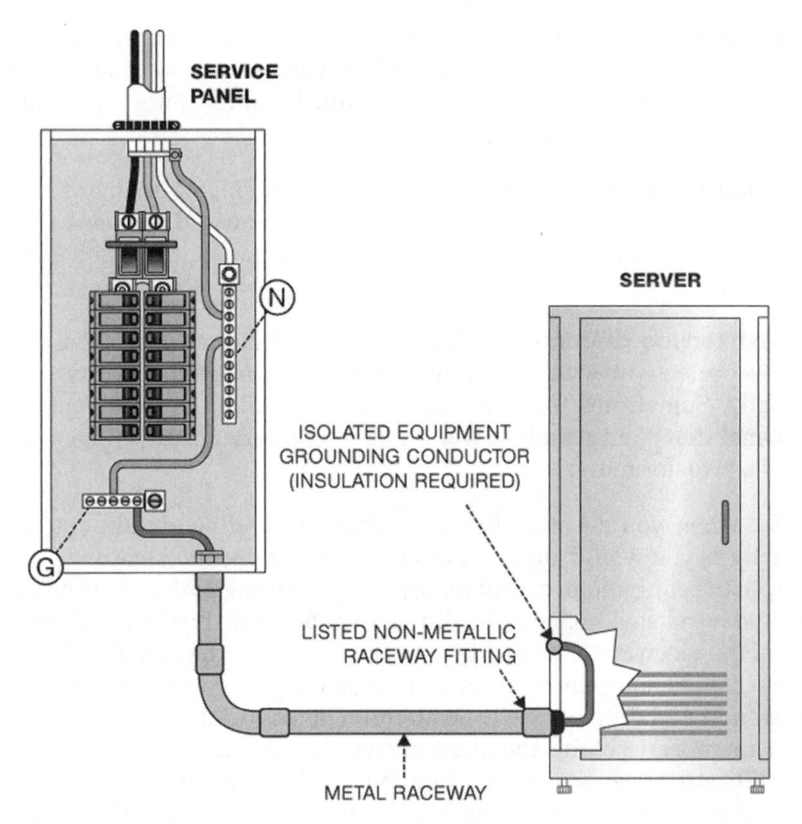

Figure 8.3 Isolated grounding for a server rack.

rack, preventing objectionable electrical noise from coming through the metal conduit and interfering with data communications.

The basic premise of isolated grounding circuits has to do with the reduction of electromagnetic interference, or electrical noise, to a given piece of sensitive equipment. Your typical electrical circuit has two ground-fault paths: one through the green-wire ground (third wire) and one through the metal raceway. The metal raceway (conduit) is part of a massive interconnected parallel steel network and is electrically very noisy. The green-wire ground (third wire) is an insulated copper wire that is tied only to the ground bar in the main electrical cabinet and should have relatively low levels of electrical noise.

This section of the Code simply allows you to place a small isolating barrier (a nonmetallic raceway fitting) between the equipment and the raceway, so as to eliminate the metal raceway as a source of electrical noise that could interfere with sensitive equipment.

Raceway An enclosed channel of metal or nonmetallic material designed expressly for holding wires, cables, bus bars, and so on. Raceways include all types of conduit, flexible tubing, raceways, floor raceways, busways, and so on.

Listed or listed item An item, product, or device that has been specifically developed for a particular function and has been independently evaluated by a product-testing organization, such as the UL or CE.

First service disconnect The very first electrical panel that will disconnect (turn off) the power coming in from the utility company. Sometimes this term is used to describe the first electrical panel that can turn off power on the secondary or low-voltage side of a transformer.

So, when you are installing an isolated grounding circuit, you run it exactly as you would any other circuit, with bonded metal raceway all the way to the equipment, except for the very last fitting in the raceway, which should be a listed nonmetallic fitting, so as to break the electrical continuity of the raceway to the sensitive enclosure. Thus, the metal enclosure of the sensitive equipment has only a single copper ground wire bonding it to ground and does not have a path through the metal raceway.

Remember that the metal raceways containing isolated grounding circuits still must be properly bonded per Art. 250.96(A). Additionally, the raceway must meet the requirements found in Art. 250.118. If your raceway does not meet the requirements in Art. 250.118, you must install a second ground wire in the circuit to bond the metal raceway and enclosures housing the isolated grounding circuit. See Art. 250.146(D) for more information.

In fact, many inspectors will simply mandate that you run two insulated green-wire grounds for any isolated grounding circuit, even if your raceway meets Art. 250.118. This is particularly true for isolated ground-type receptacles. The first ground wire goes directly to the ground lug on the back of the receptacle, and the second ground wire should be bonded directly to the metal enclosure housing the receptacle. This is certainly good practice, and you should plan on doing this for all isolated ground circuits. See Art. 250.146(D) for more information.

If the electrical circuit servicing your sensitive equipment comes from a subpanel, you are allowed to route the green ground wire directly from the first service disconnect and through the subpanel without landing it, as permitted in Art. 408.40. Figure 8.4 is a schematic showing how to properly install an isolated equipment grounding conductor to piece of equipment, such as a server rack, when being fed from a subpanel. Note that the equipment grounding conductor originates in the first service disconnect and is routed all the way to the server,

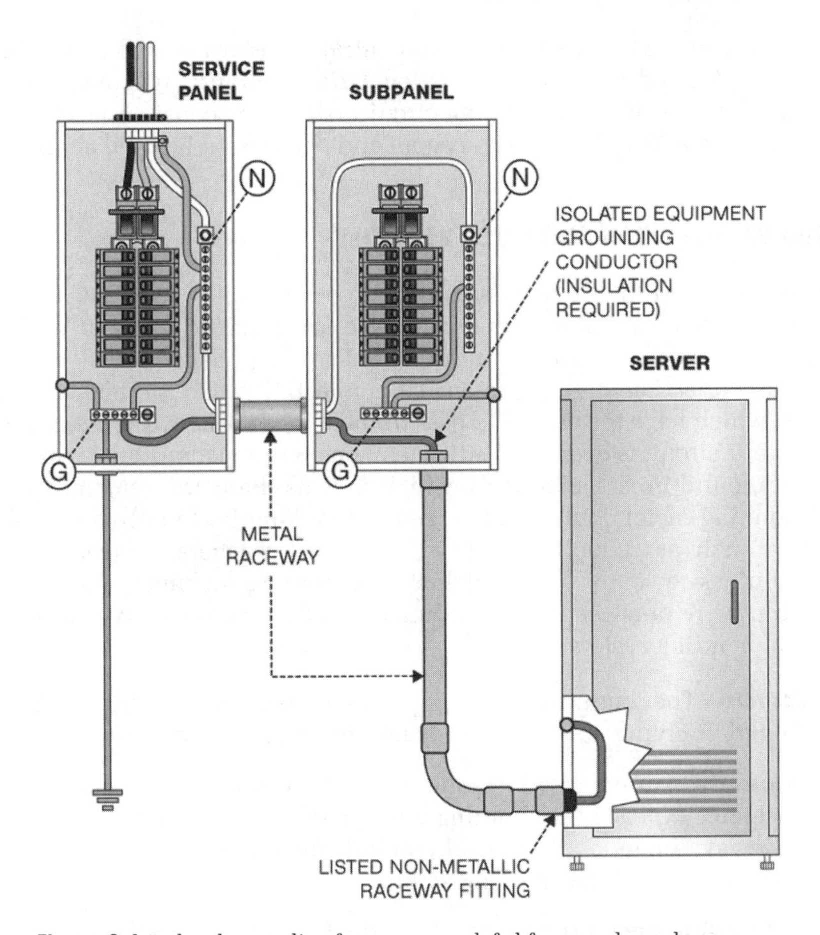

Figure 8.4 Isolated grounding for a server rack fed from a subpanel.

through and bypassing the subpanel, even though the electrical feed is supplied by the subpanel. Also, pay special attention to the listed non-metallic raceway fitting that is used to connect the metal conduit to the server rack. This non-metallic fitting provides electrical isolation for the server rack, preventing objectionable electrical noise from coming through the metal conduit and interfering with data communications.

But let us be honest here. If your equipment is so sensitive as to need an isolated grounding circuit, should you really be running it out of a subpanel? You should probably install a small dedicated isolation transformer and generate a new first service disconnect, and then route your isolated grounding circuit from there. Frankly, isolation transformers and/or uninterruptible power supplies (UPS) are a good idea for all isolated grounding circuits. It is very hard for electrical noise to pass through isolation transformers, and UPS systems will correct almost any electrical noise issue, including voltage drops.

Also, you must have a dedicated neutral wire for isolated grounding circuits. Shared neutrals, also called *Edison circuits*, will negate the benefit of an isolated grounding circuit. All things considered, Edison circuits are a bad idea for any system and should not be used at all.

250.97. BONDING FOR OVER 250 VOLTS

This section of the Code tells us that if your circuit is over 250 V, you actually need less bonding than you need for circuits under 250 V. This may seem odd and counterintuitive, but lower voltages actually need better connection strength, to ensure connectivity, than do higher voltages, which have the energy to pass through more resistive connections.

If your circuit is over 250 V, the metal raceway must meet the requirements found in Art. 250.92(B), which tells us there are some methods of bonding metal components together that simply do not work well. Items such as reducing washers, oversized washers, concentric or eccentric knockouts, standard locknuts, sealing locknuts, and metal bushings are not acceptable bonding methods; you must have an additional bonding system.

Service The conductors and equipment for delivering electric energy from the serving utility to the wiring system of the premises.

Raceway An enclosed channel of metal or nonmetallic material designed expressly for holding wires, cables, bus bars, and so on. Raceways include all types of conduit, flexible tubing, raceways, floor raceways, busways, and so on.

Supply-side bonding jumper A conductor installed on the supply (utility) side of a service, or within a service equipment enclosure(s), or for a separately derived system, that ensures the required electrical conductivity between metal parts required to be electrically connected; sometimes called a *main bonding jumper*. A conductor, screw, or strap that bonds the neutral (grounded) conductor to the service equipment enclosures on the supply side. The term *supply-side bonding jumper* is used to distinguish the bond at the supply (utility) side of the service from the *system bonding jumper*, which is used on the load side of the service (at the first service disconnect). For more details, see Arts. 250.24(A)(4), 250.24(B), 250.28, 250.28(D)(2), and 408.3(C).

Note *The NEC occasionally uses this term to indicate an equipment-grounding conductor (see Art. 250.30).*

First service disconnect The very first electrical panel that will disconnect (turn off) the power coming in from the utility company.

Sometimes this term is used to describe the first electrical panel that can turn off power on the secondary or low-voltage side of a transformer.

However, if the enclosure housing a circuit over 250 V has been tested and is listed as suitable for bonding, you do not need to have the supply-side equipment-bonding jumper connecting a bonding-type locknut to ground, given that one of the following is used:

1. Threadless couplings and connectors for cables with metal sheaths
2. Rigid or intermediate metal conduit with two locknuts, one inside and one outside the enclosure and made tight.
3. Fittings with shoulders that seat firmly against the enclosure, with one a locknut on the inside of the enclosure and made tight.
4. Listed fittings made for this purpose.

Listed connection/bonding methods are acceptable for circuits over 250 V, if the enclosure is listed for such bonding, without the use of a supply-side bonding jumper.

250.98. BONDING LOOSELY JOINTED METAL RACEWAYS

If you have a loose-fitting joint in your raceway due to the use of expansion fittings, telescoping sections of raceway, or for any other reason, you must install a bonding jumper to ensure that the raceway is electrically continuous.

Raceway An enclosed channel of metal or nonmetallic material designed expressly for holding wires, cables, bus bars, and so on. Raceways include all types of conduit, flexible tubing, raceways, floor raceways, busways, and so on.

Bonding jumper A reliable conductor sized per Art. 250 to ensure electrical conductivity between metal parts of the electrical installation.

250.100. BONDING IN HAZARDOUS (CLASSIFIED) LOCATIONS

Articles 500.5, 505.5, and 506.5 of the NEC list locations that are especially hazardous and may require special precautions to make electrical circuits safe. This section of the Code tells us that no matter what the voltage or what the classification, division, group, or zone may be (e.g., Class II, Division 1), you must ensure that the bonding requirements found in Art. 250.92(B)(2) to (B)(4) are met, whether wire-type grounding conductor are installed or not.

Remember that in hazardous locations bonding becomes even more important. Please see Arts. 501.30, 502.30, 503.30, 505.25, and 506.25 for specific bonding requirements.

The 2014 edition of the Code added the requirement to reference 505.5 and 506.5 to this section and an informational note in regard to special bonding requirements found in Arts. 501.30, 502.30, 503.30, 505.25, and 506.25.

250.102. BONDING CONDUCTORS AND JUMPERS

This section of the Code tells us all about the proper material(s), attachment, size, installation, and protection required for a variety of different bonding jumpers.

Bonding (bond, bonded) The permanent joining of metallic parts together to form an electrically conductive path. This path must have the capacity to safely conduct any fault current likely to be imposed on it.

Bonding jumper A reliable conductor sized per Art. 250 to ensure electrical conductivity between metal parts of the electrical installation.

Equipment-grounding (bonding) conductor The low-impedance fault-current path that is typically run with or encloses the circuit conductors, used to connect the non-current-carrying metal parts of equipment, raceways, and other enclosures to the grounded (neutral) conductor and equipment-grounding (bonding) conductor at service equipment or at the source of a separately derived system. Often called the *green-wire* or *third-wire* ground conductor.

Equipment-bonding jumper or load-side bonding jumper A conductor, screw, or strap that bonds the equipment-grounded conductor (neutral) to the equipment enclosures that are on the load side of the first service disconnect enclosure(s), in accordance with Art. 250.102(D). The term is also used to indicate a conductor, screw, or strap that bonds the equipment-grounding conductor of a circuit to the enclosure or raceway of that circuit, in accordance with Art. 250.146.

Main bonding jumper A conductor, screw, or strap that bonds the neutral (grounded) conductor to a grounding conductor of some type. The term *main bonding jumper* is confusingly used in the Code in various places to indicate both a *supply-side bonding jumper* and a *system bonding jumper* (load side). For more details, see Arts. 250.24(A)(4), 250.24(B), 250.28, and 408.3(C).

Supply-side bonding jumper A conductor installed on the supply (utility) side of a service, or within a service equipment enclosure(s), or for a separately derived system, that ensures the required electrical conductivity between metal parts required to be electrically connected; sometimes called a *main bonding jumper.* A conductor, screw, or strap that bonds the neutral (grounded) conductor to the service equipment enclosures on the supply side. The term *supply-side bonding jumper* is used to distinguish the bond at the supply (utility) side of the service from the *system bonding jumper,* which is used on the load side of the service (at the first service disconnect). For more details, see Arts. 250.24(A)(4), 250.24(B), 250.28, 250.28(D)(2), and 408.3(C).

Note *The NEC occasionally uses this term to indicate an equipment-grounding conductor (see Art. 250.30).*

System bonding jumper A conductor, screw, or strap that bonds the equipment-grounding conductor to the neutral (grounded) conductor; sometimes called the *main bonding jumper.* The term *system bonding jumper* is used to distinguish the bond at the load side of the service (at the first service disconnect) from the *supply-side bonding jumper,* which is the bonding used on the utility or supply side of the service. For more details, see Arts. 250.24(A)(4), 250.24(B), 250.28, and 408.3(C).

250.102(A). Material. This section tells us that bonding jumpers must be made of copper or other corrosion-resistant material and that bonding jumpers may be a wire, bus, screw, or similar suitable conductor.
250.102(B). Attachment. This section says that bonding jumpers must be attached in accordance with Art. 250.8 for circuits/equipment and Art. 250.70 for electrodes.
250.102(C). Size-Supply-Side Bonding Jumper. This section tells us how to properly size the supply-side bonding jumper.

Supply-side bonding jumper A conductor installed on the supply (utility) side of a service, or within a service equipment enclosure(s), or for a separately derived system, that ensures the required electrical conductivity between metal parts required to be electrically connected; sometimes called a *main bonding jumper.* A conductor, screw, or strap that bonds the neutral (grounded) conductor to the service equipment enclosures on the supply side. The term *supply-side bonding jumper* is used to distinguish the bond at the supply (utility) side of the service from the *system bonding jumper,* which is used on the load side of the service (at the first service disconnect). For more details, see Arts. 250.24(A)(4), 250.24(B), 250.28, 250.28(D)(2), and 408.3(C).

Note *The NEC occasionally uses this term to indicate an equipment-grounding conductor (see Art. 250.30).*

Bonding jumper A reliable conductor sized per Art. 250 to ensure electrical conductivity between metal parts of the electrical installation.

Main bonding jumper A conductor, screw, or strap that bonds the neutral (grounded) conductor to a grounding conductor of some type. The term *main bonding jumper* is confusingly used in the Code in various places to indicate both a *supply-side bonding jumper* and a *system bonding jumper* (load side). For more details, see Arts. 250.24(A)(4), 250.24(B), 250.28, and 408.3(C).

The following table is new to the 2014 Code, and roughly matches Table 250.66 (see Appendix B).

Table 250.102(C) Grounded Conductor, Main Bonding Jumper, System Bonding Jumper, and Supply-Side Bonding Jumper for Alternating-Current Systems

Size of largest phase (ungrounded) conductor for service in AWG/kcmil		Required size of grounded conductor or bonding jumper in AWG/kcmil	
Copper	Aluminum or copper-clad aluminum	Copper	Aluminum or copper-clad aluminum
2 or smaller	1/0 or smaller	8	6
1 or 1/0	2/0 or 3/0	6	4
2/0 or 3/0	4/0 or 250	4	2
Over 3/0 to 350	Over 250 to 500	2	1/0
Over 350 to 600	Over 500 to 900	1/0	3/0
Over 600 to 1100	Over 900 to 1750	2/0	4/0
Over 1100	Over 1750	See notes 1–4	See notes 1–4

Table 250.120(C) for Bonding Jumpers

1. If your phase conductors are larger than 1100 kcmil copper or 1750 kcmil aluminum, the neutral (grounded conductor) or bonding jumper shall be at least 12½ percent of the circular mil area of the largest phase conductor (ungrounded conductor). The neutral conductor is not required to be larger than the phase conductors.
2. If your system has a mix of both copper and aluminum conductors, the neutral (grounded conductor) or bonding jumper must be sized according to the highest ampacity conductor.
3. If your system has a mix of both copper and aluminum service entrance conductors, as permitted in Art. 230.40, Exception No. 2, or if you have multiple sets of supply conductors from separately derived sources, the equivalent size of the largest phase conductor must be determined by adding the total area of all the corresponding conductors in each set.
4. If you have no service entrance conductors (the utility has yet to provide them), you must assume that the largest allowable conductor for the service will be used. [Author's note: Frankly, you should be doing this anyway. If the utility brings larger conductors in to your service in the future, will someone remember to increase the size of the bonding jumper(s)?]

Table 250.120(C) uses the term *bonding jumper* to indicate any of the following: neutral conductor (grounded conductor), main bonding jumper, system bonding jumper, and supply-side bonding jumper.

The 2014 edition of the Code added Table 250.120(C) to decrease confusion regarding the sizing of bonding jumpers.

250.102(C)(1). Size for Supply Conductors in a Single Raceway or Cable. This section of the Code is very similar to Art. 250.28(D)(1) and simply states that the supply-side bonding jumpers (this is also true for the main bonding jumpers and system bonding jumpers) must be sized in accordance with Table 250.102(C). Remember that when the supply conductors (phase conductors) are larger than 1100 kcmil copper or 1750 kcmil aluminum, the jumpers shall be at least 12½ percent of the circular mil area of the largest phase conductor (ungrounded conductor). See Arts. 250.24(A)(4), 250.24(B), 250.28, 250.28(D)(1), and 250.102(C) for more information.

250.102(C)(2). Size for Parallel Conductor Installations. This section of the Code is very similar to Art. 250.28(D)(2) and states that when you have a grounded electrical service that feeds more than one first service disconnect (or enclosure), you must have a supply-side bonding jumper (the same is true for main bonding jumpers and system bonding jumpers) in each of the enclosures. It further tells us that the individual jumpers must be sized according to Table 250.102(C) to match the phase wires that feed each individual enclosure. If a single supply-side bonding jumper is used to bond two or more raceways or cables, it must be sized according to Art. 250.102(C)(1). Please see Art. 250.28(D)(2) for more information and examples.

250.102(C)(3). Different Materials. If your incoming phase wires are a different material than the supply-side bonding jumper (one is aluminum and the other copper), you must size the jumper to match the highest current rating.

For example, if your phase wires are a 500 MCM aluminum, but you want to use a copper jumper, you would simply look up the 500 MCM aluminum conductor on Table 250.66 (see Appendix B) to see that the supply-side bonding jumper needs to be a 1/0 copper.

250.102(D). Size: Equipment-Bonding Jumper on the Load Side of an Overcurrent Device. This section tells us that bonding jumpers located on the load side of an electrical system (below the main overcurrent protection device [OCPD]), shall be sized in accordance with Table 250.122.

Equipment-grounding (bonding) conductor The low-impedance fault-current path that is typically run with or encloses the circuit conductors, used to connect the non-current-carrying metal parts of equipment, raceways, and other enclosures to the grounded (neutral) conductor and equipment-grounding (bonding) conductor at service equipment or at the source of a separately derived system. Often called the *green-wire* or *third-wire* ground conductor.

Bonding jumper A reliable conductor sized per Art. 250 to ensure electrical conductivity between metal parts of the electrical installation.

Overcurrent device Usually a fuse or circuit breaker, a device that limits the maximum amount of current that can flow in a circuit. Sometimes called an *overcurrent protection device* or *OCPD*.

You are allowed to use a single continuous bonding jumper to connect two or more raceways or cables if the bonding jumper is sized per Table 250.122 for the largest overcurrent device therein.

250.102(E). Installation. You may install bonding jumpers and/or equipment-bonding jumpers inside or on the outside metal raceways and enclosures.

Bonding jumper A reliable conductor sized per Art. 250 to ensure electrical conductivity between metal parts of the electrical installation.

Equipment-grounding (bonding) conductor The low-impedance fault-current path that is typically run with or encloses the circuit conductors, used to connect the non-current-carrying metal parts of equipment, raceways, and other enclosures to the grounded (neutral) conductor and equipment-grounding (bonding) conductor at service equipment or at the source of a separately derived system. Often called the *green-wire* or *third-wire* ground conductor.

Listed or listed item An item, product, or device that has been specifically developed for a particular function and has been independently evaluated by a product-testing organization, such as the UL or CE.

250.102(E)(1). Inside a Raceway or an Enclosure. If you install the bonding jumpers and/or equipment-bonding jumpers inside a metal raceway or enclosure, they must comply with Arts. 250.119 and 250.148.

- Article 250.119 tells us that ground wires can be bare or insulated. If they are insulated, they must be colored green or green with a yellow stripe.
- Article 250.148 tells us that the bonding jumpers must be connected using a listed screw or other listed device and that you may not solder the connections.

250.102(E)(2). Outside a Raceway or an Enclosure. If you install the bonding jumpers and/or equipment-bonding jumpers outside a metal raceway or enclosure, the jumper conductor may not be longer than 6 ft and must be routed with the raceway or enclosure (see Fig. 8.5).

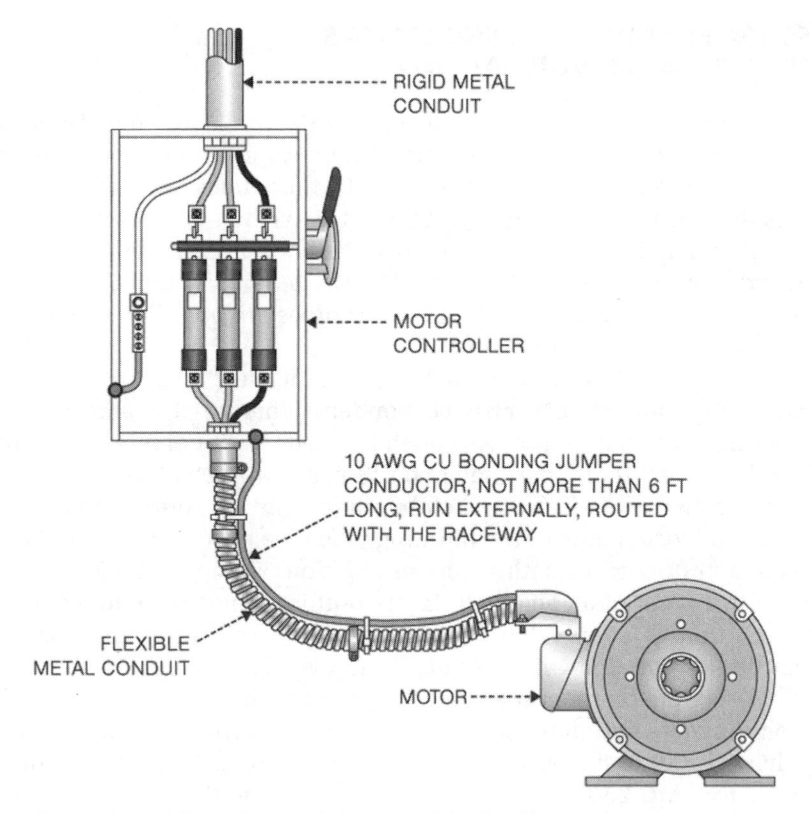

Figure 8.5 Additional grounding required for flexible metal conduit.

A good example would be flexible conduit running to a 30-A motor on vibration mounts. The flexible conduit may not provide an effective and reliable ground-fault path, so an additional equipment-bonding jumper is needed. Referring to Table 250.122, we see that a 10 AWG copper conductor is required. By strapping a 10 AWG insulated green wire on the outside of the conduit that is no longer than 6 ft in length, you can be assured of an effective ground path.

An exception exists for the 6-ft length rule: you may use longer lengths when bonding outside pole locations (power poles) and when bonding various grounding electrodes together.

250.102(E)(3). Protection. Bonding jumpers and/or equipment-bonding jumpers shall be installed in such a manner as to protect them from physical damage in accordance with Art. 250.64(A) and (B). Those sections tell us that aluminum may not come within 18 in. of the earth and which conductor sizes need to be routed in conduit and which ones do not. Please see Art. 250.64(A) and (B) for more information.

250.104. BONDING OF PIPING SYSTEMS AND EXPOSED STRUCTURAL STEEL

This part of the Code is a virtual repeat of several previous sections. That is because water pipe and structural steel are sometimes considered to be grounding electrodes and at other times they are simply items that must be bonded; Art. 250.104 is written to close any loopholes that may have resulted from differentiation of the two functions (electrode vs. bonding). What Art. 250.104 tells us is that it does not matter if your building's steel and water pipes are qualified electrodes or not, you must bond to them.

What Art. 250.104 does not state, is that the steel rebar in your concrete foundation should also be bonded, whether it qualifies as a grounding electrode or not. Although the Code does not require it, you should still do it. In particular, buildings with sensitive electronic equipment will benefit from such bonding. And, of course, it is much safer for the occupants of the building when we ensure that there is an effective fault-current path in the foundation, so circuit breakers will trip in the event of an accident. In my opinion, the failure to mandate the bonding of steel rebar in concrete in every building/structure, regardless of age, is a gross and negligent oversight by the Code.

This section of the Code recognizes that plastic water piping is becoming more and more common and your building or structure may not have the 10 ft of metal water pipe in direct contact with the earth as required in Art. 250.52(A)(1), so you may not use the water pipe as a grounding electrode. However, you must still bond your water pipe to the common ground for the building/structure.

As discussed in the introductory section to Part V above, there is a difference between bonding an object to ground and using it as a grounding electrode. You must always bond your water pipe to ground and under certain conditions listed in Arts. 250.52(A)(1) and 250.53(D), you may sometimes qualify it as an electrode.

In other words if you have a plastic water main from the city coming into your building/structure, but are using metal pipes inside the building, you may not use the city's plastic water pipe as an electrode, and you must bond your metal water pipes inside the building to ground.

This section tells you that you must bond your metal water pipe(s), whether they qualify as an electrode or not.

250.104(A). Metal Water Piping. This section states that metal water pipes will be bonded as required in the three sections discussed below: Art. 250.104(A)(1) to (A)(3), and that the bonding jumpers must be in accordance with Art. 250.64(A), (B), and (E).

Bonding jumper A reliable conductor sized per Art. 250 to ensure electrical conductivity between metal parts of the electrical installation.

Chapter Fifteen

CHANGES TO THE 2014 NATIONAL ELECTRICAL CODE

There were a number of changes made to Art. 250 in the 2014 edition of the National Electrical Code (NEC). Most of the changes were either global issues, such as the change made across the board moving the 600-V rule to 1000 V or the changes that involved adding clarification to better explain existing code. But in a few areas, we see some fairly big changes.

In particular, we see the addition of a new table, Table 250.102(C) (see Appendix B), in the 2014 Code. This new table is intended to help in determining the proper size of all types of bonding jumpers, and is very similar to Table 250.66 (see Appendix B). We see in Part VI of Art. 250 a significant expansion of Art. 250.122(F) Conductors in Parallel, in an attempt to help clarify a fairly confusing section of the Code. In Part VIII, we see a brand-new section of the Code, Art. 250.167, which adds the requirement for ground-fault detection systems on certain direct current (DC) systems.

In Part X, for systems operating at over 1000 V, we actually have two entirely new sections of code. The 2011 version of Art. 250.186 gets renumbered as Art. 250.187, giving us a new Art. 250.186 for 2014. The new Art. 250.186 is quite large and extensive and deals with new regulations regarding the neutral conductor. Additionally, this same section sees an entirely new area mandating the bonding of metal fences for electrical systems operating at 1000 V and higher in Art. 250.194.

Here is a list of each of the changes to Art. 250 by part. Only a few of the minor wording changes regarding the global 1000-V change are mentioned below, as in most cases this change had little impact on Art. 250.

PART I

Article 250.10 Protection of Ground Clamps and Fittings
There is minor simplification in wording.

PART II

Article 250.20(B) Alternating Current Systems of 50 Volts to 1000 Volts
There is a small change in this section to clarify the wording regarding the 50- to 1000-V specification.

Article 250.20(C) Alternating Current Systems of Over 1000 Volts
There is a small change in this section to clarify the wording regarding the 50- to 1000-V specification.

Article 250.21(A) General
There is a small change in this section to clarify the wording regarding the 50- to 1000-V specification.

Article 250.21(B) Ground Detectors
There is a small change in this section to clarify the wording regarding the 50- to 1000-V specification.

Article 250.21(C) Marking
The Code added the required wording to be marked on ungrounded systems.

Article 250.24(A)(1) General
The Code added overhead and underground service conductors to the existing requirements.

Article 250.24(C) Grounded Conductor Brought to Service Equipment
There is a small change in this section to clarify the wording regarding the 1000-V or less specification.

Article 250.24(C)(1) Sizing for a Single Raceway
This change referenced the new table: Table 250.102(C).

Article 250.24(D) Grounding Electrode Conductor
The Code removed a reference to Art. 250.24(A).

Article 250.24(E) Ungrounded System-grounding Connections
The Code added overhead and underground service conductors to the existing requirements.

Article 250.26 Conductor to Be Grounded—Alternating Current Systems
The Code changed the term “common conductor” to “neutral conductor.”

Article 250.28(D)(1) General
This change referenced the new table: Table 250.102(C).

You will recall that Art. 250.64(A) says aluminum may not come within 18 in. of the earth/soil, Art. 250.64(B) requires the bonding jumper be protected from damage, and Art. 250.64(E) gives us very specific instructions on how to bond the conduit enclosing the jumper so as to prevent an impedance choke. Please see Art. 250.64(A), (B), and (E) for more information.

250.104(A)(1). General. This section tells you that you must bond your building's/structure's metal water pipe at your first service disconnect or to your grounding electrode conductor system. You also must size the bonding jumper according to Table 250.66 (see Appendix B), except where permitted in Art. 250.104(A)(2) and (A)(3) below.

First service disconnect The very first electrical panel that will disconnect (turn off) the power coming in from the utility company. Sometimes this term is used to describe the first electrical panel that can turn off power on the secondary or low-voltage side of a transformer.

Grounding electrode conductor system The network of above-grade conductors used to connect building steel, cold-water pipe, concrete-encased electrodes (steel rebar in the foundation), ground rings, ground rods, and so on to the first service disconnect, per Art. 250.66.

The Code does not clearly state in this section where on the water pipe you must bond. Technically, you may bond to any convenient place along the metal pipe. However, most electrical inspectors will only allow the bond to occur within 5 ft of the building entrance.

Remember, the *NEC Handbook* clearly states that you must bond both hot and cold piping, which can add some confusion as to where to make these bonds, as the hot water is often quite far from the water main entrance to the building. The best rule of thumb is to bond as close as possible to the source of the water, either hot or cold.

For information on bonding of all metal piping systems, including the hot-water pipes, please see Art. 250.104(B).

250.104(A)(2). Buildings of Multiple Occupancy. Again, this section of the Code is dealing with the increased use of plastic pipes. For apartment buildings and the like (one single building with multiple apartments), it is becoming common to see plastic water pipe routed all the way to each individual apartment, where it connects to metal pipe inside the apartment.

When this occurs, there is no single metal water pipe location. You have multiple isolated water pipe systems unique to each apartment. You could not possibly run a conductor from the building's main electrical panel to each individual apartment. This section allows you to bond the water pipes in each individual unit (apartment) to the ground in the electrical subpanel for that specific apartment.

In this situation, the bonding jumper for the apartment's isolated metal water pipe must be sized according to Table 250.122.

250.104(A)(3). Multiple Buildings or Structures Supplied by a Feeder(s) or Branch Circuit(s). This section is very similar to Art. 250.104(A)(2), in that it is dealing with plastic main water pipes feeding multiple separated buildings that have isolated metal piping inside each building. You can imagine a campus of individual buildings all being fed by branch circuits from the same electrical source, with a plastic water main to each building. This is common for farms or large homes with guesthouses, separate garages, or pool buildings.

Just as in Art. 250.104(A)(2) above, when this is your situation, you do not need to run a bonding jumper from each isolated water pipe back to the primary electrical service. You simply must bond each building's individual metal water pipe to that building's electrical subpanel. However, the bonding jumper in this case must be sized according to Table 250.66 (see Appendix B).

The bonding jumper does not need to be larger than the largest phase wire feeding the building.

250.104(B). Other Metal Piping. This "catchall" section of the Code tells you that all metal piping, no matter what it is used for, must be bonded to ground. As discussed in the introductory section to Part V above, there is a difference between bonding an object to ground and using it as a grounding electrode.

For many types of metal pipes, it is strictly forbidden to use them as grounding electrodes, with gas pipes being a prime example. This section of the Code specifically tells us that gas piping must be bonded to ground, while Art. 250.52(B) forbids using gas pipe as an electrode. These are not contradictory requirements. You must install an electrically isolating pipe section or a listed electrically isolating device in your gas pipe at the point where it enters the earth, so as to prevent fault currents from traveling down the metal gas pipe and out to earth/soil. This way, you can bond your gas pipe to ensure that there are no hazardous differences in potential and cannot use it as an earthing electrode. For additional information on gas piping systems, please see NFPA 54-2009 National Fuel Gas Code, Section 7.13.

Please remember that gas stoves, gas clothes dryers, and gas water heaters bring your electrical service and your gas lines together in one spot. Proper bonding of the gas piping is critical for safety and should be rigorously inspected.

Another common question involves fire risers and fire sprinkler systems. Article 250.104(B) tells us that all metal piping, including fire risers and fire sprinklers, must be bonded. Again, an electrically isolating pipe section or a listed electrically isolating device in the fire pipe(s) at the point where it enters the earth will prevent fault currents from traveling down the metal fire pipe and out to earth/soil. For additional information on fire risers and fire sprinkler systems, please see the NFPA 13 Standard for the Installation of Sprinkler Systems, Section 10.6.8.1.

This section tells you that you must bond these other metal pipes at your first service disconnect or to your grounding electrode conductor system. You also must size the bonding jumper according to Table 250.122.

> **Bonding jumper** A reliable conductor sized per Art. 250 to ensure electrical conductivity between metal parts of the electrical installation.

The Code makes a special note to state that you should also bond metal air ducts for additional safety. Bonding of metal air ducts and other air-handling equipment is standard protocol for several IEEE, ANSI, TIA, EIA, Motorola, and other industrial standards. You can be sure that in a future version of the NEC, the bonding of air ducts and air-handling equipment will be mandatory.

250.104(C). Structural Metal. This section of the Code is very similar to what we see for water pipes in Art. 250.104(A). If your building has a concrete foundation with a vapor barrier or uses coated rebar, fiberglass structural support, and so on, the concrete foundation may not be used as a qualified electrode. Please see Art. 250.52(A)(2) and (3) for more information. When this occurs, you still must bond all exposed metal framework to ground.

As discussed in the introductory section to Part V above, there is a difference between bonding an object to ground and using it as a grounding electrode. You must always bond your building's structural steel to ground and under certain conditions listed in Art. 250.52(A)(2) and (3), you may sometimes qualify it as an electrode.

In other words, if you have a concrete foundation that is not electrically conductive, you may not use it as an electrode, and you still must bond your building's structural steel to ground.

You must bond your building's structural steel to ground at your first service disconnect or to your grounding electrode conductor system. You also must size the bonding jumper according to Table 250.66 (see Appendix B) and in accordance with Art. 250.64(A), (B), and (E).

> **Bonding jumper** A reliable conductor sized per Art. 250 to ensure electrical conductivity between metal parts of the electrical installation.

You will recall that Art. 250.64(A) says aluminum may not come within 18 in. of the earth/soil, Art. 250.64(B) requires the bonding jumper be protected from damage, and Art. 250.64(E) gives us very specific instructions on how to bond the conduit enclosing the jumper so as to prevent an impedance choke. Please see Art. 250.64(A), (B), and (E) for more information.

This section tells you that you must bond all exposed metal framework to ground whether it is a qualified electrode or not.

250.104(D). Separately Derived Systems. If you have a separately derived system, such as an emergency generator, wind turbine, solar panels, and so on, you must bond its ground system to any adjacent water pipes, structural metal, and/or the building's/structure's common grounding electrode conductor whether they are qualified grounding electrodes or not.

Article 250.104(D) has very similar requirements to those found in Art. 250.30. Please see Art. 250.30 for detailed instructions on properly bonding separately derived systems.

250.104(D)(1). Metal Water Piping System(s). The neutral (grounded conductor) of each separately derived system must be bonded to the nearest available point of the metal water piping system and at the same point where the grounding electrode conductor is connected. The bonding jumper should be sized according to Table 250.66 (see Appendix B).

Grounding (earthing) electrode conductor The conductor used to connect the grounding (earthing) electrode(s) to the equipment-grounding (bonding) conductor, to the neutral (grounded) conductor, or to both in accordance with Arts. 250.66 and 250.142.

Bonding jumper A reliable conductor sized per Art. 250 to ensure electrical conductivity between metal parts of the electrical installation.

There are two exceptions to this rule:

1. If the metal water piping qualifies as a grounding electrode, then only the requirements found in Arts. 250.52(A)(1) and 250.53(D) are required. You do not need a second bonding jumper to meet Art. 250.104(D)(1).
2. If the building's structural metal qualifies as a grounding electrode, and the water pipe is already properly bonded to the structural steel, then only the requirements found in Art. 250.52(A)(2) and (3) are required. You do not need a second bonding jumper to meet Art. 250.104(D)(1).

250.104(D)(2). Structural Metal. The neutral (grounded conductor) of each separately derived system must be bonded to the nearest available point of the building's structural steel frame and at the same point where the grounding electrode conductor is connected. The bonding jumper shall be sized according to Table 250.66 (see Appendix B).

Grounding (earthing) electrode conductor The conductor used to connect the grounding (earthing) electrode(s) to the equipment-grounding (bonding) conductor, to the neutral (grounded) conductor, or to both in accordance with Arts. 250.66 and 250.142.

Bonding jumper A reliable conductor sized per Art. 250 to ensure electrical conductivity between metal parts of the electrical installation.

There are two exceptions to this rule:

1. If the building's structural metal qualifies as a grounding electrode, then only the requirements found in Art. 250.52(A)(2) and (3) are required. You do not need a second bonding jumper to meet Art. 250.104(D)(2).
2. If the metal water piping qualifies as a grounding electrode, and the water pipe is already properly bonded to the structural steel, then only the requirements found in Arts. 250.52(A)(1) and 250.53(D) are required. You do not need a second bonding jumper to meet Art. 250.104(D)(2).

250.104(D)(3). Common Grounding Electrode Conductor. If you have a common grounding electrode conductor installed for multiple separately derived systems in accordance with Art. 250.30(A)(6), you simply need to bond the metal piping and building steel to the common grounding electrode conductor.

Grounding (earthing) electrode conductor The conductor used to connect the grounding (earthing) electrode(s) to the equipment-grounding (bonding) conductor, to the neutral (grounded) conductor, or to both in accordance with Arts. 250.66 and 250.142.

There is one exception to this rule, which simply states that if you have already bonded building steel and water pipe to the common grounding electrode conductor under one of the other articles (Arts. 250.52, 250.30, etc.), you do not need a second connection to meet the requirements of Art. 250.104.

250.106. LIGHTNING PROTECTION SYSTEMS

This simple section clearly states that any lightning protection system installed on your building or structure must be bonded to the grounding electrode system.

Grounding (earthing) electrode system Two or more grounding electrodes that are bonded together to form a single system. For a typical building/structure, a grounding electrode system should consist of: two ground rods, water pipe, steel frame of the building, the steel rebar in the concrete foundation, and any ground rings.

There are two informational notes in the Code:

1. You are asked to reference Art. 250.60, which states that you may not use the lightning protection system's grounding electrodes in lieu of installing grounding electrodes specifically for the building's electrical system. It also asks you to reference NFPA 780-2011 Standard for the Installation of Lightning Protection Systems for more information.
2. The second note carries on somewhat from the first note, which discusses the electrical phenomenon called *side flash* (arc flash) that can occur when a building suffers a lightning strike. NFPA 780-2011 Standard for the Installation of Lightning Protection Systems has very specific requirements for bonding of metal objects in proximity to the lightning protection system. Based on spacing and/or distance an object is from a given lightning protection system conductor, it may be required to be bonded to prevent side flash.

Chapter Nine

NEC 250 PART VI: EQUIPMENT GROUNDING AND EQUIPMENT-GROUNDING CONDUCTORS

The equipment-grounding conductor is the green-wire or third-wire ground that runs from your first service disconnect through the branch circuits to the final load. It is the downstream grounding conductor that ties the individual circuit breakers to the equipment being protected by the breaker. In fact, without the equipment-grounding conductor, your circuit breakers may not function properly.

The equipment-grounding conductor is used to connect non-current-carrying metal parts of equipment, raceways, or other enclosures to the system neutral (grounded conductor) and/or grounding electrode conductor at the service entrance (first service disconnect) or at the source of a separately derived system.

The Code can be very confusing, because it divides the grounding requirements up into four sections and uses different names for each one. Here is a simple breakdown:

1. Above-grade grounding inside the first service disconnect (neutral-to-ground bonds). Terms used: *main bonding jumper, system bonding jumper, supply-side bonding jumper.*
2. Above-grade grounding below the first service disconnect to the various metal objects in the structure and to the electrodes. Term used: *grounding electrode conductor.*
3. Below-grade grounding/earthing, which connects the electrical service to the earth/soil. Term used: *grounding electrode.*
4. Grounding via the branch circuits to the load equipment. Term used: *equipment-grounding conductor.*

Unfortunately, the Code actually makes understanding these definitions worse on occasion, as it mixes up the terms in a few spots. For example, if you look in the *National Electrical Code (NEC) Handbook* at Exhibit 250.21, you can see that the Code calls a grounding electrode conductor an equipment-grounding conductor. This is just one of several examples of naming errors within the Code. Hopefully, the above description will help to get these terms straight, and they will be corrected in future editions of the Code.

The bottom line is that the equipment-grounding conductor is the primary safety component in your electrical system. This is the ground fault–current path between the overcurrent protection device (OCPD) and the final equipment that is using the system. Without an equipment-grounding conductor, you will have no way to trip a circuit breaker and turn off power should a fault occur. Without an equipment-grounding conductor, ground-fault currents (short circuit to ground) would have to take an unpredictable return path through the building's metal components (and earth) as they travel back to the source. This can be very dangerous to personnel if the ground-fault currents travel through metal parts of the building the personnel may contact. It is also possible at ground-fault currents may even have a return path, but the impedance of the path is high enough to keep the circuit breaker from tripping. Both conditions are extremely dangerous.

Part VI of Art. 250 of the Code will give you instructions on how to properly connect/install the equipment-grounding conductor, what must be connected, what qualifies as an approved equipment-grounding conductor, and how to properly size the conductor, so it can safely handle the likely fault currents.

Here are some definitions to familiarize yourself with:

Equipment-grounding (bonding) conductor The low-impedance fault-current path that is typically run with or encloses the circuit conductors, used to connect the non-current-carrying metal parts of equipment, raceways, and other enclosures to the grounded (neutral) conductor and equipment-grounding (bonding) conductor at service equipment or at the source of a separately derived system. Often called the *green-wire* or *third-wire* ground conductor.

Bonding (bond, bonded) The permanent joining of metallic parts together to form an electrically conductive path. This path must have the capacity to safely conduct any fault current likely to be imposed on it.

Circuit or branch circuit The electrical system downstream from the last OCPD. The outlet you plug your computer into would be part of a circuit.

Load The electrical energy consumed by a component, circuit, device, piece of equipment, or system that is connected to a source of electric power to perform its function. This term often refers to all electrical components and equipment that consume electrical energy downstream from the first service disconnect. It can also be called an *electric load.*

First service disconnect The very first electrical panel that will disconnect (turn off) the power coming in from the utility company. Sometimes this term is used to describe the first electrical panel that can turn off power on the secondary or low-voltage side of a transformer.

Main bonding jumper A conductor, screw, or strap that bonds the neutral (grounded) conductor to a grounding conductor of some type. The term *main bonding jumper* is confusingly used in the Code in various places to indicate both a *supply-side bonding jumper* and a *system bonding jumper* (load side). For more details, see Arts. 250.24(A)(4), 250.24(B), 250.28, and 408.3(C).

Supply-side bonding jumper A conductor installed on the supply (utility) side of a service, or within a service equipment enclosure(s), or for a separately derived system, that ensures the required electrical conductivity between metal parts required to be electrically connected; sometimes called a *main bonding jumper.* A conductor, screw, or strap that bonds the neutral (grounded) conductor to the service equipment enclosures on the supply side. The term *supply-side bonding jumper* is used to distinguish the bond at the supply (utility) side of the service from the *system bonding jumper*, which is used on the load side of the service (at the first service disconnect). For more details, see Arts. 250.24(A)(4), 250.24(B), 250.28, 250.28(D)(2), and 408.3(C).

Note *The NEC occasionally uses this term to indicate an equipment-grounding conductor (see Art. 250.30).*

System bonding jumper A conductor, screw, or strap that bonds the equipment-grounding conductor to the neutral (grounded) conductor; sometimes called the *main bonding jumper.* The term *system bonding jumper* is used to distinguish the bond at the load side of the service (at the first service disconnect) from the *supply-side bonding jumper*, which is the bonding used on the utility or supply side of the service. For more details, see Arts. 250.24(A)(4), 250.24(B), 250.28, and 408.3(C).

Ground fault An unintentional connection between an energized conductor and earth or metallic parts of enclosures, raceways, or equipment.

Ground fault-current path An electrically conductive path from the point of a ground fault on a wiring system through normally non-current-carrying conductors, equipment, or the earth to the electrical supply source. A ground fault–current path may not use the earth/soil as a conductor or current path, as it is a violation of the NEC Arts. 250.4(A)(5) and 250.4(B)(4). This term is used for both intentional and unintentional current paths.

Overcurrent device Usually a fuse or circuit breaker, a device that limits the maximum amount of current that can flow in a circuit. Sometimes called an *overcurrent protection device* or *OCPD*.

Please remember that when operating any electrical item of any kind in wet environments (or dry ones for that matter), regardless of the voltage, the safety of persons is best served with a grounded electrical system (wye power), a properly bonded and sized equipment-grounding conductor, with power supplied on a ground-fault circuit interrupter (GFCI) circuit.

250.110. EQUIPMENT FASTENED IN PLACE (FIXED) OR CONNECTED BY PERMANENT WIRING METHODS

This section tells us that certain exposed metal parts mounted on fixed equipment do not typically need to be bonded to ground. Items such as nameplates, bolts, screws, and so on, are considered to be normally non-current-carrying metal parts that are not likely to become energized and therefore do not need to be bonded to ground (bonded to an equipment-grounding conductor). Please see the *NEC Handbook*, which gives examples of the intended items under this code.

However, there are a few circumstances where even these items (nameplates, bolts, etc.) must be bonded to ground. Those conditions are:

1. When the items are within 8 ft (2.5 m) vertically or 5 ft (1.5 m) horizontally from the earth or from a grounded metal object that comes into contact with persons.
2. When the items are not isolated and are located in a wet or damp location.
3. When the items are in electrical contact with metal (just make sure the bond is efficient in such cases).
4. When the items are in a classified or hazardous location as covered by NEC Arts. 500 to 517.
5. When the items have a grounding conductor supplied to them, with an exception as covered in Art. 250.86 for small metal pieces providing support for cable assemblies.
6. When the items are on equipment that operates with any terminal over 150 V.

There are three exceptions to this rule:

Exception no. 1: If you have an electrically heated appliance that has a frame that is permanently and effectively insulated from ground, and the appliance has special permission exempting it from bonding, then you do not have to ground items such as nameplates, bolts, screws, and so on.

Exception no. 2: Transformers and other distribution gear mounted at least 8 ft above grade and on wooden poles do not have to ground items such as nameplates, bolts, screws, and so on.

Exception no. 3: Listed equipment that has a system of double insulation does not need to be connected to an equipment-grounding conductor, as long as the equipment is distinctly marked indicating its ungrounded status.

250.112. SPECIFIC EQUIPMENT FASTENED IN PLACE (FIXED) OR CONNECTED BY PERMANENT WIRING METHODS

This section of the Code tells us that an equipment-grounding conductor must be installed on the normally non-current-carrying metal parts and/or enclosures for certain items regardless of the voltage. The list of items is broken down into two types: one list of items that have metal parts that must be bonded and another list of items that have enclosures that must be bonded.

Equipment-grounding (bonding) conductor The low-impedance fault-current path that is typically run with or encloses the circuit conductors, used to connect the non-current-carrying metal parts of equipment, raceways, and other enclosures to the grounded (neutral) conductor and equipment-grounding (bonding) conductor at service equipment or at the source of a separately derived system. Often called the *green-wire* or *third-wire* ground conductor.

Bonding (bond, bonded) The permanent joining of metallic parts together to form an electrically conductive path. This path must have the capacity to safely conduct any fault current likely to be imposed on it.

Here is the list of items that have metal parts that must be bonded to ground using an equipment-grounding conductor:

- Article 250.112(A) Switchboard or Switchgear Frames and Structures**

- Article 250.112(B) Pipe Organs**
- Article 250.112(C) Motor Frames
- Article 250.112(D) Enclosures for Motor Controllers**
- Article 250.112(E) Elevators and Cranes
- Article 250.112(F) Garages, Theaters, and Motion Picture Studios*
- Article 250.112(G) Electric Signs
- Article 250.112(H) Motion Picture Projection Equipment
- Article 250.112(I) Remote-control, Signaling, and Fire Alarm Circuits*
- Article 250.112(J) Luminaries
- Article 250.112(K) Skid-mounted Equipment

Here is the list of items that have metal enclosures that must be bonded to ground using an equipment-grounding conductor:

- Article 250.112(L) Motor-operated Water Pumps
- Article 250.112(M) Metal Well Casings

*There are actually quite a few exceptions to this rule as you will see when you read the individual items; however, Art. 250.122 only points out two exceptions and those are for:

- Article 250.112(F) Garages, Theaters, and Motion Picture Studios: pendant lamp holders operating at less than 150 V do not need an equipment-grounding conductor.
- Article 250.112(I) Remote-Control, Signaling, and Fire Alarm Circuits: Class 1 circuits operating at less than 50 V do not need an equipment-grounding conductor.

**Here is a list of the other three exceptions to the rule that the Code does not point out in 250.122:

- Article 250.112(A) Switchboard Frames and Structures: Switchboard frames of two-wire direct current (DC) and insulated from ground do not need an equipment-grounding conductor.
- Article 250.112(B) Pipe Organs: If the organ has been effectively insulated from ground and from the motor driving it, you do not need an equipment-grounding conductor.
- Article 250.112(D) Enclosures for Motor Controllers: If the motor controller enclosure is attached to portable equipment, you do not need an equipment-grounding conductor.

More information on each specific item follows.

250.112(A). Switchboard or Switchgear Frames and Structures. You must install an equipment-grounding conductor on all switchboard frames, switchgear frames, and structures supporting switching equipment.

You do not need an equipment-grounding conductor for two-wire DC switchboards or switchgear that are effectively insulated from ground. But, of course, you can and should install an equipment-grounding conductor to ensure there is an effective ground fault–current path.

The 2014 edition of the Code added switchgear to this section in addition to switchboards.

250.112(B). Pipe Organs. You must install an equipment-grounding conductor on all generator and motor frames for pipe organs, unless they have been effectively insulated from ground and the motors driving them.

250.112(C). Motor Frames. You must install an equipment-grounding conductor on all motor frames as defined in Art. 430.242 (you must bond the metal raceway and/or cable armor to the terminal housing on the motor).

250.112(D). Enclosures for Motor Controllers. You must install an equipment-grounding conductor on all motor controller enclosures, unless they are attached to ungrounded portable equipment.

250.112(E). Elevators and Cranes. You must install an equipment-grounding conductor on all electrical equipment involved with the operation of elevators and/or cranes.

250.112(F). Garages, Theaters, and Motion Picture Studios. You must install an equipment-grounding conductor on all electrical equipment in commercial garages, theaters, and motion picture studios, except for pendant lamp holders operating at less than 150 V.

250.112(G). Electric Signs. You must install an equipment-grounding conductor on all electrical equipment involved in the operation of electric signs and/or outline lighting, as defined in Art. 600.7 (you must have an equipment-grounding conductor sized in accordance with Art. 250.122). Please see NEC Art. 600.7 for more information.

250.112(H). Motion Picture Projection Equipment. You must install an equipment-grounding conductor on all electrical equipment involved in the operation of motion picture projection equipment.

250.112(I). Remote-control, Signaling, and Fire Alarm Circuits. You must install an equipment-grounding conductor on all electrical equipment involved in the operation of remote-control circuits, signaling circuits, and fire alarm circuits for the following:

- Class 1 circuits over 50 V
- Class 1 power-limited circuits
- Class 2 and class 3 remote-control and signaling circuits
- All fire alarm circuits

There are two exceptions to this rule:

1. Class 1 circuits less than 50 V do not need an equipment-grounding conductor.
2. Fire alarm circuits using two-wire DC of less than 50 V do not need an equipment-grounding conductor. Please see the exceptions in Art. 250.162(A) for more information.

For information on class 1, 2, and 3 circuits, please see NEC Arts. 501, 502, and 503, respectively.

250.112(J). Luminaries. You must install an equipment-grounding conductor on all electrical equipment involved in the operation of luminaries and lighting equipment as defined in Part V of Art. 410 (you must have an equipment-grounding conductor as specified in Art. 250.118 and sized in accordance with Art. 250.122). Please see NEC Arts. 410.44 and 410.46 for more information.

250.112(K). Skid-mounted Equipment. You must install an equipment-grounding conductor sized in accordance with Art. 250.122 on all electrical equipment involved in the operation of permanently mounted equipment and skids.

250.112(L). Motor-operated Water Pumps. You must install an equipment-grounding conductor on the enclosure of all motor-operated water pumps, including the submersible type(s).

This requirement is necessary to reduce the risks of shock hazards caused by stray voltage, differences in potential, and/or ground faults.

250.112(M). Metal Well Casings. You must install an equipment-grounding conductor on all metal well casings where a submersible motor-operated water pump is in use. The equipment-grounding conductor must be connected to the pump's circuit equipment-grounding conductor.

This requirement is necessary to reduce the risks of shock hazards caused by stray voltage, differences in potential, and/or ground faults.

250.114. EQUIPMENT CONNECTED BY CORD AND PLUG

This section of the Code tells us that an equipment-grounding conductor must be installed on the normally non-current-carrying metal parts and enclosures for any cord-and-plug equipment (you must have a third-wire green ground) for any of the following conditions:

1. Cord-and-plug equipment in hazardous (classified) locations per Arts. 500 through 517
2. Cord-and-plug equipment operating at over 150 V (there are three exceptions to this rule given below)
3. Cord-and-plug equipment operating in residential occupancies:
    a. Refrigerators, freezers, and air conditioners
    b. Clothes washers, clothes dryers, dishwashers, ranges/ovens, kitchen waste disposals
    c. Information technology equipment such as computers, printers, and the like
    d. Sump pumps and aquarium equipment
    e. Handheld, stationary, fixed motor, and/or light industrial motor operated tools of all kinds, including but not limited to:
        i. Hedge clippers, lawn mowers, and snow blowers
        ii. Wet scrubbers
        iii. Drills, saws, sanders, grinders, and other power tools
    f. Portable corded hand lamps

4. Cord-and-plug equipment operating in any other location:
   a. Refrigerators, freezers, and air conditioners
   b. Clothes washers, clothes dryers, dishwashers, ranges/ovens, kitchen waste disposals
   c. Information technology equipment such as computers, printers, and the like
   d. Sump pumps and aquarium equipment
   e. Handheld, stationary, fixed motor, and/or light industrial motor-operated tools of all kinds, including but not limited to:
      i. Hedge clippers, lawn mowers, snow blowers
      ii. Wet scrubbers
      iii. Drills, saws, sanders, grinders, and other power tools
   f. Portable corded hand lamps
   g. Any other tool or appliance that may come in contact with wet or damp surfaces and/or is used by persons standing outside on the ground and/or by persons working on metal surfaces or inside metal tanks/boilers

Basically, everything with a cord needs an equipment-grounding conductor (third-wire green ground).

Equipment-grounding (bonding) conductor The low-impedance fault-current path that is typically run with or encloses the circuit conductors, used to connect the non-current-carrying metal parts of equipment, raceways, and other enclosures to the grounded (neutral) conductor and equipment-grounding (bonding) conductor at service equipment or at the source of a separately derived system. Often called the *green-wire* or *third-wire* ground conductor.

Bonding (bond, bonded) The permanent joining of metallic parts together to form an electrically conductive path. This path must have the capacity to safely conduct any fault current likely to be imposed on it.

There are three exceptions to this rule, given certain requirements are met. In most cases hand tools that have an exception must be double-insulated and listed by a qualified electrical testing laboratory and distinctly marked. The exceptions are:

1. If a motor operating at over 150 V is guarded to prevent contact while in operation, it does not need an equipment-grounding conductor.
2. If you have an electrically heated appliance with a frame that is permanently and effectively insulated from ground, and the appliance has special permission exempting it from bonding, then you do not need an equipment-grounding conductor.
3. Corded tools and portable hand lamps operating at less than 50 V that are likely to be used in wet locations and have power supplied

through an isolation transformer with an ungrounded secondary do not need an equipment-grounding conductor. Note: the intent of electrical items such as this is to limit the amount of electrical energy that could possibly enter the wet area by using a low-power, often only 6- or 12-V, isolation transformer that is double-insulated and integral with the cord. A GFCI circuit should be used when operating equipment in wet environments.

250.116. NONELECTRICAL EQUIPMENT

The Code requires bonding of certain metal parts, even on equipment that is nonelectrical. While the Code specifies some very specific items that must be bonded, it does not limit you from bonding other metal objects deemed likely to become energized. You are encouraged to bond any metal object to the ground, even if it falls outside the scope of the NEC, if you feel it will be safer (swimming pools often have extensive grounding systems that bond metal door frames, gates, fences, and other items beyond what is required by the Code).

The following lists objects that are required to have an equipment-grounding conductor:

1. Frames and tracks of electrically operated cranes and hoists
2. Frames of nonelectrically driven elevator cars to which electrical conductors are attached (think of a hydraulic or cable elevator with lights and control buttons)
3. Hand-operated metal shifting ropes or cables of electric elevators

Special note *While the Code cannot mandate it, because it falls outside the scope of the NEC, it is highly recommended that the metal sidings of buildings be bonded to ground if they have electrical lights, luminaries, signs, and/or receptacles.*

This part of the NEC is a commonsense section, and it does not limit your ability to bond metal objects that are deemed a potential shock hazard should they become energized. In fact, an informational note in the text of the Code encourages you to bond metal objects that could become accidentally energized (due to lightning, overhead power lines, electrical fault, etc.) to ground for additional safety. Just because a metal object is not on the list above does not mean that you cannot bond it to ground and make it safer.

Bonding (bond, bonded) The permanent joining of metallic parts together to form an electrically conductive path. This path must have the capacity to safely conduct any fault current likely to be imposed on it.

Equipment-grounding (bonding) conductor The low-impedance fault-current path that is typically run with or encloses the circuit conductors, used to connect the non-current-carrying metal parts of equipment, raceways, and other enclosures to the grounded (neutral) conductor and equipment-grounding (bonding) conductor at service equipment or at the source of a separately derived system. Often called the *green-wire* or *third-wire* ground conductor.

250.118. TYPES OF EQUIPMENT-GROUNDING CONDUCTORS

This important and commonly referenced section gives a list of items that are acceptable for use as equipment-grounding conductors. It is not just copper wire with green insulation that qualifies (although copper wire is the best); there are a number of different types of conduits, raceways, bus bars, and cable trays that are acceptable if they are made to be electrically continuous.

The Code only requires one equipment-grounding conductor, and it can be any of the types listed in 1 to 14 below, with the exception of health-care facilities. Article 517.13(B) requires that the receptacles in patient-care areas of hospitals and health-care facilities have a wire-type equipment-grounding conductor (item 1 in the list below) in addition to a properly bonded raceway.

Please note that in today's high-tech environment many industrial codes (IEEE, ANSI, EIA, TIA, etc.) are requiring a copper conductor in addition to items in the list below. Even in residences, the proliferation of big-screen televisions, sound systems, computers, and surge-protection devices are only increasing the need for copper grounding conductors (shared neutrals or "Edison circuits" are always a bad idea, but are especially bad for circuits supplying power for sensitive electronics and GFCI circuits).

Ferrous metals (e.g., steel) are at least twelve times less conductive than copper, meaning steel is a poor conductor in comparison. However, the biggest issue with steel is that it is at least 250 times more magnetic than copper, meaning that steel will hold a magnetic field very well. Copper, on the other hand, is diamagnetic, as it actually helps to collapse the magnetic fields that are the source of so much electromagnetic interference.

While the Code may allow you to use steel conduit as a ground conductor, the equipment you are supplying the power for may not. The best single-phase electrical systems supply one phase wire (hot), one neutral wire, and one ground wire per circuit breaker.

Equipment-grounding (bonding) conductor The low-impedance fault-current path that is typically run with or encloses the circuit conductors, used to connect the non-current-carrying metal parts of equipment, raceways, and other enclosures to the grounded (neutral) conductor and equipment-grounding (bonding) conductor at service equipment or at the source of a separately derived system. Often called the *green-wire* or *third-wire* ground conductor.

Listed or listed item An item, product, or device that has been specifically developed for a particular function and has been independently evaluated by a product-testing organization, such as the UL or CE.

Effective ground fault-current path An intentionally constructed, low-impedance, electrically conductive path designed and intended to carry current under ground-fault conditions from the point of a ground fault on a wiring system to the electrical supply source and that facilitates the operation of the OCPD or ground-fault detectors.

The following list of equipment-grounding conductors are approved for use by the NEC. This part of the Code is discussing equipment-grounding conductors that are run with or enclosing circuit conductors (phase wires).

A valid equipment-grounding conductor must have an effective ground fault–current path and can be one or more or a combination of the following:

1. A copper, aluminum, or copper-clad aluminum conductor. This conductor may be bare or insulated, stranded or solid, and formed as a wire or as a bus bar of any shape.
2. Rigid metal conduit.
3. Intermediate metal conduit.
4. Electrical metallic tubing.
5. Listed flexible metal conduit meeting ALL of the following:
   a. The flexible conduit is terminated using listed fittings.
   b. The circuit in the flexible conduit has a circuit breaker of 20 A or less.
   c. The flexible conduit is 6 ft in length or less.
   d. The equipment needs the flexible conduit for normal operation due to vibration or where movement is required (objects that remain stationary do not qualify).
6. Listed liquid-tight flexible metal conduit meeting ALL of the following:
   a. The liquid-tight flexible conduit is terminated using listed fittings.

b. The circuit in the liquid-tight flexible conduit with trade sizes 3/8–½ in. (metric sizes 12–16) has a circuit breaker of 20 A or less.
c. The circuit in the liquid-tight flexible conduit with trade sizes ¾–1¼ in. (metric sizes 21–35) has a circuit breaker of 60 A or less.
d. The liquid-tight flexible conduit is 6 ft in length or less.
e. The equipment needs the liquid-tight flexible conduit for normal operation due to vibration or where movement is required (objects that remain stationary do not qualify).

7. Flexible metallic tubing meeting ALL of the following:
   a. The flexible tubing is terminated using listed fittings.
   b. The circuit in the flexible tubing has a circuit breaker of 20 A or less.
   c. The flexible tubing is 6 ft in length or less.
8. Armor of Type AC cable per Art. 320.108.
9. The copper sheath of mineral-insulated, metal-sheathed cable Type MI.
10. Type MC cable that has an effective ground-fault path meeting any one of the following:
    a. Contains an equipment-grounding conductor (insulated or bare) per Art. 250.118(1).
    b. The combined metallic sheath and bare equipment-grounding conductor of interlocked metal tape–type MC cable that is listed and identified as having an equipment-grounding conductor (i.e., an MC cable marked as having an equipment-grounding conductor).
    c. The metal sheath of the combined metallic sheath and equipment-grounding conductors of the smooth or corrugated tube-type MC cable that is listed and identified as having an equipment-grounding conductor (i.e., an MC cable marked as having an equipment-grounding conductor).
11. Cable trays per Arts. 392.10 and 392.60 (must be electrically continuous or made to be continuous).
12. Cable bus framework per Art. 370.3 (must be electrically continuous or made to be continuous).
13. Other listed electrically continuous metal raceways and listed auxiliary gutters.
14. Surface metal raceways listed for grounding.

In short, just make sure that your raceway, conduit, or other metal object is listed as being an acceptable equipment-grounding conductor. Again, we do not recommend using conduit or other raceways for your equipment-grounding conductors, because there are many joints and/or bonding jumpers required to make a complete path back to the panel. Each of these joints or bonds is a potential point of failure. The conduit should be bonded together to form a continuous conductive path, but it should be considered a backup equipment-grounding conductor and not your sole (primary) ground fault–current path.

The 2014 edition of the Code added Type MI identification to line item 9, regarding the mineral-insulated, metal-sheathed cable, and changed the location of the definition of effective ground fault–current path to Art. 100 definitions.

250.119. IDENTIFICATION OF EQUIPMENT-GROUNDING CONDUCTORS

An equipment-grounding conductor may be bare, covered, or insulated. When it is insulated, it must have a continuous finish that is green in color or green with a yellow stripe(s). You may not use a green colored wire for anything other than grounding.

You may not place different colored electrical tape on a green wire and then use it as a phase wire or neutral. This applies to 8 AWG conductors and smaller; see Art. 250.119(A).

> Exception no. 1: Alarm and signaling circuits under 50 V, as permitted in Art. 250.112(I). They may use green and/or green with yellow stripes for something other than grounding.
>
> Exception no. 2: Flexible cords may have an outside finish that is colored green, even if they do not have an equipment-grounding conductor inside the cord. An example would be a green-colored holiday tree light cord with only two wires (Type SPT-2 cord).
>
> Exception no. 3: Traffic signaling and traffic control systems may use a wire-type equipment-grounding conductor that is bare or has insulation that is colored green with a yellow stripe, in order to allow the use of a solid green insulated conductor as one of the signaling control-phase conductors. This green-colored signaling control-phase conductor must be installed between the output terminals of the control equipment. This is similar to the alarm and signaling rules in Art. 250.112(I), except that there is no voltage limit and a wire-type equipment-grounding conductor (that is bare or has insulation that is green with a yellow stripe) is still required to be routed in accordance with Art. 250.118.

The 2014 edition of the Code added two new exceptions: Exceptions nos. 2 and 3.

250.119(A). Conductors Larger than 4 AWG. Green insulated conductors larger than 4 AWG can be difficult to find, although they are more readily available today. The Code recognizes that it may not be possible to get green-colored conductors for 4 AWG and larger. Equipment-grounding conductors larger than 4 AWG not covered in green colored insulation must comply with the following:

1. You must identify the non-green conductor as being for grounding at both ends of the conductor and at any point where the conductor is accessible, such as a junction box or enclosure.
2. You must identify the grounding conductor by encircling it by one of the following methods:
   a. Stripping the insulation off the entire exposed length of the conductor.
   b. Coloring the insulation green at the points of termination and as required in item 1 above.
   c. Marking it with green electrical tape or green adhesive labels at the points of termination and as required in item 1 above.

Exception no. 1: For conductors 4 AWG and larger, if the junction box is simply a pull box, and there are no splices, joints, or connections of any kind, you do not have to identify the grounding conductor.

The 2014 edition of the Code increased the size of the conductor in this rule from 6 AWG to 4 AWG.

250.119(B). Multiconductor Cable. For multiconductor cables that can only be accessed by maintenance and qualified personnel, non-green conductors within the cable may be permanently marked as being ground conductors by one of the following methods:

1. Stripping the insulation off the entire exposed length of the conductor
2. Coloring the insulation green for entire exposed length of the conductor
3. Marking the exposed sections with green electrical tape or green adhesive labels

250.119(C). Flexible Cord. In flexible cords, the equipment-grounding conductor may be bare. But if it is insulated, it must be green or green with a yellow stripe(s).

250.120. EQUIPMENT-GROUNDING CONDUCTOR INSTALLATION

This simple section is primarily a repeat of early sections of the Code, except that Art. 250.120(C) mandates that conductors smaller than 6 AWG must be protected from physical damage by installing them in raceways or in the hollow spaces of framing members.

250.120(A). Raceway, Cable Trays, Cable Armor, Cable Bus, or Cable Sheaths. You must install all raceways, cable trays, cable armor, cable buses, and/or cable sheath in compliance with the Code by using listed fittings. All joints, connections, and fittings must be made tight.

If you are using FHIT systems, please see Underwriters Laboratory Standard ANSI/UL 2196 and the IEEE 848 (1996), "IEEE Standard Procedure for the Determination of the Ampacity Derating of Fire-Protected Cables" for more information.

250.120(B). Aluminum and Copper-clad Aluminum Conductors. This is a repeat of the 18-in. aluminum rule in Art. 250.64(A). Aluminum and/or copper-clad aluminum may not come within 18 in. (450 mm) of the earth/soil.

250.120(C). Equipment-grounding Conductors Smaller Than 6 AWG. This section of the Code mandates that conductors smaller than 6 AWG must be protected from physical damage by installing them in raceways or in the hollow spaces of framing members. Please see Arts. 250.130(C) and 250.134(B) for more information and an exception.

250.121. USE OF EQUIPMENT-GROUNDING CONDUCTORS

You may not use an equipment-grounding conductor as a grounding electrode conductor. In other words, one conductor cannot be used to do two jobs. Also, grounding electrode conductors must be a wire or bus bar, per Art. 250.62, and may not be a raceway.

Equipment-grounding (bonding) conductor The low-impedance fault-current path that is typically run with or encloses the circuit conductors, used to connect the non-current-carrying metal parts of equipment, raceways, and other enclosures to the grounded (neutral) conductor and equipment-grounding (bonding) conductor at service equipment or at the source of a separately derived system. Often called the *green-wire* or *third-wire* ground conductor.

Grounding (earthing) electrode conductor The conductor used to connect the grounding (earthing) electrode(s) to the equipment-grounding (bonding) conductor, to the neutral (grounded) conductor, or to both in accordance with Arts. 250.66 and 250.142.

250.122. SIZE OF EQUIPMENT-GROUNDING CONDUCTORS

Equipment-grounding conductors must be sized to be able to handle the likely fault currents over the distances those currents must travel.

The following parts of Art. 250.122 give instruction on sizing your equipment-grounding conductor, how to increase the size when required, grounding for motor circuits, and how to deal with multiple and parallel circuits.

Overcurrent device Usually a fuse or circuit breaker, a device that limits the maximum amount of current that can flow in a circuit. Sometimes called an *overcurrent protection device* or *OCPD*.

The general rule for the proper sizing of your equipment-grounding conductor is to look at the OCPD (circuit breaker) protecting your circuit and size it to the requirements of Table 250.122.

250.122(A). General. Equipment-grounding conductors, whether they are copper, aluminum, or copper-clad aluminum, must be sized as shown in Table 250.122 or larger. You may need to increase the size from what is shown in Table 250.122 to account for the increase in impedance caused by long conductor runs. You are never required to install an equipment-grounding conductor that is larger than the corresponding phase wires.

Equipment-grounding (bonding) conductor The low-impedance fault-current path that is typically run with or encloses the circuit conductors, used to connect the non-current-carrying metal parts of equipment, raceways, and other enclosures to the grounded (neutral) conductor and equipment-grounding (bonding) conductor at service equipment or at the source of a separately derived system. Often called the *green-wire* or *third-wire* ground conductor.

If you are using raceway, cable tray, or other listed objects as your equipment-grounding conductor, it must meet the Code per Art. 250.118 and comply with Art. 250.4(A)(5) or (B)(4), which tells us that you must have a low-impedance ground-fault path and that you may not use the earth as a ground fault–current path.

If your equipment-grounding conductor is located inside a multiconductor cable as provided by Art. 250.119(B), you are allowed to use several conductors within the cable in order to meet the circular mils requirement of Table 250.122. This is common in variable-speed drive motor cables, where three smaller grounds in a symmetrical pattern add up to the necessary sized equipment-grounding conductor. This configuration is used to help cancel harmonics created by the variable-speed drives.

Table 250.122 lists the circular mils of common conductors and the required equipment-grounding conductor based on material (copper or aluminum) and rating of the OCPD, such as circuit breakers or fuses.

Table 250.122 Minimum Size of Equipment-Grounding Conductors for Grounding Raceway and Equipment

Rating of OCPD not exceeding:	Copper		Aluminum or copper-clad aluminum	
	Size (AWG or kcmil)	Circular mils	Size (AWG or kcmil)	Circular mils
15 A	14	4,107	12	6,530
20 A	12	6,530	10	10,383
60 A	10	10,383	8	16,509
100 A	8	16,509	6	26,251
200 A	6	26,251	4	41,740
300 A	4	41,740	2	66,369
400 A	3	52,633	1	83,680
500 A	2	66,369	1/0	105,518
600 A	1	83,680	2/0	133,056
800 A	1/0	105,518	3/0	167,780
1,000 A	2/0	133,056	4/0	211,566
1,200 A	3/0	167,780	250	250,000
1,600 A	4/0	211,566	350	350,000
2,000 A	250	250,000	400	400,000
2,500 A	350	350,000	600	600,000
3,000 A	400	400,000	600	600,000
4,000 A	500	500,000	750	750,000
5,000 A	700	700,000	1200	1,200,000
6,000 A	800	800,000	1200	1,200,000

Circular Mils and Required Equipment-Grounding Conductor

250.122(B). Increased in Size. If your electrical system has incoming phase (ungrounded) conductors that are larger than what the service needs (larger than the main circuit breaker), you must size the equipment-grounding conductor per these larger phase wires. Installing larger service-feed conductors is a common practice of many utilities to accommodate predicted future increases in power consumption. In other words, you cannot run oversized phase conductors (in case you need to increase the breaker size later) and not run proportionally sized equipment-grounding conductors.

The 2014 edition of the Code made some additional statements to clarify this section of the Code.

250.122(C). Multiple Circuits. The Code allows you to have only a single equipment-grounding conductor in any given raceway, even when the raceway houses multiple circuits. When this occurs, the equipment-grounding conductor must be sized to match the circuit with the highest amperage rating (highest OCPD). You do not need to size the equipment-grounding conductor to match the total current levels in the raceway.

Please remember that the requirements in this section of the Code are designed to ensure that circuit breakers function properly. Sensitive electronic equipment may require individual equipment-grounding conductors for proper operation.

If your equipment-grounding conductor is installed in a cable tray in an industrial environment, it must be at least a 1/0 AWG or larger to meet the requirements of 392.10(B)(1)(c).

250.122(D). Motor Circuits. The equipment-grounding conductor for circuits that operate motors must be sized according to Table 250.122 and based on the current rating of the motors OCPD (branch-circuit short-circuit, or ground-fault protective device, instantaneous-trip circuit breaker, or a motor short-circuit protector).

Overcurrent device Usually a fuse or circuit breaker, a device that limits the maximum amount of current that can flow in a circuit. Sometimes called an *overcurrent protection device* or *OCPD.*

Equipment-grounding (bonding) conductor The low-impedance fault-current path that is typically run with or encloses the circuit conductors, used to connect the non-current-carrying metal parts of equipment, raceways, and other enclosures to the grounded (neutral) conductor and equipment-grounding (bonding) conductor at service equipment or at the source of a separately derived system. Often called the *green-wire* or *third-wire* ground conductor.

250.122(D)(1). General. The equipment-grounding conductor for circuits that operate motors must be sized according to Art. 250.122(A) based on the amperage rating of the branch-circuit, short-circuit, and ground-fault protective device.

250.122(D)(2). Instantaneous-trip Circuit Breaker and Motor Short-circuit Protector. When your motor has an instantaneous-trip circuit breaker or a motor short-circuit protector, the equipment-grounding conductor must be sized according to Art. 250.122(A), based on the amperage ratings derived from Art. 430.52(C)(1) and 430.52(C)(1) Exception no. 1.

Parts IV and V of Art. 430 have extensive lists, tables, and instructions related to the proper installation and setup of OCPDs required for motors. Please see Arts. 430.51 to 430.63 for more information.

250.122(E). Flexible Cord and Fixture Wire. Flexible cord and fixture wire are the type of cords designed for installations requiring lots of movement and flexibility. Think of elevator wires, lamps on pendants, and so on. These useful wires are great, due to their flexibility, but they are often relatively fragile and require extra precautions.

This confusing section of the Code breaks down flexible cord and fixture wire into two types: those that have phase conductors of 10 AWG or smaller and those that are 8 AWG or larger. Here is the breakdown:

- Flexible cords and fixture wires with phase conductors 10 AWG or smaller must have an equipment-grounding conductor between 18 AWG and 10 AWG, based on the requirements of Art. 240.5.

- Flexible cords and fixture wires with phase conductors 8 AWG or larger must have an equipment-grounding conductor based on Table 250.122.

Please reference Art. 240.5, as it has very extensive tables and rules regarding the proper sizing of the flexible cord and/or fixture wire based on the lengths, types of cords, and the planned applications. The instructions from Art. 240.5 will determine the size of the equipment-grounding conductor needed for your situation.

250.122(F). Conductors in Parallel. Phase conductors that are installed in parallel must have equipment-grounding conductors installed in accordance with Art. 250.122 (based on the amperage rating of the circuit breaker and not the size of the individual conductors) and the following:

1. If the parallel phase conductors are installed in multiple raceways or cables (as allowed in Art. 310.10(H)), the equipment-grounding conductors must be wire type and installed in the parallel raceways or cables with the phase conductors. The equipment-grounding conductor does not need to be larger than the largest phase conductors.
2. If the parallel phase conductors are installed in the same raceway, cable, or cable tray (as allowed in Art. 310.10(H)), you may use a single equipment-grounding conductor (instead of parallel conductors matching the phase wires). However, if routed in a cable tray, this single equipment-grounding conductor must be at least 4 AWG and may be insulated, covered, or bare.

Exception: This exception applies to Art. 250.122(F)(1). If under engineering supervision (designed by an electrical engineer) and only for electrical system in industrial locations, you may use smaller equipment-grounding conductors sized to match the phase conductors in each raceway, as long as the total of all the equipment-grounding conductors match the circular mils required in Art. 250.122, instead of using multiple larger equipment-grounding conductors. In other words, an electrical engineer can design a system for industrial areas that allows parallel phase conductors to be routed with parallel equipment-grounding conductors that are sized to match the phase conductors they are routed with, instead of the size of the circuit breaker.

Figure 9.1 is a schematic showing the proper ground wire sizing when routing phase conductors in parallel conduits. In the schematic we see that the main feed is a 3/0 AWG conductor capable of handling 200 amps and is routed in a single conduit. The full 200 amps is then routed out of the enclosure in two conduits via a pair of 100A rated #3 AWG conductors, to another 200A enclosure. What the schematic shows is that while the phase conductors in the parallel runs need only be rated for 100 amps (two conductors at 100A = 200A), the individual ground wires may NOT be downsized. The individual ground wire that

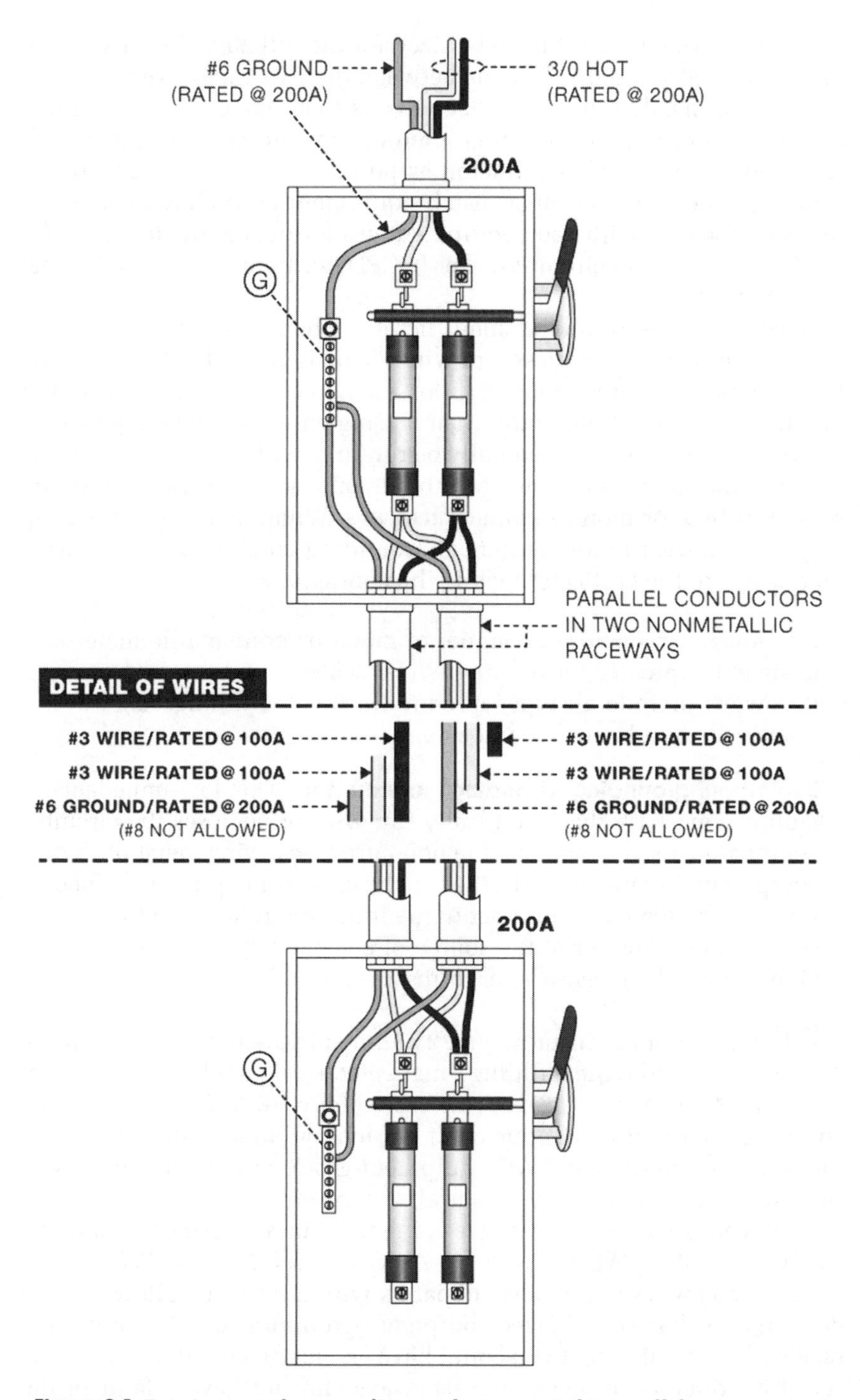

Figure 9.1 Requirements for grounding conductors routed in parallel raceways.

is routed in each conduit must be sized for the full 200A load (two conductors at 200A = 400A) flowing between the two enclosures.

For most installations, the Code tells us that when you run conductors in parallel, even if they are routed in separate raceways, it considers them to be a single conductor, as permitted under Art. 310.10(H). When this occurs, you must install equipment-grounding conductors in each raceway, with each equipment-grounding conductor sized for the total load of the circuit breaker (OCPD) and not per the individual phase conductors.

What the Code is talking about here is when you split the current load between two or more sets of wires. This is generally done because the amperage required is too high for a single conductor, or it is easier to route multiple smaller wires than one big wire. When conductors are run in parallel, they will generally be run in one of two ways: either the parallel conductors are routed together in one large raceway, or they are routed in two (or more) separate raceways. When using separate raceways, you must run an equipment-grounding conductor that is based on the size of the OCPD protecting the conductors.

Raceway An enclosed channel of metal or nonmetallic material designed expressly for holding wires, cables, bus bars, and so on. Raceways include all types of conduit, flexible tubing, raceways, floor raceways, busways, and so on.

Equipment-grounding (bonding) conductor The low-impedance fault-current path that is typically run with or encloses the circuit conductors, used to connect the non-current-carrying metal parts of equipment, raceways, and other enclosures to the grounded (neutral) conductor and equipment-grounding (bonding) conductor at service equipment or at the source of a separately derived system. Often called the *green-wire* or *third-wire* ground conductor.

Let us say you are installing a 225-amp subpanel and need to route 4/0 AWG wire from the building's main electrical panel to the new subpanel. That means you need to run a set of four 4/0 AWG wires plus the equipment-grounding conductor. If we look at Table 250.122, we see that this feed needs a 4 AWG equipment-grounding conductor. If you run this set in one raceway, this is all you need.

However, if you determine it is going to be much easier to route two smaller sets of 2 AWG wires (2 × 2 AWG Wires = 1 × 4/0 AWG) in two separate raceways to the new subpanel, which the Code allows you to do, you must have a full-sized equipment-grounding conductor in each raceway. Remember that you could have an electrical fault occur in one set of conductors in one of the raceways and not have a fault in the other. Because you only have one circuit breaker protecting both sets of conductor runs, the equipment-grounding conductor in both raceways must be sized to carry the full load back to that 225-A breaker. In other words, you now need two 4 AWG equipment-grounding conductors, one wire for each raceway. You may NOT downgrade the equipment-grounding conductor to two 6 AWG, because a 6 AWG wire is not rated to handle a 225-A circuit breaker.

If you are running your parallel conductors in a cable tray, you must use a minimum size 4 AWG conductor or the required conductor from Table 250.122, whichever is greater.

The 2014 edition of the Code expanded this section to improve clarity and added the exception allowing design changes by electrical engineers.

250.122(G). Feeder Taps. Equipment-grounding conductor(s) that are run with feeder taps must be sized according to Table 120.122 and based on the OCPD protecting the feeder taps. To be specific, the equipment-grounding conductor must be sized according to the circuit breaker on the supply side of the feeder. You may not downsize the ground wire to match the circuit that is being fed by the tap.

Feeder circuit All circuit conductors between the first service disconnect and the final OCPD. The circuits after the final OCPD are called the *branch circuits*.

Feeder taps The conductor that is tapped into a feeder circuit.

Equipment-grounding (bonding) conductor The low-impedance fault-current path that is typically run with or encloses the circuit conductors, used to connect the non-current-carrying metal parts of equipment, raceways, and other enclosures to the grounded (neutral) conductor and equipment-grounding (bonding) conductor at service equipment or at the source of a separately derived system. Often called the *green-wire* or *third-wire* ground conductor.

For example, if your feeder circuit has a 1200-A circuit breaker and it feeds three 400-A subpanels, the equipment-grounding conductor to each subpanel must match Table 120.122 for the 1200-A circuit breaker. Of course, the equipment-grounding conductor is still not required to be larger than the tap conductors.

250.124. EQUIPMENT-GROUNDING CONDUCTOR CONTINUITY

This section of the Code is in place to ensure that the equipment-grounding conductor is always in place before electrical energy is applied to a circuit.

250.124(A). Separable Connections. This section of the Code is in place to ensure that when using plug-and-cord connections, the equipment-grounding conductor is connected before the phase wires. This means that plugs, receptacles, and connections of all types, must use a first-make, last-break–type connection to ensure the equipment-grounding conductor is always in contact before any electrical energy is applied to a circuit.

First-make, last-break A type of electrical connector with pins of staggered lengths. The long pins make contact first during connector mating and break last during unmating.

The first-make, last-break connection is very common. Just look at any three-wire plug and you will see that the prong for the ground wire is slightly longer than the phase (hot) and neutral wire. When this style of connector is plugged into a receptacle, the longer ground prong will make contact first, and will be the last to leave contact when the plug is removed.

First-make, last-break connections are only required for separable connections that require an equipment-grounding conductor.

250.124(B). Switches. You may not install switches or cutoffs of any kind on an equipment-grounding conductor, unless that switch or cutoff also shuts off all sources of electrical energy at the same time.

250.126. IDENTIFICATION OF WIRING DEVICE TERMINALS

Terminals involved in the connection of equipment-grounding conductors must be properly identified. Here is the list of acceptable identifications:

1. A green, not readily removable terminal screw with a hexagonal head.
2. A green, not readily removable terminal nut.
3. A green pressure-wire connector. If the terminal of the pressure wire connector is not visible, the entrance hole for the conductor or the area around it must be marked to indicate that it is for a ground conductor. The following are acceptable marks:
   a. The letter "G"
   b. The letters "GR"
   c. The word "Ground"
   d. The word "Green"
   e. The universal electrical symbol for ground
   f. Colored green

Chapter Ten

NEC 250 PART VII: METHODS OF EQUIPMENT GROUNDING

Part VII of Art. 250 gives us a number of different methods we are allowed to use for grounding our equipment. But primarily the Code requires that we have at least one ground fault–current path that is continuous from our first service disconnect to the equipment and is made of a properly sized wire-type equipment-grounding conductor.

Since 1996, the National Electrical Code (NEC) has required that the third-wire or "green" ground be routed with every circuit. This wire is called the *equipment-grounding conductor*, and it is mandatory.

When circuits are encased in metal raceway (see NEC Art. 300 for rules governing raceway usage), the metal raceway must also provide a continuous electrical path.

This section of the Code also discusses the difference between how the neutral wire is to be handled in regard to the supply side of an electrical system and the load side. The *supply side* of an electrical service typically comprises the electrical components and systems that connect the power from the utility company to your building/structure, commonly thought of as the system above the meter and/or the first service disconnect. The *load side* refers to the electrical components and systems downstream from the meter and first service disconnect panel.

There are different rules as to how the neutral wire is bonded or not bonded, depending on what side of the electrical system it is located. In general, when on the supply side, the neutral wire is both the neutral and the ground wire at the same time. When on the load side of a service, the two must be separate conductors and kept apart at all times.

This section also tells us how to install the equipment-grounding conductors for an isolated receptacle. Isolated receptacles do NOT have the internal connection from the yoke to the equipment-grounding conductor and require two equipment-grounding conductors, one wire type for the grounding terminal on the receptacle, per Art. 250.146(A), and one for the receptacle's metal enclosure (box), per Art. 250.118.

Ground fault-current path An electrically conductive path from the point of a ground fault on a wiring system through normally non-current-carrying conductors, equipment, or the earth to the electrical supply source. A ground fault–current path may not use the earth/soil as a conductor or current path, as it is a violation of the NEC Arts. 250.4(A)(5) and 250.4(B)(4). This term is used for both intentional and unintentional current paths.

Equipment-grounding (bonding) conductor The low-impedance fault-current path that is typically run with or encloses the circuit conductors, used to connect the non-current-carrying metal parts of equipment, raceways, and other enclosures to the grounded (neutral) conductor and equipment-grounding (bonding) conductor at service equipment or at the source of a separately derived system. Often called the *green-wire* or *third-wire* ground conductor.

First service disconnect The very first electrical panel that will disconnect (turn off) the power coming in from the utility company. Sometimes this term is used to describe the first electrical panel that can turn off power on the secondary or low-voltage side of a transformer.

Load The electrical energy consumed by a component, circuit, device, piece of equipment, or system that is connected to a source of electric power to perform its function. This term often refers to all electrical components and equipment that consume electrical energy downstream from the first service disconnect. It can also be called an *electric load.*

Service The conductors and equipment for delivering electric energy from the serving utility to the wiring system of the premises.

Yoke or contact yoke A set of circular metal tabs found on both ends of a standard three-wire, 120-V outlet, often called "Mickey Mouse ears" due to their similarity in shape. These metal tabs provide a metal-to-metal contact between the circuit's (receptacles) equipment-grounding conductor and the receptacle's enclosure via an internal connection integral to the receptacle, as required under Art. 250.146(A) and (B). An isolated grounding receptacle does NOT have the internal connection from the yoke to the equipment-grounding conductor and requires two equipment-grounding conductors, one wire type for the grounding terminal on the receptacle per Art. 250.146(D), and one for the receptacle's metal enclosure (box) per Art. 250.118.

250.130. EQUIPMENT-GROUNDING CONDUCTOR CONNECTIONS

This section of the Code actually covers a number of different issues: how to connect the equipment-grounding conductor to a separately derived system, how to connect to a service entrance, and how to add a ground to a two-wire (hot and neutral only) branch circuit, which is now mandatory. Here are the four areas discussed in this section:

1. When connecting the required equipment-grounding conductor to separately derived systems (transformers, generators, solar panels, etc.), you must make sure to also comply with Art. 250.30(A)(1) regarding the neutral-to-ground connection. Primarily, you do not want to inadvertently connect your equipment-grounding conductor to a generator and form an unintentional neutral-to-ground connection. Please refer to Art. 250.30(A)(1) for more information.
2. When connecting the required equipment-grounding conductor at a service entrance, the connection must be made in accordance with Art. 250.130(A) or (B) below.
3. When replacing nongrounding-type receptacles (two-wire receptacles) with grounding-type receptacles (three-wire receptacles), if you do not have an equipment-grounding conductor in your circuit (you only have the hot and neutral wires, and no third green-wire ground), you must make the connection according to Art. 250.130(C).
4. When extending an existing branch circuit and you do not have an equipment-grounding conductor in your circuit (you only have the hot and neutral wires, and no third green-wire ground), you must make the connection according to Art. 250.130(C).

The Code has required for some time now that you must have a third-wire or green-wire ground in all circuits, regardless of amperage. In the past, the Code had excluded 15- and/or 20-A, 120-V circuits from having an equipment-grounding conductor, but now you must upgrade branch circuits whenever you are replacing receptacles or extending existing branch circuits.

Remember that Art. 250.20(B) requires circuits between 50 to 1,000 V to meet the requirements of NEC Art. 250 and to be grounded.

Equipment-grounding (bonding) conductor The low-impedance fault-current path that is typically run with or encloses the circuit conductors, used to connect the non-current-carrying metal parts of equipment, raceways, and other enclosures to the grounded (neutral) conductor and equipment-grounding (bonding) conductor at service equipment or at the source of a separately derived system. Often called the *green-wire* or *third-wire* ground conductor.

Separately derived system(s) A premises wiring system whose power is derived from a source of electrical energy other than a (utility) service. Examples include solar panels, generators, and wind turbines that have no direct connection to another electrical source. Specifically, the neutral conductor must not be solidly bonded to the neutrals of other sources. This term is also used in regard to isolation and/or step-down transformers that generate a distinct power source separate from the utility.

Neutral-to-ground connection (bond) is where a grounded current-carrying conductor, the neutral, is intentionally bonded to earth/ground. In general, this should only occur at the transformer and at the first service disconnect(s). The Code uses several names for neutral-to-ground connections: system bonding jumper, main bonding jumper, and supply-side bonding jumper.

Service The conductors and equipment for delivering electric energy from the serving utility to the wiring system of the premises.

Circuit or branch circuit The electrical system downstream from the last OCPD. The outlet you plug your computer into would be part of a circuit.

250.130(A). For Grounded Systems. The equipment-grounding conductor must be connected to the neutral (grounded conductor) from the service and to the grounding electrode conductor, for grounded electrical systems.

Grounded system An electrical service that has one of the current-carrying conductors intentionally grounded (i.e., the neutral wire). Floating or delta-type electrical systems are not grounded systems, whereas wye-type systems are grounded.

Equipment-grounding (bonding) conductor The low-impedance fault-current path that is typically run with or encloses the circuit conductors, used to connect the non-current-carrying metal parts of equipment, raceways, and other enclosures to the grounded (neutral) conductor and equipment-grounding (bonding) conductor at service equipment or at the source of a separately derived system. Often called the *green-wire* or *third-wire* ground conductor.

First service disconnect The very first electrical panel that will disconnect (turn off) the power coming in from the utility company. Sometimes this term is used to describe the first electrical panel that can turn off power on the secondary or low-voltage side of a transformer.

What this means is that you must tie the equipment-grounding conductor to the ground bar in the electrical cabinet that has the circuit breaker matched to your ground wire. That ground bar must in turn be tied back to the utility's incoming neutral wire via the neutral-to-ground bond, as discussed at great length in Part II of Art. 250.

If your circuit breaker is in a subpanel, you must bond the equipment-grounding conductor to the ground bar in the subpanel. The ground bar in the subpanel must be bonded back to the first service disconnect via an equipment-grounding conductor, in most cases. See Art. 250.32 for additional information.

250.130(B). For Ungrounded Systems. The equipment-grounding conductor must be connected to the grounding electrode conductor in the panel with the overcurrent protection device (OCPD) matched for the circuit, for ungrounded electrical systems.

Ungrounded system An electrical service that does not have any of the current-carrying conductors intentionally grounded (i.e., no neutral wire). Delta-type electrical systems and some standby generators are examples of ungrounded systems.

Overcurrent device Usually a fuse or circuit breaker, a device that limits the maximum amount of current that can flow in a circuit. Sometimes called an *overcurrent protection device* or *OCPD*.

250.130(C). Nongrounding Receptacle Replacement or Branch Circuit Extensions. This section applies to both grounded and ungrounded electrical systems. It governs the extension of existing branch circuits that do not have an equipment-grounding conductor (two-wire systems) and the replacement of an existing nongrounding receptacle (two-wire type) by a grounded-type receptacle (three-wire type).

What this section tells us is that you must establish an equipment-grounding conductor for your receptacles and branch circuits via any of the following methods:

1. Connecting to the building's/structure's grounding electrode system at any accessible point as described in Art. 250.50.
2. Connecting to the building's/structure's grounding electrode conductor at any accessible point.
3. Connecting to the grounding terminal bar within the enclosure from which the circuit originates (route a ground wire from the electrical cabinet to the circuit/receptacle).
4. For grounded electrical systems, connect to the neutral terminal bar (grounded conductor) within the enclosure from which the circuit originates (route a ground wire from the electrical cabinet to the circuit/receptacle), as long as it is a first service disconnect and not a subpanel. See Art. 250.30(A)(1) regarding neutral-to-ground connections.

5. For ungrounded electrical systems, connect to the grounding terminal bar (tied to the grounding electrode conductor) within the enclosure from which the circuit originates (route a ground wire from the electrical cabinet to the circuit/receptacle).

The Code has required for some time now that you have a third-wire or green-wire ground in all circuits regardless of amperage. In the past, the Code had excluded 15- and/or 20-A, 120-V circuits from having an equipment-grounding conductor, but now you must upgrade branch circuits whenever you are replacing receptacles or extending existing branch circuits.

Remember that Art. 250.20(B) requires circuits between 50 to 1000 V to meet the requirements of NEC Art. 250 and to be grounded.

Grounding (earthing) electrode system Two or more grounding electrodes that are bonded together to form a single system. For a typical building/structure, a grounding electrode system should consist of: two ground rods, water pipe, steel frame of the building, the steel rebar in the concrete foundation, and any ground rings.

Grounding (earthing) electrode conductor The conductor used to connect the grounding (earthing) electrode(s) to the equipment-grounding (bonding) conductor, to the neutral (grounded) conductor, or to both in accordance with Arts. 250.66 and 250.142.

In summary, this section of the Code requires that your branch circuits and receptacles must have an equipment-grounding conductor connected from the grounding terminal on the receptacle to any accessible point on the building's/structure's:

- Grounding electrode systems
- Grounding electrode conductor
- Grounding terminal bar within the service enclosure
- Neutral bar in the service enclosure, if it does not violate the neutral-to-ground bonding rules in Art. 250.30(A)(1)

Note *You may NOT connect any equipment-grounding conductor to the water piping system beyond 5 ft from where water supply enters the building. See Arts. 250.68(C)(1), 250.52(A)(1), 250.53(D), and 250.104(A) for more information on bonding/connecting to water pipe.*

The 2014 edition of the Code added wording to clarify the requirements in Art. 250.130(C)(4).

250.132. SHORT SECTIONS OF RACEWAY

If you have an isolated section of metal raceway, a metal enclosure, or cable armor, you must connect it to an equipment-grounding conductor in accordance with Art. 250.134. See Art. 250.80 for an exception regarding buried metal raceway elbows.

Raceway An enclosed channel of metal or nonmetallic material designed expressly for holding wires, cables, bus bars, and so on. Raceways include all types of conduit, flexible tubing, raceways, floor raceways, busways, and so on.

What this section tells us is that if you have a metal receptacle box (or any other metal raceway/enclosure) that is isolated (not connected to metal conduit), you must bond it to the circuit's equipment-grounding conductor.

250.134. EQUIPMENT FASTENED IN PLACE OR CONNECTED BY PERMANENT WIRING METHODS (FIXED)—GROUNDING

This section of the Code is in place primarily to eliminate a conflict that arose between Art. 250.32(B)(1), which requires an equipment-grounding conductor to be routed in the same raceway as the service feed, and some exceptions that are in place for certain appliances, as found in Arts. 250.140 and 250.142.

This section tells us that you must have an equipment-grounding conductor bonding the non-current-carrying metal parts of equipment, raceways, and enclosures, unless it meets the rules in Arts. 250.140 and 250.142.

250.134(A). Equipment-grounding Conductor Types. As expected, the types of acceptable equipment-grounding conductors are as permitted in Art. 250.118.

250.134(B). With Circuit Conductors. You must route the equipment-grounding conductor inside the same raceway or cable as the circuit conductors. Keeping the phase (hot), neutral, and ground wires together in the same raceway (or cable) keeps the impedance on the equipment-grounding conductor to a minimum.

The phase (hot) wire will develop a magnetic field as the current travels down the conductor to the load. The same is true with regard to the returning current on the neutral wire. Forming these magnetic fields takes time and energy that can be measured in an overall increase in impedance. Because the strength of the magnetic field (magnetic flux) is inversely proportional to the square of the distance between the two conductors, cross-canceling these magnetic fields is best done by close proximity of another low-impedance conductor that is either out of phase (i.e., the neutral wire) or has no current on it (i.e., the ground wire) or both.

During normal operations, the hot and neutral wires will cross-cancel the majority of each other's magnetic fields. The equipment-grounding conductor will gather the rest of the magnetic flux, reducing heat and objectionable currents from forming in the surrounding metal raceways.

During fault conditions, the only viable ground-fault path is the equipment-grounding conductor. Keeping the ground wire with the circuit wires ensures that the cancellation of the magnetic flux will still occur, thereby keeping the impedance low.

If the ground wire is routed separately from the circuit conductors, during normal operation, the levels of heat and objectionable current induced onto the surrounding raceways will be somewhat higher than when the ground wire is included.

If the ground wire is routed separately from the circuit conductors, during fault conditions, the cancellation of the magnetic flux will be negligible, thereby increasing the impedance of the ground fault–current path. This could result in the failure of OCPDs such as circuit breakers and fuses to operate properly. Please see Art. 250.4(A)(5) and (B)(4) for more information regarding ground fault–current paths.

Ground fault An unintentional connection between an energized conductor and earth or metallic parts of enclosures, raceways, or equipment.

Ground fault-current path An electrically conductive path from the point of a ground fault on a wiring system through normally non-current-carrying conductors, equipment, or the earth to the electrical supply source. A ground fault–current path may not use the earth/soil as a conductor or current path, as it is a violation of the NEC Arts. 250.4(A)(5) and 250.4(B)(4). This term is used for both intentional and unintentional current paths.

Raceway An enclosed channel of metal or nonmetallic material designed expressly for holding wires, cables, bus bars, and so on. Raceways include all types of conduit, flexible tubing, raceways, floor raceways, busways, and so on.

There are a few exceptions to the rule found in Art. 250.134(B). They are as follows:

Exception no. 1: When installing new receptacle or branch circuits as in Art. 250.130(C), the equipment-grounding conductor does not need to be in the raceway but may be run separately from the circuit conductors, if you are installing a ground-type receptacle on an existing circuit where there is no ground.

Exception no. 2: For direct current (DC) circuits, the equipment-grounding conductor does not need to be routed in the raceway. Direct current circuits do not form the magnetic fields that alternating current (AC) systems have, and therefore do not have the impedance issues discussed above.

Please see Arts. 250.102 and 250.168 for more information regarding equipment-bonding jumpers. Please see Art. 400.7 for more information regarding cords for fixed equipment.

250.136. EQUIPMENT CONSIDERED GROUNDED

This section tells us that there are certain non-current-carrying metal parts that are considered grounded under certain conditions. See Art. 250.136(A) and (B).

250.136(A). Equipment Secured to Grounded Metal Supports. This section tell us that if you have a structural metal rack that is designed to support electrical equipment that has been bolted securely to its frame, you do not need to run an individual equipment-grounding conductor to each piece of equipment. You only need to securely bond an equipment-grounding conductor to the metal frame of the rack, in accordance with Arts. 250.118 and 250.134.

Just because the Code allows you to do this, it does not mean that it is a good idea. In fact, this exception should be removed from the Code. Many equipment manufacturers, especially for sensitive electronic systems, will require an equipment-grounding conductor to each individual piece of equipment, regardless of what the Code says. Keep in mind that the physical separation between the equipment-grounding conductor and the circuit conductors increases the impedance of the ground fault–current path and may in fact violate Arts. 250.134(B) and 300.3(B).

Ground fault An unintentional connection between an energized conductor and earth or metallic parts of enclosures, raceways, or equipment.

Ground fault-current path An electrically conductive path from the point of a ground fault on a wiring system through normally non-current-carrying conductors, equipment, or the earth to the electrical supply source. A ground fault–current path may not use the earth/soil as a conductor or current path, as it is a violation of the NEC Arts. 250.4(A)(5) and 250.4(B)(4). This term is used for both intentional and unintentional current paths.

Equipment-grounding (bonding) conductor The low-impedance fault-current path that is typically run with or encloses the circuit conductors, used to connect the non-current-carrying metal parts of equipment, raceways, and other enclosures to the grounded (neutral) conductor and equipment-grounding (bonding) conductor at service equipment or at the source of a separately derived system. Often called the *green-wire* or *third-wire* ground conductor.

Additionally, you must ensure that the metal frame is also tied to ground via an equipment-grounding conductor. See Section V of Art. 250 for more information regarding bonding requirements.

250.136(B). Metal Car Frames. A metal car frame is considered to be grounded if it is supported by metal hoisting cables that meet all of the following requirements:

- The metal hoisting cables must be attached to or running over metal sheaves or drums of an elevator.
- The metal sheaves or drums of the elevator must be bonded/connected to an equipment-grounding conductor that meets the requirements of Art. 250.134.

This means that if you are hoisting cars around with a crane, you can consider the cars grounded if the metal hoisting cables are grounded via the metal sheaves or drums of the crane (assuming the sheaves or drums are properly grounded).

250.138. CORD AND PLUG-CONNECTED EQUIPMENT

All metal non-current-carrying components of equipment that are connected via a cord and plug must be bonded to the equipment-grounding conductor. See Art. 250.138 (A) and (B) for more information.

250.138(A). By Means of an Equipment-grounding Conductor. This section states that if your cord-and-plug equipment may an equipment-grounding conductor routed inside the cable assembly, then it must have a fixed grounding contact, such as a pin for attachment. Think of any three-wire extension cord and you have the type discussed here. Please see Art. 250.114 for additional information regarding cord-and-plug equipment.

There is a small exception to this rule that involves a particular type of grounding contact pole on movable, self-restoring type, plug-in ground fault circuit interrupter (GFCI) systems. This exception only involves the manufacturers. All you need to know is to use listed GFCI systems.

Listed or listed item An item, product, or device that has been specifically developed for a particular function and has been independently evaluated by a product-testing organization, such as the Underwriters Laboratory (UL) or Conformité Européenne (CE).

Equipment-grounding (bonding) conductor The low-impedance fault-current path that is typically run with or encloses the circuit conductors, used to connect the non-current-carrying metal parts of equipment, raceways, and other enclosures to the grounded (neutral) conductor and equipment-grounding (bonding) conductor at service equipment or at the source of a separately derived system. Often called the *green-wire* or *third-wire* ground conductor.

250.138(B). By Means of a Separate Flexible Wire or Strap. This section states that cord-and-plug equipment may have an equipment-grounding conductor that is made by means of a separate strap of flexible wire that is insulated or bare, as long as it is protected as well as practicable against physical damage and is part of the equipment.

250.140. FRAMES OF RANGES AND CLOTHES DRYERS

This section tells us that the frames of ranges and clothes dryers must be connected to an equipment-grounding conductor as specified in Art. 250.134 or 250.138. This includes electric ovens/ranges, wall-mounted ovens/ranges, counter-mounted cooking units, clothes dryers, and outlets or junction boxes that are part of the circuit supplying power for the appliances.

There is an exception to this rule that applies only to existing electrical circuits that do not have an equipment-grounding conductor. Prior to 1996, the Code allowed appliances to be grounded using the neutral wire, and many such old circuits still exist. Newly installed circuits may not use this rule. When installing an appliance to an old circuit, if you choose to use this exception, all of the following must be met:

1. If you have a single-phase, 240-V, three-wire appliance, you may use the neutral wire as the equipment-grounding conductor. The same is true if it is a single-phase, 208-V circuit derived from a three-phase system.

2. The neutral wire (grounded conductor) must be at least a 10 AWG copper or an 8 AWG aluminum conductor; it may not be smaller.
3. The neutral wire must be insulated, or it may be part of a Type SE service entrance cable with the branch service originating at the first service disconnect.
4. The grounding contacts of any receptacles furnished as part of the equipment (appliance) are bonded to the frame of the equipment.

While the Code allows appliances to be connected this way, the safety of the people using this equipment is best served with an equipment-grounding conductor (and a GFCI system, where possible). A new circuit should be run if at all possible that includes a wire-type equipment-grounding conductor.

Equipment-grounding (bonding) conductor The low-impedance fault-current path that is typically run with or encloses the circuit conductors, used to connect the non-current-carrying metal parts of equipment, raceways, and other enclosures to the grounded (neutral) conductor and equipment-grounding (bonding) conductor at service equipment or at the source of a separately derived system. Often called the *green-wire* or *third-wire* ground conductor.

First service disconnect The very first electrical panel that will disconnect (turn off) the power coming in from the utility company. Sometimes this term is used to describe the first electrical panel that can turn off power on the secondary or low-voltage side of a transformer.

Neutral-to-ground connection (bond) is where a grounded current-carrying conductor, the neutral, is intentionally bonded to earth/ground. In general, this should only occur at the transformer and at the first service disconnect(s). The Code uses several names for neutral-to-ground connections: system bonding jumper, main bonding jumper, and supply-side bonding jumper.

Special caution should be taken when connecting an older appliance to a new circuit. The older appliance may have an internal and integral bond connecting the neutral to the chassis of the appliance. You must remove this neutral-to-ground bond inside the old appliance prior to installing it to the new electrical circuit. (A four-wire circuit/cord will carry two phase conductors: a neutral and an equipment-grounding conductor. When this is installed, the neutral-to-ground bond will now be located in the first service disconnect and therefore may NOT also be in the appliance.)

250.142. USE OF GROUNDED CIRCUIT CONDUCTOR FOR GROUNDING EQUIPMENT

This section tells us that under certain circumstances, you may use the neutral (grounded conductor) to provide grounding for non-current-carrying metal parts of equipment, raceway, and enclosures. These circumstances are for the supply side of your service, for separately derived systems such as generators, and for certain building/structure installations being supplied by a branch circuit.

250.142(A). Supply-side Equipment. This section tells us that when dealing with the supply side of your service (above the meter), you may use the neutral (grounded conductor) to provide grounding for non-current-carrying metal parts of equipment, raceways, and enclosures. This applies to any of the following conditions:

1. Equipment, raceways, and enclosures that are on the supply side of your first service disconnect (AC systems). This is the utility company side that is above the meter at your service.
2. Inside the enclosure of the main disconnect for a separate building/structure supplied by a feeder or branch circuit as allowed in Art. 250.32(B)(1). This would be for a building or structure that is electrically isolated from the supply building, has a master circuit breaker/disconnect, and has its own neutral-to-ground bond. Please see Art. 250.32(B) for more information.
3. Inside the enclosure of the main disconnect for a separately derived system (transformer, generator, solar, etc.) as allowed in Arts. 250.30(A)(1) and 250.32(B)(2).

The utility company will only provide phase wires and a neutral wire to your service. There will be no ground wire brought to your service.

First service disconnect The very first electrical panel that will disconnect (turn off) the power coming in from the utility company. Sometimes this term is used to describe the first electrical panel that can turn off power on the secondary or low-voltage side of a transformer.

Service The conductors and equipment for delivering electric energy from the serving utility to the wiring system of the premises.

Neutral-to-ground connection (bond) is where a grounded current-carrying conductor, the neutral, is intentionally bonded to earth/ground. In general, this should only occur at the transformer and at the first service disconnect(s). The Code uses several names for neutral-to-ground connections: system bonding jumper, main bonding jumper, and supply-side bonding jumper.

Note *The process allowed in Art. 250.32(B)(1) and listed in item 2 above should be banned. Even if a building or structure is electrically isolated, branch circuits and feeders should not be allowed to form a second neutral-to-ground bond. There is simply no way of knowing what future systems will be installed in a building/structure. A new metal pipe, a new cable television line, and/or a new data wire is all it takes to cause a major problem and a potential fire hazard.*

250.142(B). Load-side Equipment. This section states that you may NOT use a neutral wire (grounded conductor) to ground non-current-carrying metal parts on the load side (below the meter) of your service. There are a few exceptions to this rule, primarily Arts. 250.30(A)(1) and 250.32(B), which discuss the bonding jumpers that are used to form the neutral-to-ground bond in the first service disconnect of your building/structure. Here are the other exceptions:

Exception no. 1: Frames of certain appliances listed in Art. 250.140

Exception no. 2: You may ground meter enclosures using the neutral wire (grounded conductor) when the meter enclosure is adjacent to the first service disconnect when ALL of the following apply:

1. There are no ground-fault protection devices installed.
2. All meter enclosures are located immediately adjacent to the service disconnection means.
3. The size of the neutral wire (grounded conductor) is in accordance with Table 250.122 (see Appendix B).

Exception no. 3: Direct current systems may be grounded on the load side of the disconnection means in accordance with Art. 250.164.

Exception no. 4: Electrode-type boilers operating at over 1000 V must be grounded as required in Arts. 490.72(E)(1) and 490.74 (bond the neutral conductor to the pressure vessel).

The reason that it is important to prevent the neutral wire (grounded conductor) from being bonded to normally non-current-carrying metal parts on the load side of the service is that this establishes a second and parallel path on by which returning currents may travel. In other words, using the neutral-to-ground metal parts on the load side allows electrical energy to return to the transformer via those metal parts during normal operations. This is a serious electrical hazard and must be avoided.

Testing your electrical service for improper neutral-to-ground bonds is relatively easy; the amount of current on the neutral of a single-phase circuit must be equal (within a few milliamps) to the amount of current on the hot (phase) wire. You should also have very low amperages on your ground wires, less than half an amp in most cases. A thermal camera can easily spot amperages on ground conductors. If your electrical system has an imbalance between the neutrals and their corresponding phase wires (three-phase, and single-phase 208/240-V three-wire systems will have low or even zero neutral currents), or you have excessive currents on your ground wires, corrective action is required immediately.

The 2014 edition of the Code globally changed the 600-V rule to 1000 V; this was changed in Exception no. 4 above.

250.144. MULTIPLE CIRCUIT CONNECTIONS

If you have a piece of equipment that is fed by multiple circuits, you must have an equipment-grounding conductor for each of those circuits, as specified in Arts. 250.134 and 250.138.

You can think of a very large and complex equipment-processing system located inside a factory. There are often multiple circuits used to feed this equipment, including three-phase circuits to power motors and single-phase systems for computers and Programmable Logic Controllers (PLCs). You must have an equipment-grounding conductor for each circuit, and it must be located in the same raceway as its corresponding phase conductors.

250.146. CONNECTING RECEPTACLE-GROUNDING TERMINAL TO BOX

This section simply tells us that you must install an equipment-bonding jumper from the ground terminal of a receptacle to the enclosure box of the receptacle.

> **Equipment-bonding jumper or load-side bonding jumper** A conductor, screw, or strap that bonds the equipment-grounded conductor (neutral) to the equipment enclosures that are on the load side of the first service disconnect enclosure(s), in accordance with Art. 250.102(D). The term is also used to indicate a conductor, screw, or strap that bonds the equipment-grounding conductor of a circuit to the enclosure or raceway of that circuit, in accordance with Art. 250.146.

There are several ways to do this. You can use a physical wire that comes from the ground terminal on a receptacle and is properly connected to the box per Art. 250.8 and sized per Table 250.122 (see Appendix B), or it can be a device that is integral to the receptacle and establishes a metal-to-metal contact from the receptacle to the box, as is discussed in Art. 250.146(A) and (B).

This rule applies to grounded metal boxes and to grounding-type receptacles.

250.146(A). Surface-mounted Boxes. When installing your typical 120-V receptacle (outlet), it comes standard with "Mickey Mouse ears" at the top and the bottom of the outlet and a set of screws pre-inserted. The "Mickey Mouse ears" are actually called a yoke or contact yoke, and they provide a direct metal-to-metal contact between the receptacle and the box or enclosure. Internal to the receptacle is a connection between the yoke and the receptacle's ground terminal, fulfilling the requirement in Art. 250.146.

> **Yoke or contact yoke** A set of circular metal tabs found on both ends of a standard three-wire, 120-V outlet, often called "Mickey Mouse ears" due to their similarity in shape. These metal tabs provide a metal-to-metal contact between the circuit's (receptacles) equipment-grounding conductor and the receptacle's enclosure via an internal connection integral to the receptacle, as required under Arts. 250.146(A) and (B). An isolated grounding receptacle does NOT have the internal connection from the yoke to the equipment-grounding conductor and requires two equipment-grounding conductors, one wire type for the grounding terminal on the receptacle per Art. 250.146(D), and one for the receptacle's metal enclosure (box) per Art. 250.118.

> **Listed or listed item** An item, product, or device that has been specifically developed for a particular function and has been independently evaluated by a product-testing organization, such as the UL or CE.

When you examine the receptacle, you will notice that there are often plastic or fiber retention washers (often square shaped) installed on the metal screw to prevent the screw from falling off the receptacle during handling and shipping. The Code requires you to remove at least one of these plastic retention washers, although you should remove both, prior to the installation of the receptacle.

There is an exception to the rule regarding the removal of the plastic square washers if you are installing the receptacle using a conduit-style 4-in.^{2} metal cover box that requires two screws per receptacle for mounting. In this case, you can technically leave the plastic square retaining washers on, but you really should just remove them.

The best method is to always remove all of the plastic retaining washers from your receptacles.

And, of course, you must use a listed receptacle.

250.146(B). Contact Devices or Yokes. Receptacles that are listed as self-grounding shall be permitted as long as the support screws are connected per Art. 250.146(A).

Listed or listed item An item, product, or device that has been specifically developed for a particular function and has been independently evaluated by a product-testing organization, such as the UL or CE.

250.146(C). Floor Boxes. Floor boxes that house receptacles qualify as establishing an equipment-bonding jumper between the ground terminal of a receptacle and the floor box, as long as the floor box is listed and designed for such use.

Listed or listed item An item, product, or device that has been specifically developed for a particular function and has been independently evaluated by a product-testing organization, such as the UL or CE.

Equipment-bonding jumper or load-side bonding jumper A conductor, screw, or strap that bonds the equipment-grounded conductor (neutral) to the equipment enclosures that are on the load side of the first service disconnect enclosure(s), in accordance with Art. 250.102(D). The term is also used to indicate a conductor, screw, or strap that bonds the equipment-grounding conductor of a circuit to the enclosure or raceway of that circuit, in accordance with Art. 250.146.

250.146(D). Isolated Receptacles. Isolated receptacles are typically installed for reducing electrical noise, harmonics, stray currents, and other objectionable currents from a given electrical circuit. On an isolated receptacle, the grounding terminal or post is purposely insulated from having contact with the metal mounting gear of the receptacle (the yoke). Isolated receptacles are typically visibly marked with an orange triangle (and sometimes an orange triangle with a colored dot) on the face of the receptacle.

The Code states that when using an isolated receptacle, you may not use the conduit as a grounding electrode conductor, as in Art. 250.118, but that you must use an insulated wire-type equipment-grounding conductor run from the first service disconnect to the ground terminal on the isolated receptacle, sized per Table 250.122 (see Appendix B). In Fig. 10.1A, we see a wiring schematic for a typical 120V receptacle (outlet) with the equipment grounding conductor routed from the grounding bar in the electrical cabinet, through the metal conduit, and on to the grounding lug (screw) on the receptacle. The yoke of the receptacle is internally bonded to the grounding lug. When properly installed the yoke will be in direct contact with the metal enclosure, thereby providing a physical connection between the equipment grounding conductor and the metal enclosure housing the receptacle.

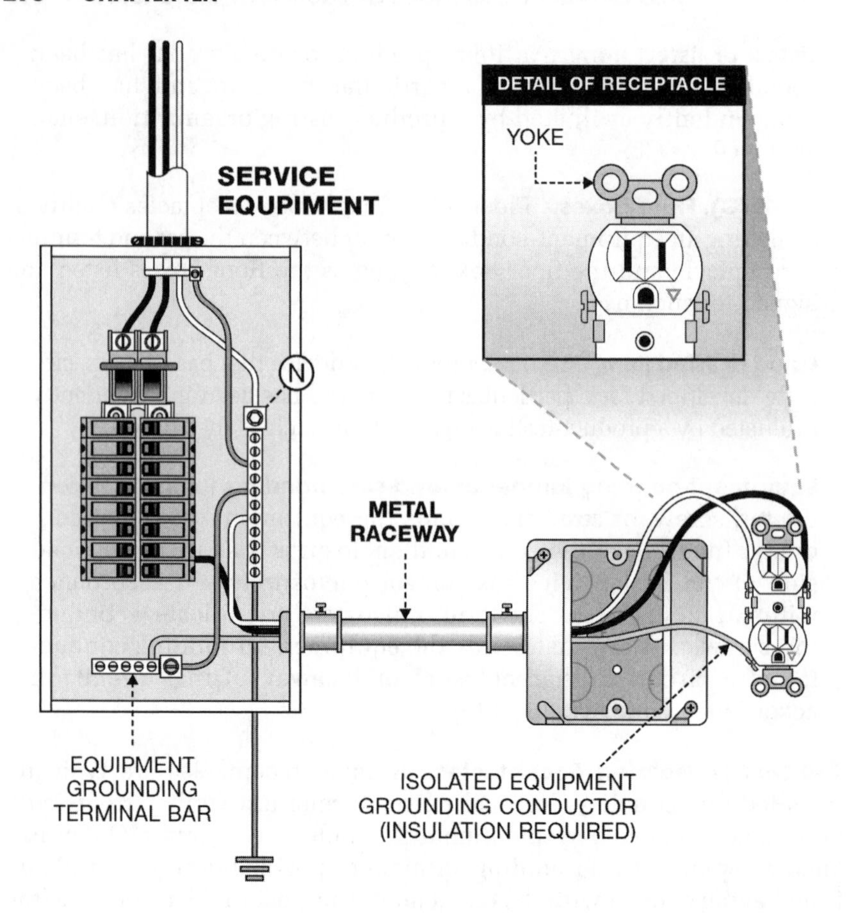

Figure 10.1A Isolated ground-type receptacle installed with continuous metal raceway.

The metal raceway and/or enclosures of the isolated receptacle's circuit must be properly bonded to ground per Art. 250.118, even when using an isolated receptacle. If you cannot ensure that the raceway (conduit) qualifies as an equipment-grounding conductor per Art. 250.118 (flexible metal conduit, PVC, etc.), you must run a second wire-type equipment-grounding conductor just for the metal enclosure of the isolated receptacle. You may find that within certain municipalities the inspector will even mandate that you route two green insulated wires when using isolated receptacles. In fact, it is simply good engineering practice to route two equipment-grounding conductors for each isolated receptacle. In Fig. 10.1B, we see a wiring schematic for an isolated ground-type 120V receptacle (outlet), as can be noted by the triangle on the face of the outlet. The circuit has two equipment grounding conductors routed from the grounding bar in the electrical cabinet and through the metal conduit. One of the equipment grounding conductors

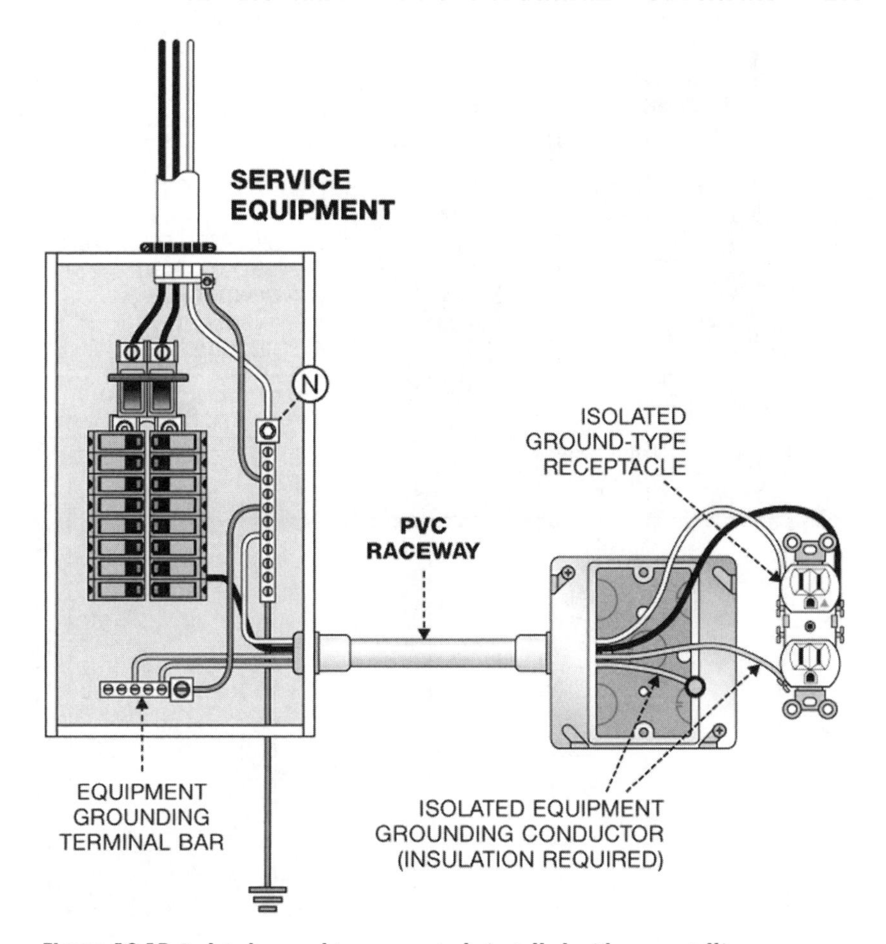

Figure 10.1B Isolated ground-type receptacle installed with nonmetallic raceway.

is bonded to the metal enclosure housing the receptacle. The other equipment grounding conductor is landed on the grounding lug (screw) on the isolated ground-type receptacle. The yoke of the isolated ground-type receptacle is NOT internally bonded to the grounding lug. When properly installed the isolated yoke will NOT provide a physical barrier to block any electrical noise that may be on the metal enclosure housing the receptacle, from entering the protected circuit.

Now, if your isolated receptacle is fed from a subpanel and not from the first service disconnect, you are allowed to route the wire-type equipment-grounding conductor directly from the first service disconnect and bypass the subpanel; see Art. 250.96(B) for more information. This is the best practice for isolated receptacles and should be done wherever possible. To be specific, the ground wire connected to the metal enclosure of the receptacle would go to the subpanel. The ground wire connected to the terminal on the isolated receptacle would be

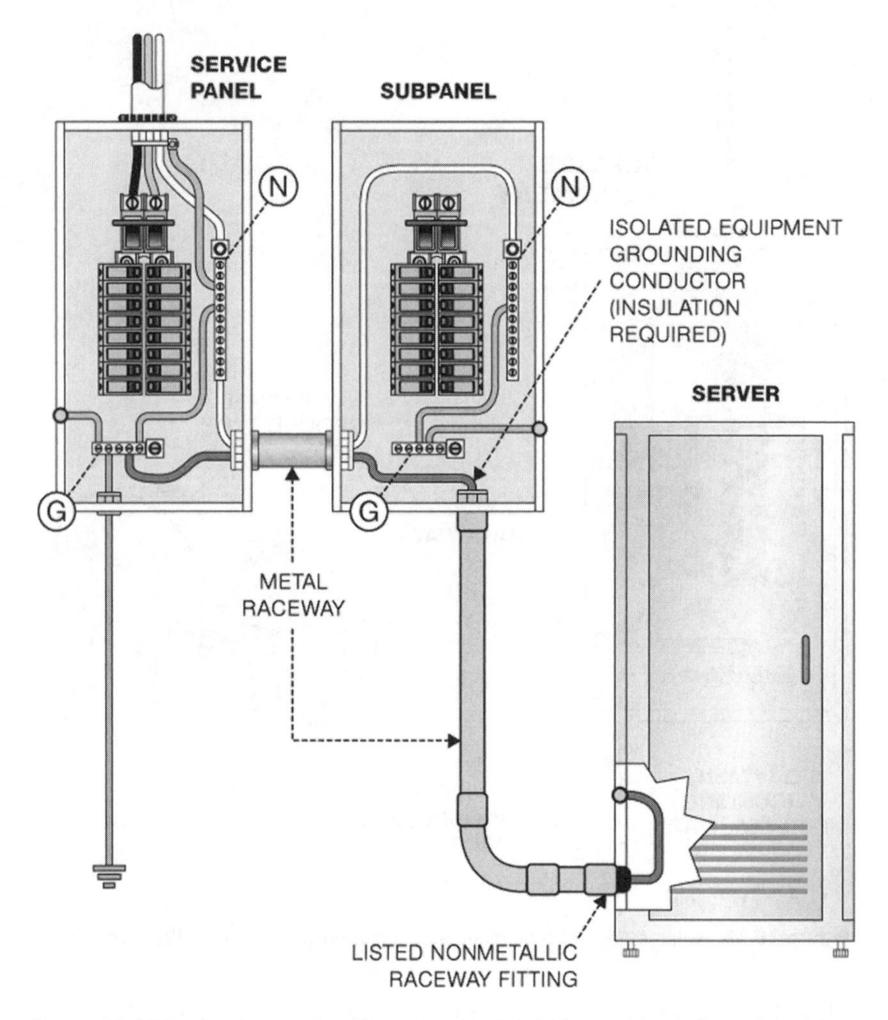

Figure 10.1C Isolated grounding for a server rack fed from a subpanel.

routed through the subpanel and all the way back to the first service disconnect.

Figure 10.1C is a schematic showing how to properly install an isolated equipment grounding conductor to piece of equipment, such as a server rack, when being fed from a subpanel. Note that the equipment grounding conductor originates in the first service disconnect and is routed all the way to the server, through and bypassing the subpanel, even though the electrical feed is supplied by the subpanel. Also, pay special attention to the listed non-metallic raceway fitting that is used to connect the metal conduit to the server rack. This non-metallic fitting provides electrical isolation for the server rack, preventing objectionable electrical noise from coming through the metal conduit and interfering with data communications.

However, if your sensitive equipment is in need of isolated receptacles due to electrical noise, and your circuit is to be fed from a subpanel, you really should consider using an isolation transformer to form a new first service disconnect.

Certain IEEE, ANSI, EIA, and other industrial standards actually require that sensitive equipment have a second equipotential grounding system installed that is specifically designed for removing objectionable currents, harmonics, transients, and other "noise" from the chassis and exposed metal components of equipment and other metal objects in the vicinity of the sensitive equipment. Consider a computer server room, a cellular base station, or radio transmission equipment as examples of the types of equipment that require additional grounding systems.

Yoke or contact yoke A set of circular metal tabs found on both ends of a standard three-wire, 120-V outlet, often called "Mickey Mouse ears" due to their similarity in shape. These metal tabs provide a metal-to-metal contact between the circuit's (receptacles) equipment-grounding conductor and the receptacle's enclosure via an internal connection integral to the receptacle, as required under Art. 250.146(A) and (B). An isolated grounding receptacle does NOT have the internal connection from the yoke to the equipment-grounding conductor and requires two equipment-grounding conductors, one wire type for the grounding terminal on the receptacle per Art. 250.146(D), and one for the receptacle's metal enclosure (box) per Art. 250.118.

Listed or listed item An item, product, or device that has been specifically developed for a particular function and has been independently evaluated by a product-testing organization, such as the UL or CE.

First service disconnect The very first electrical panel that will disconnect (turn off) the power coming in from the utility company. Sometimes this term is used to describe the first electrical panel that can turn off power on the secondary or low-voltage side of a transformer.

Replacing existing receptacles with isolated receptacles is generally considered to be unacceptable. Isolated receptacles should only be utilized for new branch circuits, and they should never be used with a shared neutral or even on a circuit with multiple receptacles. Isolated receptacles should have their own dedicated circuit all the way back to the first service disconnect. However, if you pull new conductors to the outlet with two green-wire grounds, one for the enclosure and one for the ground terminal on the receptacle, you will have a proper system.

250.148. CONTINUITY AND ATTACHMENT OF EQUIPMENT-GROUNDING CONDUCTORS TO BOXES

This section tells us that when wire-type equipment-grounding conductors are spliced within a box or terminated on a lug, there are certain methods, devices, and systems that are acceptable, and ones that are not.

> **Equipment-grounding (bonding) conductor** The low-impedance fault-current path that is typically run with or encloses the circuit conductors, used to connect the non-current-carrying metal parts of equipment, raceways, and other enclosures to the grounded (neutral) conductor and equipment-grounding (bonding) conductor at service equipment or at the source of a separately derived system. Often called the *green-wire* or *third-wire* ground conductor.

See Art. 250.146(D) for an exception when installing isolated receptacles and the required wire-type equipment-grounding conductor.

250.148(A). Connections. When you splice the wire-type equipment-grounding conductor, you must make that splice in accordance with NEC Art. 110.14(B), except that insulation is not required. NEC Art. 110.14(B) says that you must use listed connection devices or brazing or welding. Please see NEC Art. 110.14(B) for more information.

> **Listed or listed item** An item, product, or device that has been specifically developed for a particular function and has been independently evaluated by a product-testing organization, such as the UL or CE.

250.148(B). Grounding Continuity. This section tells us that the metal box or raceway housing a receptacle or luminaire must have a connection to an equipment-grounding conductor that is not broken (disconnected) when the receptacle, luminaire, or the device fed from the circuit is removed.

> **Luminaire** Complete lighting unit, consisting of one or more lamps (bulbs or tubes that emit light), along with the socket and other parts that hold the lamp in place and protect it, wiring that connects the lamp to a power source, and a reflector that helps direct and distribute the light.

> **Equipment-grounding (bonding) conductor** The low-impedance fault-current path that is typically run with or encloses the circuit conductors, used to connect the non-current-carrying metal parts of equipment, raceways, and other enclosures to the grounded (neutral) conductor and equipment-grounding (bonding) conductor at service equipment or at the source of a separately derived system. Often called the *green-wire* or *third-wire* ground conductor.

Raceway An enclosed channel of metal or nonmetallic material designed expressly for holding wires, cables, bus bars, and so on. Raceways include all types of conduit, flexible tubing, raceways, floor raceways, busways, and so on.

In other words, if you cannot ensure that the raceway and enclosures are properly bonded to ground per Art. 250.118, you must bond the wire-type equipment-grounding conductor to the box first, and then to the receptacle/luminaire.

250.148(C). Metal Boxes. This section tells us that metal boxes must be connected to at least one equipment-grounding conductor (wire type and/or metal raceway), via a listed grounding screw, grounding device, and/or grounding equipment that is used for no other purpose.

What this section tells us is that when installing a metal box in a raceway, it must have a connection to an equipment-grounding conductor that is in accordance with Art. 250.118. This could be a continuous metal raceway or a wire-type equipment-grounding conductor.

So, if you install a metal box as a pull section and there are NO splices in any of the wire-type conductors, you can treat the metal box as if it were an "L"- or a "T"-style conduit body, and no bond is required.

However, if you install a metal box as a pull section and there ARE splices in the conductors, you must bond all of the equipment-grounding conductor(s) associated with those circuit(s) to the metal box, as required by Art. 250.96(A).

250.148(D). Nonmetallic Boxes. When pulling a circuit through a nonmetallic outlet box, you must ensure that at least one of the wire-type equipment-grounding conductors is routed in such a manner as to allow for a future connection(s).

250.148(C). Solder. You may not make connections that depend solely on solder. Please see NEC Arts. 114.14(B), 250.8(B), and 250.70 for more information.

Chapter Eleven

NEC 250 PART VIII: DIRECT CURRENT SYSTEMS

The grounding and bonding of direct current (DC) systems are primarily handled by the Code the same manner as for alternating current (AC) systems, with only a few differences. This makes sense, because DC systems are simply AC systems that are at 0 Hz instead of the typical 50 or 60 Hz.

Many people think that DC systems cannot hurt you. This is not true. A rule of thumb is that if a person is exposed to electrical energy for 1 s, 70 mA of 60-Hz power is lethal, whereas 300 mA of DC power is lethal. There are many industrial processes that utilize low-voltage (less than 60 V) DC electrical sources with very high amperages (over 10,000 A) that can easily kill, such as electroplating, chemical electrolysis systems, and electrolytic smelting. Unfortunately, many of these types of systems are exempt from the **National Electrical Code (NEC)** due to human safety studies that were conducted just after World War II. It is important to understand that there is absolutely no basis for assuming that DC electrical energy under 60 V is safe.

The principles for providing a safe electrical environment for people working/living near AC systems is the same as for DC systems, regardless of the voltage.

Grounding the neutral conductor to earth is just as important in DC systems as it is in AC systems. Ensuring that you do not have erroneous neutral-to-ground bonds in a DC system must be conducted as vigilantly as with AC systems. The proper bonding and grounding of exposed non-current-carrying metal parts will keep people safe in both AC and DC systems equally. And, of course, circuit breakers and other overcurrent protection devices (OCPDs) will function reliably when equipment-grounding conductors are properly installed for both AC and DC systems.

In short, you should ground and bond DC systems in the same general manner as AC systems.

250.160. GENERAL

This section of the Code simply states that DC systems must comply with Part VIII of Art. 250 and any other part of Art. 250 that is intended for DC systems. See Arts. 250.6 (E) and 250.142(B), Exception No. 3 for more information.

250.162. DIRECT CURRENT CIRCUITS AND SYSTEMS TO BE GROUNDED

All DC systems must be grounded as stated in Art. 250.162(A) and (B), which states that all three-wire DC systems must be grounded, and all two-wire DC systems operating between 60- to 300-V DC must be grounded. See Art. 250.162 (A) and (B) for more information.

250.162(A). Two-wire, Direct Current Systems. All two-wire DC systems operating between 60- to 300-V DC must be grounded.

Given the recent understanding of DC arc flash, you should expect to see a change in the rules regarding DC systems over 300 V in the future. Properly grounding the exposed non-current-carrying metal components of these systems will only help to protect personnel and equipment. See more information in the last paragraph of this section.

If your DC system is less than 60-V DC, you do not have to ground it. However, there is no provision saying that you cannot ground the system, and you probably should. The exposed metal parts and metal chassis of systems using DC at less than 60 V will only benefit from having a properly sized and routed equipment-grounding conductor.

Equipment-grounding (bonding) conductor The low-impedance fault-current path that is typically run with or encloses the circuit conductors, used to connect the non-current-carrying metal parts of equipment, raceways, and other enclosures to the grounded (neutral) conductor and equipment-grounding (bonding) conductor at service equipment or at the source of a separately derived system. Often called the *green-wire* or *third-wire* ground conductor.

If your DC system is over 300 V, the Code says that you do not have to ground the system. Again, there is no provision saying that you cannot ground the system, and you probably should. The exposed metal parts and metal chassis of systems using DC at over 300 V will only benefit from having a properly sized and routed equipment-grounding conductor. This is especially true given the hazards that DC arc flash represents for personnel. Arc flash for DC systems can actually be worse than for AC systems. The 2012 edition of the National Fire Protection

Association's Standard for Electrical Safety in the Workplace (NFPA 70E®) is just beginning to discuss these issues, as there are no clear methods of accurately calculating DC arc flash. But what is clear is that DC arc flash is a serious issue, and properly grounded non-current-carrying components will be necessary.

Additionally, many DC systems are located near AC systems and are subject to induced currents, harmonics, electrical noise, stray voltages, and other objectionable currents inherent in AC systems. Ungrounded DC systems will have no effective means of removing these stray voltages other than through the neutral conductor of the electrical system, especially in isolated DC systems. Proper grounding and bonding will only improve your electrical system.

Grounding of a DC system is in general to be conducted in the same manner as grounding an AC system.

> Exception no. 1: If your DC system is installed adjacent to or is integral to the power supply, has ground detectors, and is only used to supply industrial equipment in limited areas, it is not required to be grounded.

> Exception no. 2: If you have an AC system that is in compliance with Art. 250.20, that is, feeding power to a rectifier-derived DC system, you are not required to ground the DC system.

The 2014 edition of the Code increased the DC voltage from 50 V to 60 V, and required in Exception no. 1 that the source of power be adjacent to or integral to the system.

250.162(B). Three-wire, Direct Current Systems. All three-wire DC systems shall be grounded regardless of the voltage.

Three-wire DC systems are not much different than any standard 208-V, single-phase AC system, with the exception of the frequency (0 Hz). The neutral terminal on the rectifier on a DC system will be grounded just as a neutral wire would in an AC system.

Grounding of a DC system is in general to be conducted in the same manner as grounding of an AC system.

250.164. POINT OF CONNECTION FOR DIRECT CURRENT SYSTEMS

Although the Code does not make it clear, this section is discussing both two-wire and three-wire DC systems, which must be grounded in accordance with Art. 250.162(A) and (B).

Three-wire DC systems are not much different than any standard 208-V, single-phase AC system, with the exception of the frequency (0 Hz). The center point of the winding on an AC transformer (the XO) is 0 V, so we ground it at this point. The same is true for a DC system, where the center point is 0 V, and we will therefore ground it at this point. The wire attached to the zero point is the neutral conductor (grounded conductor), just as in an AC system.

As with AC systems, you are to have one and only one neutral-to-ground connection at the first service disconnect for the system. See NEC Art. 250 Parts II and III for more information.

Two-wire DC systems are not much different than any standard 120-V, single-phase AC system, with the exception of the frequency (0 Hz). The return conductor of the two-wire DC system is the neutral, and it must be bonded to ground in the first service disconnect, just as in an AC system.

First service disconnect The very first electrical panel that will disconnect (turn off) the power coming in from the utility company. Sometimes this term is used to describe the first electrical panel that can turn off power on the secondary or low-voltage side of a transformer.

Neutral-to-ground connection (bond) is where a grounded current-carrying conductor, the neutral, is intentionally bonded to earth/ground. In general, this should only occur at the transformer and at the first service disconnect(s). The Code uses several names for neutral-to-ground connections: system bonding jumper, main bonding jumper, and supply-side bonding jumper.

And just as in an AC system, the grounding system for a DC circuits involves connecting the non-current-carrying metal components to ground. All metal chassis, enclosures, raceways, and so on are to be bonded to ground, just as in an AC system. See NEC Art. 250 Parts IV, V, and VI for more information.

Direct current systems must also have grounding electrodes installed into the earth/soil and grounding electrode conductors, just as in AC systems. These electrodes and conductors must comply with NEC Art. 250 parts II, III, and V.

250.164(A). Off-Premises Source. This section tells us that any DC supply station(s) that is located off-premises from the building/structure it is providing power for must have a ground connection made at one or more supply station(s). A ground connection shall not be made at any individual services or premises.

Supply station, DC A battery or a series of batteries, generator, power supply, solar panel(s), or other device that provides DC electrical power to a circuit.

First service disconnect The very first electrical panel that will disconnect (turn off) the power coming in from the utility company.

Sometimes this term is used to describe the first electrical panel that can turn off power on the secondary or low-voltage side of a transformer.

250.164(B). On-premises Source. This section tells us that any DC supply station(s) that is located inside or on the building/structure it is providing power for must have a neutral conductor (grounded conductor) connected at one of the following:

1. The source (at the supply station)
2. The first service disconnect or OCPD
3. Other listed system-protection equipment

Supply station, DC A battery or a series of batteries, generator, power supply, solar panel(s), or other device that provides DC electrical power to a circuit.

First service disconnect The very first electrical panel that will disconnect (turn off) the power coming in from the utility company. Sometimes this term is used to describe the first electrical panel that can turn off power on the secondary or low-voltage side of a transformer.

Overcurrent device Usually a fuse or circuit breaker, a device that limits the maximum amount of current that can flow in a circuit. Sometimes called an *overcurrent protection device* or *OCPD.*

Listed or listed item An item, product, or device that has been specifically developed for a particular function and has been independently evaluated by a product-testing organization, such as the UL or CE.

250.166. SIZE OF DIRECT CURRENT GROUNDING ELECTRODE CONDUCTOR

This section of the Code tells us that you must size the grounding electrode conductor for DC systems in accordance with Art. 250.166(A) and (B), except as allowed in Art. 250.166(C) through (E). Additionally, the grounding electrode conductor is not required to be larger than a 3/0 copper or 250 kcmil aluminum.

Grounding (earthing) electrode conductor The conductor used to connect the grounding (earthing) electrode(s) to the equipment-grounding (bonding) conductor, to the neutral (grounded) conductor, or to both in accordance with Arts. 250.66 and 250.142.

In general, Table 250.66 (see Appendix B) can be used for proper sizing of grounding electrode conductors for DC systems.

The 2014 edition of the Code placed a maximum size of 3/0 copper or 250 kcmil aluminum for the grounding electrode conductor on DC systems.

250.166(A). Not Smaller Than the Neutral Conductor. Two-wire DC generators used in conjunction with balancer sets in order to obtain a neutral point for three-wire systems must be equipped with OCPDs that electrically disconnects the three-wire system, in accordance with NEC Art. 445.12(D). When this is the case, the grounding electrode conductor must be the same size as the neutral conductor or an 8 AMG copper (or 6 AWG aluminum) conductor, whichever is larger.

Overcurrent device Usually a fuse or circuit breaker, a device that limits the maximum amount of current that can flow in a circuit. Sometimes called an *overcurrent protection device* or *OCPD*.

If this sounds like it refers to a very small set of DC systems, you are correct. This rule only applies to DC systems with three-wire balancer sets or balancer windings and with OCPDs. Please see Art. 250.166(B) for all grounding electrode conductor requirements for all other DC systems.

250.166(B). Not Smaller Than the Largest Conductor. This section deals with grounding electrode conductor requirements of all DC systems other than those in Art. 250.166(A). The grounding electrode conductor for DC systems must be equal in size to that of the largest phase conductor or an 8 AWG copper (or 6 AWG aluminum) conductor, whichever is larger.

250.166(C). Connected to Rod, Pipe, or Plate Electrodes. The grounding electrode conductor does not need to be larger than a 6 AWG copper (or 4 AWG aluminum) when it is the sole connection to a rod, pipe, or plate electrode, as in Art. 250.52(A)(5) or (A)(7). This matches the requirements of AC systems that are connected to ground rods, as found in Art. 250.66(A).

250.166(D). Connected to a Concrete-encased Electrode. The grounding electrode conductor does not need to be larger than a 4 AWG copper when it is the sole connection to steel rebar in your concrete foundation (concrete-encased electrode), as in Art. 250.52(A)(3). This matches the requirements of AC systems that are connected to steel rebar in concrete foundations (concrete-encased electrodes), as found in Art. 250.66(B).

Remember that aluminum is not allowed to come within 18 in. of the earth or to be encased in concrete or to be used as an electrode, under Arts. 250.52(B)(1) and 250.64(A).

250.166(E). Connected to a Ground Ring. The grounding electrode conductor does not need to be larger than the size of the copper ground ring conductor when it is the sole connection to the ground ring, as in Art. 250.52(A)(4). This matches the requirements of AC systems that are connected to ground rings, as found in Art. 250.66(C).

What this section tells you is that for DC systems, you must reference Table 250.66 (see Appendix B) and size the grounding electrode conductor accordingly or match the size of the ground ring. The minimum size for a ground ring is 2 AWG copper per Art. 250.52(A)(4).

Remember that aluminum is not allowed to come within 18 in. of the earth or to be encased in concrete or to be used as an electrode, under Arts. 250.52(B)(1) and 250.64(A).

250.167. DIRECT CURRENT GROUND-FAULT DETECTION

This new section of the Code requires ground-fault monitoring and detection on ungrounded DC systems and makes it optional to install them on grounded DC systems. Proper marking is also required.

Ground fault An unintentional connection between an energized conductor and earth or metallic parts of enclosures, raceways, or equipment.

This section is new to the 2014 Code.

250.167(A). Ungrounded Systems. You are required to install ground-fault detection systems on all ungrounded DC systems.

This section is new to the 2014 Code.

250.167(B). Grounded Systems. You are permitted to install ground-fault detection systems on grounded DC systems.

This section is new to the 2014 Code.

250.167(C). Marking. You must identify the type of DC system that is in use at the first service disconnect or at the supply by legibly and durably (must be able to withstand the environment) marking it as “Grounded DC System” or “Ungrounded DC System.”

This section of code applies to all DC systems.

First service disconnect The very first electrical panel that will disconnect (turn off) the power coming in from the utility company. Sometimes this term is used to describe the first electrical panel that can turn off power on the secondary or low-voltage side of a transformer.

Grounded (earthed) Connected to earth/soil.

Ungrounded system An electrical service that does not have any of the current-carrying conductors intentionally grounded (i.e., no neutral wire). Delta-type electrical systems and some standby generators are examples of ungrounded systems.

This section is new to the 2014 Code.

250.168. DIRECT CURRENT SYSTEM BONDING JUMPER

This section tells us that the grounding electrode conductor must be connected to the neutral conductor (grounded conductor) at the first service disconnect. The system bonding jumper must be the same size as the grounding electrode conductor, as in Art. 250.166 above. You must also comply with all the main bonding jumper and system bonding jumper rules found in Art. 250.28(A), (B), and (C). Please see Art. 250.28 for more information.

System bonding jumper A conductor, screw, or strap that bonds the equipment-grounding conductor to the neutral (grounded) conductor; sometimes called the *main bonding jumper.* The term *system bonding jumper* is used to distinguish the bond at the load side of the service (at the first service disconnect) from the *supply-side bonding jumper*, which is the bonding used on the utility or supply side of the service. For more details, see Arts. 250.24(A)(4), 250.24(B), 250.28, and 408.3(C).

Neutral-to-ground connection (bond) is where a grounded current-carrying conductor, the neutral, is intentionally bonded to earth/ground. In general, this should only occur at the transformer and at the first service disconnect(s). The Code uses several names for neutral-to-ground connections: system bonding jumper, main bonding jumper, and supply-side bonding jumper.

Grounding (earthing) electrode conductor The conductor used to connect the grounding (earthing) electrode(s) to the equipment-grounding (bonding) conductor, to the neutral (grounded) conductor, or to both in accordance with Arts. 250.66 and 250.142.

First service disconnect The very first electrical panel that will disconnect (turn off) the power coming in from the utility company. Sometimes this term is used to describe the first electrical panel that can turn off power on the secondary or low-voltage side of a transformer.

This provision is of course only applicable to those DC systems that must be grounded in accordance with Art. 250.162.

250.169. UNGROUNDED DIRECT CURRENT SEPARATELY DERIVED SYSTEMS

This section tells us that ungrounded DC separately derived systems, such as stand-alone DC generators and solar panels, must have a grounding electrode conductor and a grounding electrode that complies with Part III of NEC Art. 250.

In other words, any DC stand-alone separately derived system, whether it is a grounded electrical system or not, must be connected to earth via a qualified grounding electrode.

Additionally, all non-current-carrying exposed metal parts must also be connected to ground via the grounding electrode conductor, just as in a standard AC system. This includes the first service disconnect and the metal enclosure of the separately derived DC supply source.

Separately derived system(s) A premises wiring system whose power is derived from a source of electrical energy other than a (utility) service. Examples include solar panels, generators, and wind turbines that have no direct connection to another electrical source. Specifically, the neutral conductor must not be solidly bonded to the neutrals of other sources. This term is also used in regard to isolation and/or step-down transformers that generate a distinct power source separate from the utility.

Grounding (earthing) electrode A device that establishes an electrical connection to the earth/soil.

Grounding (earthing) electrode conductor The conductor used to connect the grounding (earthing) electrode(s) to the equipment-grounding (bonding) conductor, to the neutral (grounded) conductor, or to both in accordance with Arts. 250.66 and 250.142.

First service disconnect The very first electrical panel that will disconnect (turn off) the power coming in from the utility company. Sometimes this term is used to describe the first electrical panel that can turn off power on the secondary or low-voltage side of a transformer.

Supply station, DC A battery or a series of batteries, generator, power supply, solar panel(s), or other device that provides DC electrical power to a circuit.

An exception for vehicle-mounted and portable generators is made for this rule and matches the rules found in Art. 250.34. Please see Art. 250.34 for more information.

Chapter Twelve

NEC 250 PART IX: INSTRUMENTS, METERS, AND RELAYS

This short section of the **National Electrical Code (NEC)** tells us that there are some special requirements for instruments, meters, and relays under certain circumstances regarding the grounding of these items. Many of these circumstances involve instruments that monitor circuits of over 1000 V or those circuits that actually handle 1000 V or more.

Instrument transformers are a category of high-accuracy electrical devices used to isolate or transform voltage/current levels, specifically for metering and instrumentation. The most typical usage of these transformers is to monitor high-voltage/high-current circuits, and safely isolate secondary electronic systems from high-voltage/high-current hazards. They may also be used in control systems for phase-shifting circuitry and similar electronic systems.

250.170. INSTRUMENT TRANSFORMER CIRCUITS

This section tells us that the secondary circuitry of instrumentation transformers must be connected to ground for the following instances:

- All instrument transformer circuits mounted on switchboards, regardless of voltage
- All instrument transformer circuits mounted on metal-enclosed switchgear, regardless of voltage
- Any instrument transformer circuit with primary voltage operating at 300 V to ground or more

Switchboard One or more panels accommodating control switches, indicators, and other apparatus for operating electric circuits.

There are two exceptions to this rule.

Exception no. 1: If the primary windings are connected to circuits under 1000 V and any live parts or wires are only accessible to qualified personnel.

Exception no. 2: The secondaries of current transformers connected to three-phase delta circuits do not need to be grounded.

The 2014 edition of the Code added requirements in this section to include metal-enclosed switchgear.

250.172. INSTRUMENT TRANSFORMER CASES

The cases, frames, and/or chassis of instrument transformers must be grounded using an equipment-grounding conductor per Parts IV, V, and VI of Article 250 of the NEC.

Equipment-grounding (bonding) conductor The low-impedance fault-current path that is typically run with or encloses the circuit conductors, used to connect the non-current-carrying metal parts of equipment, raceways, and other enclosures to the grounded (neutral) conductor and equipment-grounding (bonding) conductor at service equipment or at the source of a separately derived system. Often called the *green-wire* or *third-wire* ground conductor.

In other words, you must ground the chassis of instrument transformers using a properly sized equipment-grounding conductor per Table 250.122 (see Appendix B).

There is one exception:

Exception no. 1: The frames of current transformers for systems less than 150 V do not need to be grounded if they are used exclusively to supply current to meters.

250.174. CASES OF INSTRUMENTS, METERS, AND RELAYS OPERATING AT LESS THAN 1000 VOLTS

You must connect any equipment-grounding conductor to relays, meters, and instruments with windings operating at 1000 V or less, as specified in 250.174(A), (B), and (C).

Equipment-grounding (bonding) conductor The low-impedance fault-current path that is typically run with or encloses the circuit conductors, used to connect the non-current-carrying metal parts of equipment, raceways, and other enclosures to the grounded (neutral) conductor and equipment-grounding (bonding) conductor at service equipment or at the source of a separately derived system. Often called the *green-wire* or *third-wire* ground conductor.

250.174(A). Not on Switchboards or Switchgear. You must ground all instruments, relays, and meters that operate at 300 V or more to ground if they are not installed on a switchboard or switchgear. The cases, chassis, and/or exposed metal parts must be grounded. There are exceptions for these items if access is restricted to only qualified personnel. See Art. 250.170 for more information.

Switchboard One or more panels accommodating control switches, indicators, and other apparatus for operating electric circuits.

The 2014 edition of the Code added requirements in this section to include switchgear.

250.174(B). On Metal-enclosed Switchgear and Dead-front Switchboards. All instruments, meters, and/or relays installed on switchboards or metal-enclosed switchgear and having no live parts exposed on the front of the panels must be grounded via a properly sized equipment-grounding conductor, regardless of the voltage, or if they are connected directly in the circuit, or if they use current or potential transformers.

Switchboard One or more panels accommodating control switches, indicators, and other apparatus for operating electric circuits.

Equipment-grounding (bonding) conductor The low-impedance fault-current path that is typically run with or encloses the circuit conductors, used to connect the non-current-carrying metal parts of equipment, raceways, and other enclosures to the grounded (neutral) conductor and equipment-grounding (bonding) conductor at service equipment or at the source of a separately derived system. Often called the *green-wire* or *third-wire* ground conductor.

The 2014 edition of the Code added requirements in this section to include metal-enclosed switchgear.

250.174(C). On Live-front Switchboards. All instruments, meters, and/or relays installed on switchboards that DO HAVE live parts exposed on the front of the panels should NOT be grounded via the equipment-grounding conductor if the voltage exceeds 150 V, if they are connected directly in the circuit, or if they use current or potential transformers. Mats of rubber insulation or other suitable insulation materials should be installed on the floor to provide protection to personnel.

When dealing with circuits of this type, the concern involves the accidental formation of erroneous neutral-to-ground bonds, which must be avoided.

Neutral-to-ground connection (bond) is where a grounded current-carrying conductor, the neutral, is intentionally bonded to earth/ground. In general, this should only occur at the transformer and at the first service disconnect(s). The Code uses several names for neutral-to-ground connections: system bonding jumper, main bonding jumper, and supply-side bonding jumper.

250.176. CASES OF INSTRUMENTS, METERS, AND RELAYS—OPERATING VOLTAGE 1000 VOLTS AND OVER

If you have an instrument, meter, and/or relay that has electrical circuitry that carries 1000 V or over, you should *not* connect the equipment-grounding conductor to their cases. You must isolate these systems by elevation, suitable barriers, grounded metal, insulated covers, insulated guards, or other suitable means to prevent contact by persons.

This does not apply to instruments, meters, and/or relays that use secondary transformers to step voltage down (such as a 100:1 ratio transformer) from higher voltages to lower voltages for monitoring. It only applies to those systems that actually handle 1000 V or more.

250.178. INSTRUMENT EQUIPMENT-GROUNDING CONDUCTOR

The equipment-grounding conductor for instruments, relays, and meters must be sized according to Table 250.122 (see Appendix B), but may not be smaller than a 12 AWG copper or 10 AWG aluminum conductor. If the cases are mounted directly to a grounded metal switchboard or to a metal-enclosed switchgear panel, they will be considered grounded, and no additional equipment-grounding conductor is needed.

Of course, you may ground these items, and you may even be required to have an individual equipment-grounding conductor for each meter, instrument, or relay based on manufacturer requirements. This is especially true for sophisticated electronic systems that may need the additional grounding for the elimination of electrical noise and harmonics. See Art. 250.136 for more information.

The 2014 edition of the Code added requirements in this section to include metal-enclosed switchgear.

Chapter Thirteen

NEC 250 PART X: GROUNDING OF SYSTEMS AND CIRCUITS OF OVER 1 KILOVOLT

This section of the National Electrical Code (NEC) covers equipment and electrical systems that handle 1000 V or higher. With only a few exceptions, 1000-V or higher electrical systems are handled the same way as lower-voltage electrical systems.

Some of the differences for systems operating at 1,000 V or higher include:

- The allowance of using the shield on shielded cable as an equipment-grounding conductor.
- When dealing with mobile/portable equipment, such as that found in mining operations, where a long cord is trailed behind the equipment, wye-type systems must be impedance-grounded neutral systems.
- The allowance of the use of multigrounded neutral systems to supply power to single-point, grounded neutral systems.
- Fences must be grounded.

In almost all cases, systems operating at 1000 V or higher require that access be limited to qualified personnel who have received special training. If you do not have the requisite training, please do not operate, maintain, or install such gear.

Consider utility power poles and distribution to homes and commercial buildings as a good example of much of what is discussed in this section.

250.180. GENERAL

This section tells us that grounded electrical systems operating at 1000 V or higher, must comply with Part X of NEC Art. 250.

Part X contains items that are unique to systems operating over 1000 V, and only modifies the requirements found in Parts I through IX.

Grounding of electrical systems higher than 1000 V is to be done exactly the same as for any other system and should comply with all of Art. 250, with a few exceptions found in Part X.

As a general rule, Tables 250.66 and Table 250.122 (see Appendix B) apply to systems operating at over 1000 V; however, you will see that there are some special rules regarding shielded, solid, dielectric-insulated cables that are unique to Part X of the Code.

250.182. DERIVED NEUTRAL SYSTEMS

This section simply tells us that a grounded transformer operating at over 1000 V may be used for deriving a system's neutral point.

In other words, the XO terminal of the transformer must be grounded, and the conductor will become the neutral wire (grounded conductor) for your electrical system, just as in any system below 1000 V.

250.184. SOLIDLY GROUNDED NEUTRAL SYSTEMS

This section tells us that solidly grounded electrical systems operating at over 1000 V may ground/earth the neutral conductor (grounded conductor) at a single point or at multiple points.

Multigrounded neutral systems are the most common form of electrical distribution found in the United States. Undoubtedly, your and your neighbor's homes are being fed by a single utility pole-mounted transformer, with each home receiving a neutral wire that is grounded to earth at the individual services; this is a multigrounded neutral system.

> **Multigrounded/multiple-grounded system** A system of conductors in which a neutral conductor is intentionally grounded solidly at specified intervals. A multigrounded or multiple-grounded system, one of the most predominant electrical distribution systems used in the world, is one in which one transformer provides neutral (grounded conductors) wires to multiple electrical services. Consider a single pole-mounted transformer feeding several homes as an example.

There is much debate over the safety of multigrounded neutral systems. They are undoubtedly the source of much unwanted electrical noise, harmonics, stray currents, and other objectionable currents in our electrical systems. Other countries actually utilize an additional conductor (called the *protective earth* conductor) to bond the chassis of transformers to the ground systems of the services for which they provide power. In any case, this section of code allows the practice of mulitgrounded neutral systems.

Article 250.184(B) and (C) discusses two different methods of solidly grounding a neutral conductor in an electrical system over 1000 V: single point and multigrounded. Remember that single-point systems are required to have an equipment-grounding conductor run with the circuit conductors, that is, not used as a continuous line-to-neutral load conductor.

250.184(A). Neutral Conductor. The following subsection covers information regarding the neutral or grounded conductor for systems operating at over 1000 V.

> **Grounded conductor (neutral)** A system or circuit conductor that is intentionally grounded. This is a current-carrying conductor typically called the *neutral wire*.

250.184(A)(1). Insulation Level. For solidly grounded electrical systems operating at over 1000 V, the neutral conductor (grounded conductor) must have an insulation rating of 600 V.

> **Solidly grounded** The intentional electrical connection of the neutral terminal (grounded conductor) to the equipment-grounding (bonding) conductor per Art. 250.30(A)(1).

There are a few exceptions to this rule for systems operating at over 1000 V:

Exception no. 1: You may use bare copper conductors as the neutral conductor for the following:

1. Service entrance conductors may be bare copper.
2. Service laterals or underground service conductors may be bare copper.
3. Direct-buried portions of feeders may be bare copper.

Exception no. 2: Bare conductors (copper, aluminum, copper-clad aluminum) may be used as neutral conductors (grounded conductors) for overhead outdoor installations.

Exception no. 3: The grounded neutral conductor may be bare, if it is protected from physical damage and isolated from the contact with the phase conductors.

The 2014 version of the Code added underground service conductors to Exception no. 1, item 2.

250.184(A)(2). Ampacity. The neutral conductor (grounded conductor) must be of sufficient size to handle the current levels (load/ampacity) that will be imposed on the conductor. At a minimum, the neutral conductor may not have an ampacity of less than one-third the ampacity of the corresponding phase conductors.

Ampacity The maximum amount of electrical current a conductor or device can carry before sustaining immediate or progressive deterioration. Ampacity is also described as current rating or current-carrying capacity. Ampacity is the RMS electric current that a device or conductor can continuously carry while remaining within its temperature rating.

In other words, even if your neutral wire is unlikely to see any current on it (all three-phase loads), you must still size the neutral conductor to at least one-third the circular mills of the phase wire.

There is one exception to this rule:

Exception no. 1: In industrial and/or commercial premises under engineering supervision, you may size the neutral wire to only 20 percent of the ampacity of the phase conductors.

250.184(B). Single-Point Grounded Neutral System. A single-point grounded neutral system is the type of electrical system we all learned about in school. Think of a utility transformer that is supplying power to a single commercial building (although it could supply power to more than one service) as an example. The neutral of that transformer would only be grounded only in a single place.

Where you have a single-point grounded neutral system operating at 1000 V or higher, all of the following must apply:

1. Single-point grounded neutral systems may be supplied from:
   a. Separately derived systems.
   b. A multigrounded neutral system in which an equipment-grounding conductor must be connected to the neutral conductor of the multigrounded system at the source of the single-point neutral system. In other words, you may get power from a multigrounded neutral system to feed your single-point neutral system if there is an equipment-grounding conductor tied to the neutral (of the multigrounded system) at the source (of the single-point system).
2. A grounding electrode must be properly installed for the system, per Part III of Art. 250.
3. There must be a properly sized grounding electrode conductor connecting the grounding electrode to the neutral, per Table 250.66 (see Appendix B).
4. There must be a properly sized bonding jumper connecting the equipment-grounding conductor to the grounding electrode conductor (see Table 250.102(C) in Appendix B).
5. There must be a properly sized equipment-grounding conductor provided for each building, structure, and/or equipment enclosure within the system, per Table 250.122 (see Appendix B).

6. There is a neutral conductor supplied for all phase-to-neutral loads.
7. The neutral conductor is insulated and isolated from earth, except in one location.
8. There is an equipment-grounding conductor run with the phase conductors, and the following three requirements are met:
   a. Does not carry a continuous load.
   b. May be bare or insulated.
   c. Must have sufficient ampacity to carry the fault-current duty.

As you can see from the above list, 1000-V or higher electrical systems must be grounded in virtually the same way that any other electrical system is to be grounded.

In summary, single-point grounded neutral systems are the type of systems we used as examples in school. A single transformer supplying power to a single load (although it could supply more than one service). The neutrals must of course be bonded together. You must also have an equipment-grounding conductor, for each building/structure service, and it must be run with the phase wires.

Bonding jumper A reliable conductor sized per Art. 250 to ensure electrical conductivity between metal parts of the electrical installation.

Electrode: A conductor used to establish electrical contact with the earth.

Equipment-grounding (bonding) conductor The low-impedance fault-current path that is typically run with or encloses the circuit conductors, used to connect the non-current-carrying metal parts of equipment, raceways, and other enclosures to the grounded (neutral) conductor and equipment-grounding (bonding) conductor at service equipment or at the source of a separately derived system. Often called the *green-wire* or *third-wire* ground conductor.

Grounding (earthing) electrode conductor The conductor used to connect the grounding (earthing) electrode(s) to the equipment-grounding (bonding) conductor, to the neutral (grounded) conductor, or to both in accordance with Arts. 250.66 and 250.142.

Multigrounded/multiple-grounded system A system of conductors in which a neutral conductor is intentionally grounded solidly at specified intervals. A multigrounded or multiple-grounded system, one of the most predominant electrical distribution systems used in the world, is one in which one transformer provides neutral (grounded conductors) wires to multiple electrical services. Consider a single pole-mounted transformer feeding several homes as an example.

250.184(C). Multigrounded Neutral Systems. The NEC allows certain 1000-V and higher transformers to have their common neutral wire connected to ground at multiple places. Consider a series of pole-mounted transformers running down a residential street supplying power to homes. One transformer will provide power to several residences, meaning the neutral of the transformer will be grounded to earth multiple times, hence the term *multigrounded neutral.*

This section of the Code allows utility companies to bond the neutral conductor on pole-mounted transformers to ground at multiple locations for solidly grounded neutral systems. Where you have a multigrounded neutral system operating at 1000 V or higher, all of the following must apply:

1. The neutral conductor of a solidly grounded neutral system may be grounded at more than one point, if grounded at the following locations:
   a. The transformers supplying power to the building/structure.
   b. Underground circuits where a bare neutral wire is used (exposed neutral conductor).
   c. Overhead circuits installed outdoors, such as those used by the utility to provide power from pole-mounted transformers to homes.
2. The multigrounded neutral system is grounded at each transformer and/or location via a grounding electrode. See Part III of Art. 250 for proper electrode installation.
3. The neutral conductor of the multigrounded neutral system must be connected to a grounding electrode every 1300 ft (400 m).
4. Which means, of course, that the maximum distance between any two electrodes may not exceed 1300 ft (400 m).
5. For systems using multigrounded shielded cables, the shielding (grounded conductor) must be grounded at each cable joint that is exposed to contact by personnel.

Solidly grounded The intentional electrical connection of the neutral terminal (grounded conductor) to the equipment-grounding (bonding) conductor per Art. 250.30(A)(1).

Electrode A conductor used to establish electrical contact with the earth.

Multigrounded/multiple-grounded system A system of conductors in which a neutral conductor is intentionally grounded solidly at specified intervals. A multigrounded or multiple-grounded system, one of the most predominant electrical distribution systems used in the world, is one in which one transformer provides neutral (grounded conductors) wires to multiple electrical services. Consider a single pole-mounted transformer feeding several homes as an example.

In summary, multigrounded neutral systems are the type of systems used by utility companies to distribute power to multiple premises. That is a single transformer supplying power to more than one service, with each service connecting the neutral to ground. These systems are not required to have an equipment-grounding conductor.

250.186. GROUND-FAULT CIRCUIT CONDUCTOR BROUGHT TO SERVICE EQUIPMENT

This section of the Code discusses how to properly connect and install the neutral (grounded) conductor for systems operating at 1000 V and higher. In general, 1000 V and higher systems are treated the same way as systems below 1000 V, with only a few exceptions.

Author's Note *For some reason, the 2014 Code decided to call the neutral conductor a ground-fault circuit conductor. This is unfortunate, as this new name only adds confusion. Currently, the Code uses at least four names to describe the same thing: neutral conductor, grounded conductor, common conductor, and now ground-fault circuit conductor. Do not get confused: this section of the Code is simply talking about the neutral (grounded) conductor.*

250.186(A). Systems with a Grounded Conductor at the Service. If you have a grounded alternating current (AC) system operating at 1000 V or higher, the neutral (grounded) conductor must be routed with the phase conductors to each service. At each service, you must connect the neutral (grounded) conductor to the neutral terminal bus in the first service disconnect enclosure. You must also have a neutral-to-ground bond (i.e., supply-side bonding jumper or main bonding jumper) connecting the neutral (grounded) conductor to the enclosure.

You must install the neutral (grounded) conductor in accordance with Art. 250.186(A)(1) through (A)(3). The neutral (grounded) conductor must be sized in accordance with Art. 250.184 or 250.186(A)(1) or (A)(2), whichever results in the larger conductor.

Grounded conductor (neutral) A system or circuit conductor that is intentionally grounded. This is a current-carrying conductor typically called the *neutral wire*.

First service disconnect The very first electrical panel that will disconnect (turn off) the power coming in from the utility company. Sometimes this term is used to describe the first electrical panel that can turn off power on the secondary or low-voltage side of a transformer.

Neutral-to-ground connection (bond) is where a grounded current-carrying conductor, the neutral, is intentionally bonded to earth/ground. In general, this should only occur at the transformer and at the first service disconnect(s). The Code uses several names for neutral-to-ground connections: system bonding jumper, main bonding jumper, and supply-side bonding jumper.

Exception: You are allowed to connect the neutral (grounded) conductor to the common neutral terminal or bus in a service enclosure that has two or more disconnecting means.

This section is new to the 2014 Code.

250.186(A)(1). Sizing for a Single Raceway or Overhead Conductor. The neutral (grounded) conductor must be sized in accordance with Table 250.66 (see Appendix B), but does not need to be larger than the corresponding phase conductors.

If your phase conductors are larger than 1100 kcmil copper or 1750 kcmil aluminum, the neutral (grounded conductor) or bonding jumper shall be at least 12½ percent of the circular mil area of the largest phase conductor (ungrounded conductor). The neutral conductor is not required to be larger than the phase conductors.

This section is new to the 2014 Code.

250.186(A)(2). Parallel Conductors in Two or More Raceways of Overhead Conductors. You must route the neutral (grounded) conductor in parallel with the corresponding parallel routed phase conductors, whether enclosed in a raceway or installed as overhead conductors.

The size of the neutral (grounded) conductor in each branch of the parallel run must be sized according to Art. 250.186(A)(1) but not smaller than 1/0 AWG. See Table 250.66 (Appendix B).

Please see Art. 310.10(H) for more information on grounded conductors installed in parallel.

This section is new to the 2014 Code.

250.186(A)(3). Delta-connected Service. When dealing with three-phase, three-wire, delta-type systems supplying power to a service, the grounded conductor must be the same size as the phase conductors.

This section is new to the 2014 Code.

250.186(B). Systems without a Grounded Conductor at the Service Point. If you have a grounded AC system operating at 1000 V or higher that does not provide a neutral (grounded) conductor to the service point, a supply-side bonding jumper (equipment-grounding conductor) must be installed and routed with the phase conductors to each service.

Author's Note *The Code uses the term* supply-side bonding jumper, *which in actuality is an equipment-grounding conductor. Please do not get confused here: the Code is simply saying that you must have* *an equipment-grounding conductor bonding the transformer to the first service disconnect when you do not have a neutral.*

At each service, you must connect the supply-side bonding jumper (equipment-grounding conductor) to the ground terminal bus in the first service disconnect enclosure.

Exception: You are allowed to connect the supply-side bonding jumper (equipment-grounding conductor) to the common ground terminal or bus in a service enclosure that has two or more disconnecting means.

Supply-side bonding jumper A conductor installed on the supply (utility) side of a service, or within a service equipment enclosure(s), or for a separately derived system, that ensures the required electrical conductivity between metal parts required to be electrically connected; sometimes called a *main bonding jumper*. A conductor, screw, or strap that bonds the neutral (grounded) conductor to the service equipment enclosures on the supply side. The term *supply-side bonding jumper* is used to distinguish the bond at the supply (utility) side of the service from the *system bonding jumper*, which is used on the load side of the service (at the first service disconnect). For more details, see Arts. 250.24(A)(4), 250.24(B), 250.28, 250.28(D)(2), and 408.3(C).

Note *The NEC occasionally uses this term to indicate an equipment-grounding conductor (see Art. 250.30).*

Equipment-grounding (bonding) conductor The low-impedance fault-current path that is typically run with or encloses the circuit conductors, used to connect the non-current-carrying metal parts of equipment, raceways, and other enclosures to the grounded (neutral) conductor and equipment-grounding (bonding) conductor at service equipment or at the source of a separately derived system. Often called the *green-wire* or *third-wire* ground conductor.

This section is new to the 2014 Code.

250.186(B)(1). Sizing for a Single Raceway or Overhead Conductor. The supply-side bonding jumper (equipment-grounding conductor) must be sized in accordance with Table 250.66 (see Appendix B), but does not need to be larger than the corresponding phase conductors.

If your phase conductors are larger than 1100 kcmil copper or 1750 kcmil aluminum, the supply-side bonding jumper (equipment-grounding conductor) must be at least 12½ percent of the circular mil area of the largest phase conductor (ungrounded conductor). The neutral conductor is not required to be larger than the phase conductors.

This section is new to the 2014 Code.

250.186(B)(2). Parallel Conductors in Two or More Raceways or Overhead Conductors. You must route the supply-side bonding jumper (equipment-grounding conductor) in parallel with the corresponding parallel routed phase conductors, whether enclosed in a raceway or installed as overhead conductors.

The size of the supply-side bonding jumper (equipment-grounding conductor) in each branch of the parallel run must be sized according to Art. 250.186(A)(1), but not smaller than 1/0 AWG. See Table 250.66 (Appendix B).

Please see Art. 310.10(H) for more information on grounded conductors installed in parallel.

This section is new to the 2014 Code.

250.187. IMPEDANCE-GROUNDED NEUTRAL SYSTEMS

This section of the Code deals with impedance-grounded neutral systems of 1000 V or more, where a grounding resistance, usually a resistor, limits the ground-fault current. These systems are permitted when all of the following are met:

1. Only qualified maintenance and supervision personnel can service the associated equipment.
2. The system has ground-fault detectors installed.
3. There are no line-to-neutral loads served on the system.

Fault current The electrical energy released by a given electrical system during an unintentional line-to-ground or line-to-line fault.

Ground fault An unintentional connection between an energized conductor and earth or metallic parts of enclosures, raceways, or equipment.

Impedance-grounded neutral system An electrical system wherein a grounding impedance, usually a resistor or an inductor, is added between the neutral point (XO terminal) of the transformer and the grounding electrode to limit the amperage of fault currents.

Additionally, impedance-grounded neutral systems operating at 1000 V or more must comply with Art. 250.187(A) through (D). Please see Art. 250.36 for more information.

250.187(A). Location. The grounding impedance must be inserted between the grounding electrode conductor and the neutral point of the transformer and/or generator. Please see Art. 250.36(A) for more information.

Grounding (earthing) electrode conductor The conductor used to connect the grounding (earthing) electrode(s) to the equipment-grounding (bonding) conductor, to the neutral (grounded) conductor, or to both in accordance with Arts. 250.66 and 250.142.

250.187(B). Identified and Insulated. The neutral conductor of impedance-grounded neutral systems operating at 1000 V or more must be properly identified and insulated with the same insulation as the phase conductors.

250.187(C). System Neutral Conductor Connection. The neutral conductor shall not be connected to earth/ground, except through the impedance, usually a resistor. Please see Art. 250.36(C) for more information.

250.187(D). Equipment-grounding Conductors. Equipment-grounding conductors in impedance-grounded neutral systems operating at 1000 V or more may be bare. They must also be electrically connected to the ground bus and to the grounding electrode conductor.

Grounding (earthing) electrode conductor The conductor used to connect the grounding (earthing) electrode(s) to the equipment-grounding (bonding) conductor, to the neutral (grounded) conductor, or to both in accordance with Arts. 250.66 and 250.142.

250.188. GROUNDING OF SYSTEMS SUPPLYING PORTABLE OR MOBILE EQUIPMENT

Mobile or portable systems operating at 1000 V or more, other than substations, and installed on a temporary basis, must comply with Art. 250.188(A) through (F).

Remember that in the National Electrical Code (NEC), *portable* describes equipment that can be carried by a person, and *mobile* describes equipment that can be moved on wheels, skids, treads, and so on.

This section discusses electrical systems that are supplying power to mobile and/or portable equipment operating at over 1000 V and does not refer to generators. Consider mining equipment on rail tracks and being supplied by power from trailing cables as an example of the equipment being discussed in Art. 250.188.

250.188(A). Portable or Mobile Equipment. You must utilize impedance-grounded neutral electrical systems for all portable or mobile equipment operating at 1000 V or higher. If the system is operating with delta-type power at 1000 V or higher, a neutral point and neutral conductor must be derived. See Arts. 250.20(B), 250.36, and 250.186 for more information.

Impedance-grounded neutral system An electrical system wherein a grounding impedance, usually a resistor or an inductor, is added between the neutral point (XO terminal) of the transformer and the grounding electrode to limit the amperage of fault currents.

250.188(B). Exposed Non-current-carrying Metal Parts. Mobile equipment operating at 1000 V or more that has exposed, normally non-current-carrying metal parts must have an equipment-grounding conductor tied to where the system neutral impedance is grounded. This section only applies to wye-type systems and not to delta-type systems.

Equipment-grounding (bonding) conductor The low-impedance fault-current path that is typically run with or encloses the circuit conductors, used to connect the non-current-carrying metal parts of equipment, raceways, and other enclosures to the grounded (neutral) conductor and equipment-grounding (bonding) conductor at service equipment or at the source of a separately derived system. Often called the *green-wire* or *third-wire* ground conductor.

In other words, the chassis, frames, enclosures, raceways, and other metal components associated with the mobile equipment must be properly bonded and grounded to the earth at the point where the resistor or inductor of the impedance-grounded neutral system is grounded/earthed. See Art. 250.36 for more information on impedance-grounded neutral systems.

250.188(C). Ground-fault Current. The voltage between the portable/mobile equipment frame and the earth/ground may not exceed 100 V. Technically speaking, this is the voltage that develops between the mobile/portable equipment frame and the ground by the flow of the maximum ground-fault current.

Fault current The electrical energy released by a given electrical system during an unintentional line-to-ground or line-to-line fault.

Ground fault An unintentional connection between an energized conductor and earth or metallic parts of enclosures, raceways, or equipment.

250.188(D). Ground-fault Detection and Relaying. You must install monitoring equipment that can detect a ground fault and automatically disconnect power to mobile or portable systems operating at 1000 V or more. These systems must also continuously monitor the continuity of the equipment-grounding conductor and automatically de-energize the system should there be a loss of continuity.

Ground fault An unintentional connection between an energized conductor and earth or metallic parts of enclosures, raceways, or equipment.

Equipment-grounding (bonding) conductor The low-impedance fault-current path that is typically run with or encloses the circuit conductors, used to connect the non-current-carrying metal parts of equipment, raceways, and other enclosures to the grounded (neutral) conductor and equipment-grounding (bonding) conductor at service equipment or at the source of a separately derived system. Often called the *green-wire* or *third-wire* ground conductor.

250.188(E). Isolation. The grounding electrode used for the resistor in the impedance-grounded neutral system must be installed at least 20 ft (6 m) from any other grounding electrode or other buried metal object, including metal fence posts. This is because of the ground potential rise that can occur during a ground fault, which can energize metal objects through the voltage rise in the earth/soil. See Arts. 250.36 and 250.186 for more information.

Impedance-grounded neutral system An electrical system wherein a grounding impedance, usually a resistor or an inductor, is added between the neutral point (XO terminal) of the transformer and the grounding electrode to limit the amperage of fault currents.

Electrode A conductor used to establish electrical contact with the earth.

250.188(F). Trailing Cable and Couplers. All trailing cables and couplers for mobile or portable systems operating at 1000 V or more must meet the requirements of NEC Arts. 400 Part III and 490.55. Please also see Arts. 400.30 through 400.36 and 490.55 for more information.

250.190. GROUNDING OF EQUIPMENT

This section tells us how to ground equipment, shielded cables, and other apparatus operating at 1000 V or higher.

250.190(A). Equipment Grounding. You must ground all non-current-carrying metal parts of fixed, portable, or mobile equipment operating at 1000 V or higher. This includes chassis, enclosures, raceways, housings, and all associated supporting structures and fences. Please see Art. 250.194 for more information regarding the grounding of fences and other metal structures.

There is one exception to this rule:

Exception no. 1: Where equipment operating at 1000 V or higher is isolated from ground and located such that a person in contact with the ground/earth cannot contact the metal of the equipment when energized, the equipment does not need to be grounded. This exception is primarily for pole-mounted transformers and other pole-mounted apparatus. See Art. 250.110, Exception no. 2 for more information.

250.190(B). Grounding Electrode Conductor. The grounding electrode conductor for systems operating at 1000 V or higher must be sized according to Table 250.66; however, it may not be smaller than 6 AWG copper or 4 AWG aluminum wire. Please see Art. 250.66 and Table 250.66 (Appendix B) for more information.

Grounding (earthing) electrode conductor The conductor used to connect the grounding (earthing) electrode(s) to the equipment-grounding (bonding) conductor, to the neutral (grounded) conductor, or to both in accordance with Arts. 250.66 and 250.142.

Please remember the aluminum rule that states that you may not bring aluminum within 18 in. of the earth/soil. Please see Arts. 250.52(B)(1) and 250.64 (A) for more information.

250.190(C). Equipment-grounding Conductor. Equipment-grounding conductors in systems operating at 1000 V or higher must comply with Art. 250.190(C)(1) through (C)(3).

Equipment-grounding (bonding) conductor The low-impedance fault-current path that is typically run with or encloses the circuit conductors, used to connect the non-current-carrying metal parts of equipment, raceways, and other enclosures to the grounded (neutral) conductor and equipment-grounding (bonding) conductor at service equipment or at the source of a separately derived system. Often called the *green-wire* or *third-wire* ground conductor.

250.190(C)(1). General. Equipment-grounding conductors for systems operating at 1000 V or higher that are NOT part of an integral cable assembly must be sized according to Table 250.122, but must not be smaller than 6 AWG copper or 4 AWG aluminum wire. See Art. 250.122 and Table 250.122 (Appendix B) for more information.

For equipment-grounding conductors that ARE part of an integral shield cable, see Art. 250.190(C)(2). For equipment-grounding conductors that ARE part of an integral cable assembly, see Art. 250.190(C)(3).

250.190(C)(2). Shielded Cables. For electrical systems of 1000 V or higher, you may use the shield wire encircling shielded cables as the equipment-grounding conductor if it is listed and rated to be able to handle the ground-fault currents for the expected clearing time of the circuit's overcurrent protection device (OCPD; circuit breaker).

Equipment-grounding (bonding) conductor The low-impedance fault-current path that is typically run with or encloses the circuit conductors, used to connect the non-current-carrying metal parts of equipment, raceways, and other enclosures to the grounded (neutral) conductor and equipment-grounding (bonding) conductor at service equipment or at the source of a separately derived system. Often called the *green-wire* or *third-wire* ground conductor.

There are two types of shielded cables, those that are rated/listed to carry ground-fault currents and those that have shields composed of copper tape and drain wires. Article 310.10(E) details the use of shielded cables. Only those shielded cables listed/rated for use as a ground fault–current path for the required clearing time may be used.

Listed or listed item An item, product, or device that has been specifically developed for a particular function and has been independently evaluated by a product-testing organization, such as the UL or CE.

Overcurrent device Usually a fuse or circuit breaker, a device that limits the maximum amount of current that can flow in a circuit. Sometimes called an *overcurrent protection device* or *OCPD*.

Clearing time The time it takes for an OCPD to detect fault currents and de-energize a circuit. Typically expressed in cycles: one cycle = 0.0167 s for 60 Hz and 0.02 s for 50 Hz.

250.190(C)(3). Sizing. All equipment-grounding conductors for system operating at 1000 V or higher, must be sized according to Table 250.122 (see Appendix B), unless part of a listed shield cable. See Art. 250.190(C)(2) for more information.

Remember, if the equipment-grounding conductor is NOT part of an integral cable assembly, it must be sized according to Table 250.122 (see Appendix B), but must not be smaller than 6 AWG copper or 4 AWG aluminum wire. Please see Art. 250.190(C)(1) for more information.

Equipment-grounding conductors that ARE part of an integral cable need only be sized according to Table 250.122 (see Appendix B).

Please see Art. 250.122 and Table 250.122 (see Appendix B) for more information.

250.191. GROUNDING SYSTEMS AT ALTERNATING CURRENT SUBSTATIONS

All AC substations must be grounded in accordance with Part III of Art. 250 and in accordance with ANSI/IEEE Std. 80-2000, *IEEE Guide for Safety in AC Substation Grounding*.

250.194. GROUNDING AND BONDING OF FENCES AND OTHER METAL STRUCTURES

The metal fences that surround (enclose) any substation and come within 16 ft (5 m) of exposed electrical conductors and/or equipment must be bonded and grounded. This is required to limit the step-and-touch voltages, along with transfer voltages, that could form on the fence. Please see the earthing section in Chap. 2 of this book for more information on step-and-touch voltage hazards.

This section also applies to other metal structures, meaning that any metal object located within 16 ft (5 m) of the exposed electrical conductors of equipment must be bonded together to form a common grounding system.

This section is new to the 2014 Code.

250.194(A). Metal Fences. Metal fences that are located within 16 ft (5 m) of the exposed electrical conductors and/or equipment must be bonded to the common grounding electrode system using a wire-type conductor, as follows:

1. Wire-type copper conductors (bonding jumpers) must be installed at each fence corner (king post) and at support posts at a minimum of 160-ft (50-m) intervals along the fence traverse. Note IEEE and other industry standards require the bonding of fence support posts at far tighter intervals, some as tight as every other support post.
2. You must install wire-type copper conductors (bonding jumpers) to bond fence posts to the common grounding electrode system at each point where the overhead conductors cross the fence.
3. You must install wire-type copper conductors (bonding jumpers) to bond all gate support posts (king posts) to the common grounding electrode system.
4. You must install wire-type copper conductors (bonding jumpers) to bond support posts together at all fence openings. This would be a post-to-post bonding conductor.
5. You must extend the grounding grid or grounding electrode system to cover the swing-arc radius of all gates.

Note *In order to meet the requirements for personnel safety found in IEEE Std.-80, you may need to extend the ground grid 3 ft past the swing-arc radius of the gates in order to protect personnel from touch voltages.*

6. You must install wire-type copper conductors (bonding jumpers) to bond the barbed wire strands above the fence to the common grounding electrode system.

You may use alternative fence grounding and bonding designs that have been developed under engineering supervision.

Note *An engineering firm specializing in grounding and step-and-touch voltage hazard analysis should be consulted.*

Grounding (earthing) electrode system Two or more grounding electrodes that are bonded together to form a single system. For a typical building/structure, a grounding electrode system should consist of: two ground rods, water pipe, steel frame of the building, the steel rebar in the concrete foundation, and any ground rings.

Bonding (bond, bonded) The permanent joining of metallic parts together to form an electrically conductive path. This path must have the capacity to safely conduct any fault current likely to be imposed on it.

Bonding jumper A reliable conductor sized per Art. 250 to ensure electrical conductivity between metal parts of the electrical installation.

Author's Note *The Code uses the term bonding jumper to indicate a grounding electrode conductor that bonds two or more electrodes together. This is simply a properly sized, buried copper conductor that interconnects metal objects to form a common grounding electrode system.*

Informational Note 1 *Nonconductive fences or sections of fences may reduce transfer voltages by providing electrical isolation to other areas.*

Informational Note 2 *Fences should be designed grounded in accordance with ANSI/IEEE Std. 80-2000,* IEEE Guide for Safety in AC Substation Grounding.

This section is new to the 2014 Code.

250.194(B). Metal Structures. All exposed conductive metal structures that are located within 16 ft (5 m) horizontally or 8 ft (2.5 m) vertically of the exposed electrical conductors and/or equipment and that are subject to contact by persons must be bonded to the common grounding electrode system using a wire-type copper conductor.

In other words, any metal object that is within a 16-ft-wide by 8-ft-high zone must be bonded to the common grounding electrode system.

This section is new to the 2014 Code.

Chapter Fourteen

NEC ARTICLE 680: SWIMMING POOLS, FOUNTAINS, AND SIMILAR INSTALLATIONS

Swimming pools first appeared in the National Electrical Code (NEC) in the 1962 edition and had a grounding section (680-7) that was just three pages long. The 1965 NEC expanded the requirements and made several changes, including increasing the bonding conductor size from a 14 AWG to an 8 AWG. The 1968 NEC introduced the 5-ft horizontal bonding rule we still have today. This rule tells us that every metal object within 5 ft of the water must be bonded to form one continuous metal object.

This section of the Code covers virtually every type of pool imaginable, including the build-it-yourself above-grade pools that homeowners seem to always fail to get their pool permits for, to the grief of many an electrical inspector. Article 680 is applicable to all permanently installed or storable pools and tubs, including: swimming pools, wading pools, therapeutic pools, decorative pools, fountains, hot tubs, spas, and hydromassage bathtubs, and to all of the metallic auxiliary support equipment, such as pumps, filters, and so on. Additionally, Art. 680 covers the actual body of water contained within pool/tub, as the water itself is considered to be a conductive object capable of conducting hazardous electrical energy to persons.

The NEC has some very specific grounding requirements in regard to swimming pools. The primary concept that must be understood is that you must have two grounding systems: the standard grounding found in Art. 250 plus an additional equipotential bonding system unique to pools.

The typical equipotential bonding grid is installed immediately next to the steel rebar in the foundation and is completely encased in concrete with the rebar. The equipotential bonding system for pools is primarily a bare, solid 8 AWG copper wire routed throughout the pool area in a 1-ft (300-mm) on-center grid bonding the steel rebar in the concrete together. Many pool equipment providers offer a 3-ft wide copper mesh roll for use around the immediate pool perimeter to ease the construction of the mandatory equipotential bonding system.

This grid must extend out to at least 3 ft (1 m) past the water's edge, and farther in some cases. In general, the entire "pool area" where persons are typically wet, engaged in water-related activities, and/or utilizing the area involved in swimming pool operations should have all metallic objects bonded to the equipotential grid regardless of the distance the metallic object is from the body of water. While not necessarily required by the Code, single-pour concrete pads and/or pads with integrated structural steel rebar should be bonded together with a common equipotential bonding system. See Art. 680.26(B)(1)(b) for more information.

There is often some confusion regarding the Code's requirements in regard to the listed distances for grounding and bonding. The rule is that the 1-ft (300-mm) on-center 8 AWG copper equipotential bonding grid must be installed throughout the pool vessel and extends at least 3-ft (1-m) horizontally past the pool, regardless of the surface covering (although there is no rule that prevents you from going farther). All other metal objects within 5 ft (1.5 m) of the pool must have an additional grounding conductor bonding them to the equipotential grounding grid. For most electrical items, this means that they will have two grounding conductors, the third-wire or green equipment-grounding conductor, and an additional 8 AWG ground wire bonding the normally non-current-carrying metal parts to the equipotential grid.

One of the primary concerns for swimming pools is the transition point between concrete foundations that have an equipotential bonding system and the concrete slabs that do not. The difference in potential will be greatest at this transition point, generating the highest step-voltage hazard located anywhere in the pool area. Even though it is not required by the Code, if you are designing/constructing a swimming pool, you are simply better off extending the equipotential bonding grid to cover all of the concrete slabs involved in the pool/wet area. Please see the earthing section in Chap. 2 for more information regarding step-and-touch voltage hazards.

The goal is to create an electrically safe environment for persons using the pool areas. People who are immersed in a pool of water, dripping wet, or walking/lying on a wet deck, with large amounts of exposed skin, are highly susceptible and vulnerable to voltage gradients.

As discussed in previous sections of this book, steel is simply not conductive enough and has far too much magnetism to effectively remove impressed voltages/currents. The steel rebar in the concrete, metal handrails, metal parts of slides, pipes that supply water, the electric pumps that push the water, grates/drains, and so on must all be bonded to ground via an additional copper conductor (wire) network, to ensure the safety of persons. Copper is between 12 to 17 times more conductive than steel and is at least 250 times less magnetic. In fact, copper is diamagnetic, which means that it actually helps to collapse magnetic fields.

The general bonding rule for swimming pools is that any metal object, including those encased in concrete, must be bonded to the additional equipotential bonding system. This is even true for electrical items that have a green-wire (equipment-grounding conductor) ground bonded to them. For example, an underwater lighting or luminaire system that already has a third-wire green ground run with the hot and neutral wire must also have a connection to the equipotential bonding system. This means that the light case would have two ground connections: one to the green ground wire and one to the bare copper equipotential ground system.

Luminaire Complete lighting unit, consisting of one or more lamps (bulbs or tubes that emit light), along with the socket and other parts that hold the lamp in place and protect it, wiring that connects the lamp to a power source, and a reflector that helps direct and distribute the light.

The Code also requires that all electrical systems associated with your swimming pool must be bonded to the equipotential grounding system. There are ground-fault circuit interrupter (GFCI) requirements for swimming pools that must also be observed. With only a few exceptions, you must have GFCI systems for nearly all electrical circuits associated with swimming pools.

For more information regarding GFCI requirements for swimming pools, please see Arts. 680.5 General, 680.21(C) Motors, 680.22(A)(4) Receptacles, 680.22(B)(4) Luminaires, 680.23(A)(3) Underwater Luminaires, 680.32 Electrical Equipment, 680.44 Spas and Hot Tubs, 680.57(B) Signs, 680.58 Receptacles, 680.62(E) Therapeutic Tubs, 680.71 Hydromassage Bathtubs, 680.72 Other Electrical Equipment, and 682.15 Ground-Fault Circuit-Interrupter (GFCI) Protection.

While the Code may appear to be written with below-grade pools in mind, the same requirements apply to above-grade pools. For years, above-grade pools have gotten away with not providing the necessary bonding and grounding. Fortunately, city inspectors are catching up and are now mandating the installation of additional ground rods and grounding conductors designed to ensure that the metallic supports and other metallic objects on the pool are bonded and grounded in accordance with the NEC.

In the following pages, you will find a discussion about selected sections of Art. 680 that are most relevant to understanding the grounding requirements, not the entire 680 code.

680.6. GROUNDING

This section tells us that all electrical equipment shall be grounded in accordance with Art. 250 Parts V, VI, and VII, except as modified by Art. 680. Specifically, the following equipment must be grounded:

1. Through-wall lighting assemblies and underwater luminaires. However, low-voltage lighting products that are specifically listed for use without a grounding conductor are exempt.
2. All electrical equipment located within 5 ft (1.5 m) of the body of water.
3. All electrical equipment associated with the water recirculation system.
4. All junction boxes.
5. All transformers and power supply enclosures.
6. All GFCIs must be grounded
7. All panel boards and any service equipment on the same premises as the swimming pool must be properly grounded, whether they are part of the supply circuit(s) for the swimming pool or not.

Listed or listed item An item, product, or device that has been specifically developed for a particular function and has been independently evaluated by a product-testing organization, such as the Underwriters Laboratory (UL) or Conformité Européenne (CE).

Luminaire Complete lighting unit, consisting of one or more lamps (bulbs or tubes that emit light), along with the socket and other parts that hold the lamp in place and protect it, wiring that connects the lamp to a power source, and a reflector that helps direct and distribute the light.

680.21. MOTORS

680.21(A). Wiring Methods

680.21(A)(1). General. This section of the Code simply states that all pool-associated electrical motors must have a properly sized, copper insulated, equipment-grounding conductor per Table 250.122 (Appendix B), but not smaller than a 12 AWG conductor (whichever is larger).

Equipment-grounding (bonding) conductor The low-impedance fault-current path that is typically run with or encloses the circuit conductors, used to connect the non-current-carrying metal parts of equipment, raceways, and other enclosures to the grounded (neutral) conductor and equipment-grounding (bonding) conductor at service equipment or at the source of a separately derived system. Often called the *green-wire* or *third-wire* ground conductor.

Aluminum may not come within 18 in. of the earth/soil, is not allowed for use where it may come in contact with water, and is specifically not allowed in this provision. Please see Arts. 250.52(B)(1) and 250.64(A) for more information.

Also remember that pool pumps connected to single-phase, 120-V through 240-V circuits must be protected by a GFCI system, whether hardwired or connected through a receptacle. See Art. 680.21(C) for more information.

680.23. UNDERWATER LUMINAIRES

680.23(B). Wet-niche Luminaires

680.23(B)(3). Equipment-grounding Provisions for Cords. Other than for low-voltage lighting systems that are listed for swimming pool use without a grounding conductor, any wet-niche luminaire supplied by flexible cord or cable must have equipment-grounding conductors integral within the cable or cord assembly. This equipment-grounding conductor must be insulated copper and must be equal in size to the phase conductors, but not smaller than 16 AWG.

Equipment-grounding (bonding) conductor The low-impedance fault-current path that is typically run with or encloses the circuit conductors, used to connect the non-current-carrying metal parts of equipment, raceways, and other enclosures to the grounded (neutral) conductor and equipment-grounding (bonding) conductor at service equipment or at the source of a separately derived system. Often called the *green-wire* or *third-wire* ground conductor.

Aluminum may not come within 18 in. of the earth/soil, is not allowed for use where it may come in contact with water, and is specifically not allowed in this provision. Please see Arts. 250.52(B)(1) and 250.64(A) for more information.

680.23(B)(4). Luminaire Grounding Terminations. The grounding connections within a luminaire must be covered or encapsulated in a suitable potting compound so as to prevent deterioration (corrosion) and the entry of water into the luminaire.

680.23(B)(5). Luminaire Bonding. The forming shell of luminaires must utilize a locking device that bonds to the luminaire, forming a low-resistance contact between the two pieces. The locking device must require the use of a tool to remove the forming shell from the luminaire.

Listed or listed item An item, product, or device that has been specifically developed for a particular function and has been independently evaluated by a product-testing organization, such as the UL or CE.

Luminaires that are listed and do not have any non-current-carrying metal parts are exempt from this rule.

680.23(F). Branch Circuit Wiring. This section tells us that only enclosures and raceways that are listed for swimming pools may be used.

Listed or listed item An item, product, or device that has been specifically developed for a particular function and has been independently evaluated by a product-testing organization, such as the UL or CE.

680.23(F)(2). Equipment Grounding. Other than for low-voltage lighting systems that are listed for swimming pool use without a grounding conductor, all through-wall lighting systems, including wet-niche, dry-niche, or no-niche luminaires must be bonded to an insulated copper equipment-grounding conductor, in accordance with Table 250.122 (see Appendix B) but not smaller than 12 AWG.

1. If you have one or more underwater luminaires supplied by the same branch circuit, you may terminate the equipment-grounding conductor that is routed between junction boxes, transformer enclosures, field-wiring compartments, or other such enclosures onto listed grounding terminals.
2. If your underwater luminaire is supplied by a transformer, GFCI, clock-operated switch, or manual snap-switch, you may terminate the equipment-grounding conductor that is routed between those items onto listed grounding terminals.

In other words, you may use grounding terminals in listed enclosures to terminate equipment-grounding conductors from the electrical disconnect and/or when routing an equipment-grounding conductor from listed enclosures.

Please see Art. 680.23(B)(3) for provisions regarding wet-niche luminaires with power supplied by cords.

Luminaire Complete lighting unit, consisting of one or more lamps (bulbs or tubes that emit light), along with the socket and other parts that hold the lamp in place and protect it, wiring that connects the lamp to a power source, and a reflector that helps direct and distribute the light.

Listed or listed item An item, product, or device that has been specifically developed for a particular function and has been independently evaluated by a product-testing organization, such as the UL or CE.

Aluminum may not come within 18 in. of the earth/soil, is not allowed for use where it may come in contact with water, and is specifically not allowed in this provision. Please see Arts. 250.52(B)(1) and 250.64(A) for more information.

680.25. FEEDERS

680.25(B). Grounding. You must have a properly sized equipment-grounding conductor routed with the feeder conductors from your swimming pool disconnect (panel board) panel to the applicable equipment being serviced. The equipment-grounding conductor must be insulated.

Aluminum conductors may be used under this provision, as long as the general rules regarding aluminum are observed. You are probably just better off using copper when it comes to swimming pool installations.

Aluminum may not come within 18 in. of the earth/soil and is not allowed for use where it may come in contact with water. Please see Arts. 250.52(B)(1) and 250.64(A) for more information.

680.25(B)(1). Size. The equipment-grounding conductor must be sized in accordance with Table 250.122 (see Appendix B) or a minimum 12 AWG, whichever is larger. Please see Art. 680.23(F)(2) and (B)(3) for more information.

For separately derived systems, the equipment-grounding conductor must be sized in accordance with Art. 250.30(A)(3) or a minimum 8 AWG, whichever is larger.

680.25(B)(2). Separate Buildings. It is quite common to see an outdoor swimming pool that has a small dedicated building/structure to contain the associated swimming pool equipment, such as electrical pumps, disconnects, panel boards. When this occurs, the building/structure must meet the requirements of Art. 250.32(B).

680.26. EQUIPOTENTIAL BONDING

The equipotential bonding grid is a separate grounding system that is installed specifically for the function of reducing voltage gradients and is not intended to replace any other grounding systems, such as those that are designed to safely conduct fault currents.

The typical equipotential bonding grid is installed immediately next to the steel rebar in the foundation and is completely encased in concrete with the rebar.

Equipotential bonding A conductor system designed to reduce earth voltage gradients in the area around a permanently installed pool, transformer, substation, high-voltage motor, or similar installation by the use of a common bonding grid. Equipotential bonding is not intended to provide the primary low-impedance ground fault–current path to the source (which would assist in clearing a ground fault) as required by Art. 250.4(A)(3).

Bonding (bond, bonded) The permanent joining of metallic parts together to form an electrically conductive path. This path must have the capacity to safely conduct any fault current likely to be imposed on it.

Grounded (earthed) Connected to earth/soil.

680.26(A). Performance. The equipotential bonding grid is to be installed so as to reduce voltage gradients within the pool area. The equipotential bonding grid is to ensure that all metallic objects are at the same potential so as to reduce the risk of shock hazards created by stray currents in the ground/earth and/or piping systems. Chlorinated water has a low resistivity and can actually conduct electrical currents through the liquid water itself, even in pools with nonmetallic piping.

680.26(B). Bonded Parts. The components in Art. 680.26(B)(1) through (B)(7) must be bonded together using solid-copper conductors not smaller than 8 AWG. The copper conductor can be bare or insulated. You may also use ridged brass (or other listed corrosion-resistant metal) conduit as a bonding conductor. All connections/bonds must be made in accordance with Art. 250.8.

While you are required to bond remote panel boards, disconnects, and other electrical equipment to the equipotential bonding grid, you are not required to extend the grid itself out to those items. Of course, you may do so if it is practical and will help to reduce voltage gradient hazards for persons utilizing or working in the pool area.

Aluminum may not come within 18 in. of the earth/soil, is not allowed for use where it may come in contact with water, and is specifically not allowed in this provision. Please see Arts. 250.52(B)(1) and 250.64(A) for more information.

680.26(B)(1). Conductive Pool Shells. Swimming pools are constructed with a number of "shells." Often, a swimming pool will start with a layer of nonconductive material called a vapor barrier that is designed to prevent water from entering into the porous concrete from the surrounding soil. The second layer or shell is typically concrete with steel rebar that is designed to provide structural support. The third and final layer may be a preformed plastic or fiberglass shell that is placed on top of the concrete shell, or it may be a vinyl or waterproof plaster coating applied to the concrete.

The Code states that it considers the following construction items or shells to be conductive: poured concrete, pneumatically applied or sprayed concrete, and concrete block with painted or plastered coatings, and of course bare steel rebar (and any other metallic object such as metal conduit, raceway, enclosures, and metal post supports for ladders, rails, slides, diving boards, etc.).

The Code states that it considers the following construction items or shells to be nonconductive: vinyl liners, fiberglass composite shells, and steel rebar with a nonconductive epoxy coating that has had any exposed steel encapsulated in nonconductive listed compound.

However your pool is constructed, the Code requires you to install a copper equipotential bonding system, and you must bond the conductive items to the grid in each shell.

If your pool is constructed of epoxy-coated steel rebar (encapsulated reinforcing steel), you must still install a copper equipotential bonding system as required below; you just do not have to bond the epoxy-coated rebar to the equipotential grid. See Art. 680.26(B)(1)(a) for more information.

1. Structural reinforcing steel: Unencapsulated structural reinforcing steel (bare rebar prior to being poured in concrete) must be bonded together by steel tie wires or the equivalent. Where structural reinforcing steel is encapsulated in a nonconductive compound, you must install a copper equipotential bonding grid in accordance with Art. 680.26(B)(1)(b).
2. Copper conductor grid: A copper conductor grid must be constructed and must comply with the following:
   a. Be constructed of a minimum of 8 AWG bare solid-copper conductors that are bonded to each other at all points of intersection and/or crossing. The bonds must occur in accordance with Art. 250.8 or other approved means.
   b. The grid must conform to the contour of the pool.
   c. The 8 AWG bare solid-copper conductors must be arranged in 12-in. × 12-in. (300-mm × 300-mm) uniformly spaced grid, with a tolerance of plus or minus 4 in. (100 mm).
   d. The grid must be installed under the pool no more than 6 in. (150 mm) from the outer contour of the pool shell.

680.26(B)(2). Perimeter Surfaces. This section tells us that the perimeter of a pool extends horizontally 3 ft past the edge of the pool (inside walls of the pool) and includes both paved and unpaved surfaces. These perimeter surfaces must have an 8 AWG bare solid-copper conductor network formed into a 1-ft on-center grid, as required in Art. 680.26(B)(1)(b). The equipotential bonding grid extending out 3 ft on the surface must be bonded to the pool's structural steel and copper conductor grid in at least four points uniformly spaced around the pool.

Note that this code specifies both paved and unpaved surfaces. This means that an above-grade permanently installed pool must have a copper equipotential bonding grid installed, even if the surface is covered with grass. Of course, the copper grid must be installed 4 to 6 in. below grade in accordance with Art. 680.26(B)(1)(b)(4) and (B)(2)(b)(5).

If there are physical constraints, such as a wall or other physical barrier, preventing a 3-ft installation, the copper equipotential bonding grid is only required to be extended to the available area. If the physical barrier is metallic, it must be bonded to the equipotential bonding grid per Art. 680.26(B)(7).

The following is required for all equipotential bonding grids installed on horizontal perimeter surfaces:

1. Structural reinforcing steel: Structural reinforcing steel must be bonded in accordance with Art. 680.26(B)(1)(a).
2. Alternate means: Where the surface uses structural steel rebar that is encapsulated in a nonconductive compound, the copper equipotential bonding grid must meet the following requirements:
   a. At least one 8 AWG bare solid-copper conductor shall be provided.
   b. The conductor shall follow the contour of the surface.
   c. Splices must use listed devices for connection.
   d. The required conductor shall be 18 to 24 in. (450 to 600 mm) from the inside of the pool wall.
   e. The conductors are bared 4 to 6 in. (100 to 150 mm) below the subgrade.

680.26(B)(3). Metallic Components. All metallic parts of the pool structure not previously addressed in the Code must be bonded to the copper equipotential bonding grid. Structural steel rebar that is encapsulated in a nonconductive compound does not need to be bonded.

680.26(B)(4). Underwater Lighting. All metal forming shells and mounting brackets of no-niche luminaires must be bonded to the copper equipotential bonding grid.

Exception no. 1: Low-voltage lighting products that are specifically listed for use without a grounding conductor are exempt, in accordance with Art. 680.6(1).

680.26(B)(5). Metal Fittings. All metal fittings that are attached to or within the pool structure must be bonded to the equipotential bonding grid.

Small isolated metal parts that are not over 4 in. (100 mm) in any dimension and do not penetrate the pool structure by more than 1 in. (25 mm) do not need to be bonded.

680.26(B)(6). Electrical Equipment. The metal parts of electrical equipment that are associated with the pool water circulating system, including the metal parts of pump motors, pool covers, electric motors, and so on, must be bonded to the equipotential bonding grid.

Exception no. 1: Double-insulated and listed equipment is not required to be bonded, unless as stated in items a and b below:

1. Where you have a double-insulated water pump motor(s), you must still install an 8 AWG solid-copper conductor of sufficient length to allow the future installation of replacement motors. This conductor must be bonded to the pool's equipotential bonding grid. If there is no connection between the swimming pool's equipotential bonding grid and the premises' equipment-grounding system, this bonding conductor must be connected to the equipment-grounding conductor of the circuit.
2. Pool water heaters rated at 50 A or more must be bonded and grounded at only those parts designated if there are such special instructions from the manufacturer.

680.26(B)(7). Fixed Metal Parts. All fixed metal parts of the pool must be bonded to the copper equipment-bonding grid. This includes, but is not limited to: metal-sheathed cables, metal raceways, metal piping, metal fences, metal awnings, metal doors, and metal window frames.

The bonding of these items may occur in a number of ways; however, the bonding must be in accordance with Art. 250.8. You may bond the parts together in series or in parallel configurations, you may use rigid brass conduit as a bonding conductor, and, of course, you can use the copper equipotential grid. Obviously, only listed connecting devices are allowed.

Corrosion is always an issue when dealing with water and chlorine. Because of this, regular inspection of exposed connections is required to ensure the integrity of the bonds.

Exception no. 1: Metal objects separated by a physical barrier such as a wall do not need to be bonded.

Exception no. 2: Metal objects farther than 5 ft (1.5 m) horizontally from the inside walls of the pool do not need to be bonded. However, you are certainly allowed and encouraged to do so.

Exception no. 3: Metal objects farther than 12 ft (3.7 m) vertically above the maximum water level of the pool do not need to be bonded. However, you are certainly allowed and encouraged to do so. If your pool has diving boards, slides, observation stands, and so on, this requirement extends to 12 ft (3.7 m) above these objects.

680.26(C). Pool Water. This part of the Code tells us that the liquid water in the pool itself must be intentionally bonded to ground. The requirement is for a direct contact between the liquid water and at least 9 in.^{2} (5800 mm^2) of conductive surface area that is bonded to ground in accordance with Art. 680.26(B). This conductive surface must be in contact with the water at all times and protected from physical damage or dislodgment during pool activities.

Conductive parts of grounded and bonded underwater luminaires, handrails, ladders, and so on may be used to qualify for this portion of the Code. There are also specially made devices that are listed and approved for intentionally bonding the water; these may also be used.

Listed or listed item An item, product, or device that has been specifically developed for a particular function and has been independently evaluated by a product-testing organization, such as the UL or CE.

The 2014 edition of the Code expanded this section to improve the clarity of the instruction.

Article 250.30 Grounding Separately Derived Alternating Current Systems

The Code added wording to confirm that multiple separately derived systems connected in parallel must also meet the rules found in Art. 250.30.

Article 250.30(A)(1) System Bonding Jumper

The Code reworded Exception no. 2 to improve clarity.

Article 250.30(A)(2) Supply-side Bonding Jumper

The Code referenced the new table: Table 250.102(C).

Article 250.30(A)(3) Grounded Conductor

The Code referenced the new table: Table 250.102(C).

Article 250.30(A)(5) Grounding Electrode Conductor, Single Separately Derived System

There was a minor change in this section to improve the clarity of the wording.

Article 250.30(A)(6) Grounding Electrode Conductor, Multiple Separately Derived Systems

There was a minor change in this section to improve the clarity of the wording.

Article 250.32(B)(1) Supplied by a Feeder or Branch Circuit

The Code was changed in this section to improve the clarity of the wording and added Exception no. 2.

Article 250.34(A) Grounded Conductor Bonding

Article 250.34(B) Portable Generators

Article 250.34(C) Vehicle-mounted Generators

The 2014 edition of the Code changed the order of sections A, B, and C to improve clarity.

Article 250.36(F) Grounding Electrode Conductor Locations

The Code clarified that this section applies to both services and separately derived systems.

PART III

Article 250.64(D)(1) Common Grounding Electrode Conductor and Taps

There was a minor change to include the provision that the ground bar must be of sufficient length to accommodate the number of terminations necessary for the installation.

Article 250.64(E) Raceways and Enclosures for Grounding Electrode Conductors

Article 250.64(E)(1) General

Article 250.64(E)(2) Methods

Article 250.64(E)(3) Size

Article 250.64(E)(4) Wiring Methods

This section of the Code was split into four subsections for clarity.

Article 250.66(A) Connections to Rod, Pipe, or Plate Electrodes

There was a minor change in this section to clarify that a combination of differing types of electrodes or multiple electrodes of the same type could be used, but that the same rules apply.

Article 250.66(B) Connections to Concrete-encased Electrodes

There was a minor change in this section to clarify that single or multiple connections to the steel rebar in the foundation (concrete-encased electrode) could be used but that the same rules apply.

Article 250.68(C) Grounding Electrode Connections

The title of this section was changed and permission to extend rebar or conductors out and above of the concrete foundation, so as to allow an accessible location for connecting the grounding electrode conductor to the concrete-encased electrode, was added

PART V

Article 250.100 Bonding in Hazardous (Classified) Locations

The requirement to reference 505.5 and 506.5 was added to this section, and an informational note with regard to special bonding requirements found in Arts. 501.30, 502.30, 503.30, 505.25, and 506.25 was also added.

Article 250.102(C) Size—Supply-side Bonding Jumper

Table 250.102(C) was added to decrease confusion regarding the sizing of bonding jumpers.

PART VI

Article 250.112(A) Switchboard or Switchgear Frames and Structures

Switchgear was added to this section on switchboards.

Article 250.118 Types of Equipment-grounding Conductors

Type MI identification was added to line item 9, regarding the mineral-insulated, metal-sheathed cable; the location of the definition of effective ground fault–current path was moved to Article 100 definitions.

Article 250.119 Identification of Equipment-grounding Conductors

The Code added two new exceptions: Exceptions nos. 2 and 3.

Article 250.119(A) Conductors Larger Than 4 AWG

The size of the conductor in this rule was changed from 6 AWG to 4 AWG.

Article 250.122(B) Increased in Size

Statements were added to clarify this section of the Code.

Article 250.122(F) Conductors in Parallel

This section was expanded to improve clarity; the exception allowing design changes by electrical engineers was also added.

PART VII

Article 250.130(C) Nongrounding Receptacle Replacement or Branch Circuit Extensions

Wording was added to clarify the requirements in Art. 250.130(C)(4).

Article 250.142(B) Load-side Equipment

A global change to the 600-V rule to 1000 V; this was changed in Exception no. 4.

PART VIII

Article 250.162(A) Two-wire, Direct Current Systems

The Code increased the DC voltage from 50 V to 60 V and required (in Exception no. 1) that the source of power be adjacent or integral to the system.

Article 250.166 Size of Direct Current Grounding Electrode Conductor

The Code placed a maximum size of 3/0 copper or 250 kcmil aluminum for the grounding electrode conductor on DC systems.

Article 250.167 Direct Current Ground-fault Detection

Article 250.167(A) Ungrounded Systems

Article 250.167(B) Grounded Systems

Article 250.167(C) Marking

This section is new to the 2014 Code.

Article 250.162(A) Two-wire, Direct Current Systems

The Code increased the DC voltage from 50 V to 60 V and required (in Exception no. 1) that the source of power be adjacent or integral to the system.

Article 250.166 Size of Direct Current Grounding Electrode Conductor

The Code placed a maximum size of 3/0 copper or 250 kcmil aluminum for the grounding electrode conductor on DC systems.

Article 250.167 Direct Current Ground-fault Detection

Article 250.167(A) Ungrounded Systems

Article 250.167(B) Grounded Systems

Article 250.167(C) Marking

This section is new to the 2014 Code.

PART IX

Article 250.170 Instrument Transformer Circuits

The Code added the requirements in this section to include metal-enclosed switchgear.

Article 250.174(A) Not on Switchboards or Switchgear

The Code added the requirements in this section to include switchgear.

Article 250.174(B) On Metal-enclosed Switchgear and Dead-front Switchboards

The Code added the requirements in this section to include metal-enclosed switchgear.

Article 250.178 Instrument Equipment-grounding Conductor

The Code added the requirement in this section to include metal-enclosed switchgear.

PART X

Article 250.184(A)(1) Insulation Level

The Code added underground service conductors to Exception no. 1, item 2.

Article 250.186(A) Systems with a Grounded Conductor at the Service

Article 250.186(A)(1) Sizing for a Single Raceway or Overhead Conductor

Article 250.186(A)(2) Parallel Conductors in Two or More Raceways of Overhead Conductors

Article 250.186(A)(3) Delta-connected Service

Article 250.186(B) Systems without a Grounded Conductor at the Service Point

Article 250.186(B)(1) Sizing for a Single Raceway or Overhead Conductor

Article 250.186(B)(2) Parallel Conductors in Two or More Raceways or Overhead Conductors

This section is new to the 2014 Code.

Article 250.194 Grounding and Bonding of Fences and Other Metal Structures

Article 250.194(A) Metal Fences

Article 250.194(B) Metal Structures

This section is new to the 2014 Code.

Chapter Sixteen

CANADIAN ELECTRICAL CODE

The Canadian Electrical Code, CSA C22.1, also known as the CE Code, is the electrical equipment installation standard for Canada and is published by the Canadian Standards Association. The current edition was released in 2012 and is scheduled to have revisions issued every 3 years, 1 year after the National Electrical Code (NEC) is released in America. The CE Code has been adopted across Canada and is the basis for electrical regulation and enforcement.

The CE Code is divided into major sections, with each section titled and given a one- or two-digit number that is always even, between 0 and 98. The sections are then further broken down into rules that are given three-digit numbers, resulting in an orderly numbering system: "00-000." This "00-000" numbering scheme is typically referred to as a "rule," even though it provides both section and rule information. Secondary-level breakdowns are called subrules and items, and are conducted by bracketing number and letters after the rule: "00-000(1)(a)(I)(A)."

The electrical requirements found in the CE Code are very similar to those found in the NEC; however, the two codes are organized in a very different manner.

Currently, the CE Code has forty-four sections, numbered 0 through 86, plus 2 annexes. The Grounding and Bonding Section is listed as Section 10 and is broken down into twelve rules, and a number of subrules. The rules are as follows:

SECTION 10—GROUNDING AND BONDING

Rule 10-000: Scope and object
Rule 10-100: System and circuit grounding
Rule 10-200: Grounding connections for systems and circuits

Rule 10-300: Conductor enclosure bonding
Rule 10-400: Equipment bonding
Rule 10-500: Methods of grounding
Rule 10-600: Bonding methods
Rule 10-700: Grounding electrodes
Rule 10-800: Grounding and bonding conductors
Rule 10-900: Grounding and bonding conductor connections
Rule 10-1000: Lightning arresters
Rule 10-1100: Installation of neutral grounding devices

The CE Code is organized in a very different manner from the NEC. The following is a list of the CE Code rules, with the equivalent parts from Art. 250 of the NEC.
It is by no means an exact list, as it is simply designed to show the extent of the differences between the two sets of codes.

Rule 10-000: Scope and object	NEC Part I
Rule 10-100: System and circuit grounding	NEC Parts II, VIII, and IX
Rule 10-200: Grounding connections for systems and circuits	NEC Parts II, IV, and VI
Rule 10-300: Conductor enclosure bonding	NEC Parts V and X
Rule 10-400: Equipment bonding	NEC Parts III, V, VI, and IX
Rule 10-500: Methods of grounding	NEC Parts III and VII
Rule 10-600: Bonding methods	NEC Parts IV, V, VII, and X
Rule 10-700: Grounding electrodes	NEC Part III
Rule 10-800: Grounding and bonding conductors	NEC Parts III, V, VI, VII, and VIII
Rule 10-900: Grounding and bonding conductor connections	NEC Parts II, III, V, and VII
Rule 10-1000: Lightning arresters	Not covered in the NEC
Rule 10-1100: Installation of neutral grounding devices	NEC Part II

The CE Code is very different from the NEC, not only in the way it is organized, but also in the way it is written. The CE Code utilizes what is called a prescriptive model writing style to outline what electrical wiring and installation methods are acceptable, versus providing hard-and-fast rules. The prescriptive model incorporates value judgments and subjective criteria, where risks, costs, and benefits are considered subjectively to aid decision making about acceptable practices. This makes the CE Code far more informative and easier to understand than the NEC. It also gives the electrical professional far more flexibility in designing safe, efficient, and cost-effective electrical systems.

The hope of the authors of this book is that the guidelines and basic theories presented will assist users of both the NEC and those using the CE Code.

Chapter Seventeen

BONDING JUMPERS

Article 250 of the National Electrical Code (NEC) uses the term *jumper* to identify a variety of different electrical engineering functions throughout the Code. Unfortunately, the Code uses the same terms to indicate different engineering functions, and in some places uses several different terms to indicate the same engineering function; in fact, it seems that the Code uses the various terminologies interchangeably and wherever it sees fit. You will find throughout the text of this book an attempt to correct the terminology confusion.

The short summary here will not get into listing specific examples, as they are identified as you go through the chapters of the book. This will simply give you, for your informational purposes, an overview of the bewildering array of terminology that is used in the Code and how the terms are intermixed.

The confusion we talk about regarding the term jumper is similar to how Art. 250 of the Code uses four different terms to describe the neutral wire: *grounded conductor*, *neutral conductor*, *common conductor*, and *ground-fault circuit conductor*. Fortunately, the 2014 edition of the Code has removed the term *common conductor* from Art. 250.26 (2011 edition). Unfortunately, in the brand-new Art. 250.186, the Code decided to add the term ground fault circuit conductor instead of just calling it a neutral conductor (the old Art. 250.186 was changed to Art. 250.187). With that said, the differing neutral conductor terminology is fairly easy to straighten out, unlike the jumper terminology.

The 2014 edition of the Code has made no effort to correct or clean up all of the different confusing applications (and in some places downright errors) of its own terminology and has left all of the different jumper terminology in place. The following will try to reduce some of the confusion by identifying the problems.

Article 250 of the NEC uses at least eight different terms related to jumpers and bonding:

- Bonding jumper
- Main bonding jumper
- Equipment-bonding jumper
- Load-side bonding jumper
- Supply-side bonding jumper
- System bonding jumper
- Supply-side equipment-bonding jumper
- Equipment-grounding conductor

Some of these jumpers are used to describe neutral-to-ground bonds, and some of them are simply used to describe the connecting of normally non-current-carrying metal parts. Here is a chart showing the different terminology and the engineering functions they describe:

Jumpers Terms and Their Uses in the NEC

Jumpers that make neutral-to-ground connections	Jumpers that bond metal parts together
Bonding jumper	Bonding jumper
Main bonding jumper	—
System bonding jumper	—
Equipment-bonding jumper	Equipment-bonding jumper
Supply-Side Bonding Jumper	Supply-side bonding jumper
Load-Side Bonding Jumper	—
Supply-side equipment-bonding jumper	Supply-side equipment-bonding jumper
—	Equipment-grounding conductor

As you can see in the chart above, Art. 250 uses the terms bonding jumper, equipment-bonding jumper, supply-side bonding jumper, and supply-side equipment-bonding jumper, to indicate connections that perform two different engineering functions: a neutral-to-ground connection and bonding metal parts together.

For the jumpers that are intended to make neutral-to-ground bonds, the Code tries to separate them into two categories: neutral-to-ground bonds that occur on the supply side of the service and those that occur on the load side of the service. But even here, the Code just cannot seem to get it right. As you can see in the chart below, the terms bonding jumper and main bonding jumper are intermixed between both the neutral-to-ground connections that occur on the supply side of the load and those that occur on the load side, adding additional and unnecessary perplexity.

Terms Used for Neutral-to-Ground Connections

On the supply side of the service	On the load side of the service
Bonding jumper	Bonding jumper
Main bonding jumper	Main bonding jumper
System bonding jumper	—
—	Equipment-bonding jumper
Supply-side bonding jumper	—
—	Load-side bonding jumper
Supply-side equipment-bonding jumper	—

As you can see from the chart above, Art. 250 of the Code uses at least five terms to indicate supply-side, neutral-to-ground connections, and four terms to indicate load-side, neutral-to-ground connections.

I wish I could say that the terminology confusion ended here … but it does not. The Code adds additional uncertainty to the situation by using several of the same terms above to indicate connections that involve equipment-grounding conductors and grounding electrode conductors, and even uses them to describe below-grade (buried) connections between electrodes. Here is a list:

Terms used to describe connections that involve equipment-grounding conductors:

- Equipment-grounding conductor
- Bonding jumper
- Equipment-bonding jumper

Note *There is also the term* intersystem bonding, *which is used to indicate equipment-grounding conductors; however, the application is appropriate.*

Terms used to describe connections that involve grounding electrode conductors:

- Grounding electrode conductor
- Bonding jumper
- Supply-side equipment-bonding jumper (this jumper is actually installed on the load side of the service and is used to describe a connection to a grounding electrode and raceway)

The Code uses the following terms to indicate below-grade (buried) connections between grounding/earthing electrodes:

- Grounding electrode conductor
- Bonding jumper
- Equipotential bonding (Art. 680)

As you can see from the list above, and by no means is this a complete list, the terminology used within Art. 250 of the NEC is very confusing. The terminology used in the Code should really be corrected in the next edition (due out in 2017) by reducing each function to a single term.

So, how can we fix the confusion? The terminology should simply be broken down by the engineering function it is trying to describe. In general, the jumper systems can be broken down into two areas with three functions.

The two areas are: above-grade ground and below-grade grounding/earthing, and the three functions are to provide neutral-to-ground connections, to provide dedicated ground-fault paths, and to interconnect normally non-current-carrying metal parts. Here is a chart showing the proposed breakdown:

Possible Method for Improving Grounding Terminology

	Neutral-to-ground connection	Dedicated ground-fault current path	Normally non-current-carrying metal parts
Above-grade	Use the term *neutral-to-ground*	Use the term *grounding*	Use the term *bonding*
Below-grade	NA	Use the term: *earthing*	Use the term: *equipotential bonding*

The term *jumper* should be used to indicate short lengths of wire used to connect objects (current-carrying or non-current-carrying) that may or may not have dedicated terminals for the connection.

Clearly such a proposal as the one shown above should be given time for extensive debate, and there are almost certainly better terms that could be used than those suggested above. But at least this is a good start.

Hopefully, this section has helped to clarify some of the terminology confusion by identifying the issues within the Code. Perhaps it has only made things worse and has caused you to become disheartened, and for that I am sorry. But ultimately, I believe that knowledge is power and with this information you will be able to better navigate the Code. Good luck!

JUMPER DEFINITIONS

Bonding jumper A reliable conductor sized per Art. 250 to ensure electrical conductivity between metal parts of the electrical installation.

Equipment-bonding jumper or load-side bonding jumper A conductor, screw, or strap that bonds the equipment-grounded conductor (neutral) to the equipment enclosures that are on the load side of the first service disconnect enclosure(s), in accordance with Art. 250.102(D). The term is also used to indicate a conductor, screw, or strap that bonds the equipment-grounding conductor of a circuit to the enclosure or raceway of that circuit, in accordance with Art. 250.146.

Main bonding jumper A conductor, screw, or strap that bonds the neutral (grounded) conductor to a grounding conductor of some type. The term *main bonding jumper* is confusingly used in the Code in various places to indicate both a *supply-side bonding jumper* and a *system bonding jumper* (load side). For more details, see Arts. 250.24(A)(4), 250.24(B), 250.28, and 408.3(C).

Supply-side bonding jumper A conductor installed on the supply (utility) side of a service, or within a service equipment enclosure(s), or for a separately derived system, that ensures the required electrical conductivity between metal parts required to be electrically connected; sometimes called a *main bonding jumper*. A conductor, screw, or strap that bonds the neutral (grounded) conductor to the service equipment enclosures on the supply side. The term *supply-side bonding jumper* is used to distinguish the bond at the supply (utility) side of the service from the *system bonding jumper*, which is used on the load side of the service (at the first service disconnect). For more details, see Arts. 250.24(A)(4), 250.24(B), 250.28, 250.28(D)(2), and 408.3(C).

Note *The NEC occasionally uses this term to indicate an equipment-grounding conductor (see Art. 250.30).*

System bonding jumper A conductor, screw, or strap that bonds the equipment-grounding conductor to the neutral (grounded) conductor; sometimes called the *main bonding jumper*. The term *system bonding jumper* is used to distinguish the bond at the load side of the service (at the first service disconnect) from the *supply-side bonding jumper*, which is the bonding used on the utility or supply side of the service. For more details, see Arts. 250.24(A)(4), 250.24(B), 250.28, and 408.3(C).

Neutral-to-ground connection (bond) is where a grounded current-carrying conductor, the neutral, is intentionally bonded to earth/ground. In general, this should only occur at the transformer and at the first service disconnect(s). The Code uses several names for neutral-to-ground connections: system bonding jumper, main bonding jumper, and supply-side bonding jumper.

Equipment-grounding (bonding) conductor The low-impedance fault-current path that is typically run with or encloses the circuit conductors, used to connect the non-current-carrying metal parts of equipment, raceways, and other enclosures to the grounded (neutral) conductor and equipment-grounding (bonding) conductor at service equipment or at the source of a separately derived system. Often called the *green-wire* or *third-wire* ground conductor.

Grounding (earthing) electrode conductor The conductor used to connect the grounding (earthing) electrode(s) to the equipment-grounding (bonding) conductor, to the neutral (grounded) conductor, or to both in accordance with Arts. 250.66 and 250.142.

Grounding (earthing) electrode A device that establishes an electrical connection to the earth/soil.

Grounding (earthing) electrode system Two or more grounding electrodes that are bonded together to form a single system. For a typical building/structure, a grounding electrode system should consist of: two ground rods, water pipe, steel frame of the building, the steel rebar in the concrete foundation, and any ground rings.

Chapter Eighteen

GROUNDING QUESTIONS AND ANSWERS

One of the most fun and rewarding parts of my job is being one of the instructors at the grounding seminars we conduct throughout the country and even around the world. Meeting new people in our industry who want to learn about the value of grounding is truly rewarding. I am fascinated to discover people who have taken the time and commitment to improve themselves for their edification or for their companies' commitment to excellence in our field.

Being around positive, successful people makes you feel great. If you have ever attended one of our grounding seminars, you would know that there are usually two questions that I ask the attendees, and they go something like this.

If you ask any engineering student, electrical engineer, or electrician, "Is grounding important?" you will get the same answer from each and every one of them, "Absolutely, grounding is very important." Then, follow up the first question with a second question and ask them, "Why is proper grounding so important?"

The answers you get back will surprise you. Almost all will say, in one form or another, that oh, yes, grounding is important, but it is the "why" of the question, that will surprise. You will get that look, you know the look I am talking about, the "I know it's important but can't explain it" look, and we have all had that look at least one time in our lives.

That deer in the headlights look: "What, what did he say?" Or that look of "Darn, I did not know there was going to be a test today or the I did not study" look. Now, I do not ask these two questions to upset anyone, I ask them because I like to hear the various answers that people give.

It helps to stimulate the seminars, and it get people's thoughts flowing, it aids in learning retention, and it is fun to get the attendees involved in the seminar. With that said, that is one of the reasons for this book; we will try and give you the "why's" of those two questions.

So you will never again have to have that "deer in the headlights" look when talking about grounding issues.

In the electrical engineering educational system, one of the least covered topics is the grounding portion of the syllabus. Most engineers and electricians will tell you that the subject was covered but not in the depth that it should have been. It is one of the most misinterpreted and misunderstood sections in the National Electrical Code (NEC).

This next chapter should be an interesting and educational guide to most people as it is written in very plain languge. We have taken some of the most asked questions from our grounding blog, called *Ask the Experts*, and have answered them. These questions come from all types of disciplines, all parts of the country, and even from all over the world. Take a look and see if they are possibly some of the same questions you may have or may have had. As the old adage goes, "The only stupid question is the one that is not asked." So, please enjoy this chapter.

BONDING QUESTIONS AND ANSWERS

I lost my green bonding screw, can I use any color screw to replace it?

The Question: One of my junior electricians lost the green bonding screw in an electrical cabinet we are installing. Can I use a regular silver machine screw to replace it?

The Answer: No. You may not replace the green bonding screw with a silver one. NEC Art. 250.28(B) clearly states that the bonding screw (either the main bonding jumper and/or the system bonding jumper) must be green in color and that it must be visible when installed.

Remember that the main bonding jumper and system bonding jumper perform the same engineering function, in that they are the two and only two mandatory neutral-to-ground bonds that must occur in your electrical system.

The main bonding jumper is the name for the bond that occurs inside the transformer; the system bonding jumper is the name for the bond that must occur in the first service disconnect. See NEC Art. 250.28. In your case, you are talking about the system bonding jumper, which must be a visible green screw. Sorry for the bad news, but you are going to have to run to the shop and grab a few green screws! Good luck!

How do you ground plastic water pipe?

The Question: When a residence service is serviced with a plastic water line(s) and ground rods are all that is available, does the requirement change for only a 6 AWG copper conductor to the rods, as it is now the primary grounding electrode and not supplemental? What if it is a larger home, with 120/240 V, 400(320) A service?

The Answer: The latest NEC has recently made some changes to the grounding requirements for homes. You are now required to install two 10-ft ground rods at least 6 ft apart (20-ft spacing is better), whether there is a copper water main or not.

There are a few exceptions to this rule. If you can test for resistance to ground and can demonstrate that one ground rod by itself is under 25 Ω, you do not need the second rod. Also, if you are willing to dig 18-in. deep holes and completely bury your ground rods, you can use 8-ft rods (you must use connections rated for direct burial).

Keep in mind, you still MUST bond to the steel rebar in the foundation, the gas main, the water main, the cable entrance, the Telco ground, any lightning protection system (LPS), your swimming pool ground, and any other ground system, just as before. It is just that you may not count these as grounding electrodes anymore.

That said, there is an additional concern in that the grounding electrode conductor must be able to handle the likely fault currents. A 400 A service would need at least a 3/0 AWG copper phase conductor to provide the supply power, which means that according to Table 250.66 of the NEC, the grounding electrode conductor needs to be a #2 AWG copper or larger. You would need to install additional grounding electrode conductors connected to qualified grounding electrodes, until you have achieved at least enough circular mils on your copper grounding electrode conductor to equal a #2 AWG.

The Question: Can all steel rebar be exothermically welded?

The Answer: Not all rebar can be welded, particularly with the exothermic welding process. This is because the welding method must be able to heat the surface of the metal rebar to near the melting point in order to ensure adhesion. In general, only steel rebar that is "Grade W" is soft enough to be welded using the exothermic welding process.

According to Wikipedia: "The American Welding Society (AWS) D 1.4 sets out the practices for welding rebar in the U.S. Without special consideration the only rebar that is ready to weld is *W grade* (Low-alloy — A706). Rebar that is not produced to the ASTM A706 specification is generally not suitable for welding without calculating the 'carbon-equivalent.' Material with a carbon-equivalent of less than 0.55 can be welded (AWS D1.4)." Here is the link:

http://en.wikipedia.org/wiki/Rebar.

Rebar cages are normally tied together with wire, although welding of cages has been the norm in Europe for many years and is becoming more common in the United States. High-strength steels for prestressed concrete are known for being very resistant to all welding processes.

How do you bond galvanized steel and copper together and still prevent galvanic corrosion at the joint?

The Question: We have a process plant grounding system where all grade-level concrete foundation of building and package skids are provided with galvanized steel strips welded to the reinforcement bars. The foundation grounding bars (galvanized steel) are brought out at certain locations to connect to a copper earth bar. Electrical equipment is bonded to copper earth bars with PVC-insulated copper cables. Now, we are a bit skeptical about connecting the steel strips with the copper earth bar because of galvanic corrosion. Can both metals be thermowelded so the galvanic corrosion problem can be avoided?

The Answer: Galvanic corrosion is an electrochemical process in which one metal corrodes preferentially to another when both metals are in electrical contact and immersed in an electrolyte. Dissimilar metals and alloys have different electrode potentials, and when two or more come into contact in an electrolyte, a galvanic coupling is set up. Scientists have indexed materials based on the nobilities of the metals. This index is called the *anodic index* and provides a voltage for each material. A voltage differential greater than 0.15 V is considered corrosive. Copper has an anodic index of 0.35, and steel has an anodic index 0.85. This is a 0.50-V differential and can certainly be a source of galvanic corrosion, as the earth will surely act as an electrolyte. But there is another concern with regard to corrosion beyond simple nobility issues, and that is induced alternating current (AC). Please see the following blog for more information:

www.esgroundingsolutions.com/blog/407/i-am-designing-a-ground-grid-for-a-new-115kv-substation-the-civil-design-is-planning-on-utilizing-steel-piles-for-strucutral-support-we-would-like-to-utilize-the-piles-as-ground-rods-if-possible-bu.

Because of the good electrical conductivity and high stability in the face of short-circuit pulses, exothermic welds are one of the options specified by Art. 250.8 of the 2011 NEC for grounding conductors and bonding jumpers. Also, exothermic welding is not only the preferred method of bonding, it is the only acceptable means of bonding copper to galvanized cable, according to the following:

John Crisp, *Introduction to Copper Cabling*, Newnes, Oxford, UK, 2002, p. 88.
Jerry C. Whitaker, *The Electronics Handbook*, 2nd ed., CRC, Boca Raton, FL, 2005, p. 1199.

Should I bond the various grounds of my building power, Telco, generator, swimming pool, and surveillance system together?

The Question: We provided an earthing system to a building to have an earth electrode resistance to earth of less than 10 Ω using multiple earth electrodes(copper-clad steel rods) forming a close grid/mesh. The main building (1) receives a three-phase,50-Hz, 400-V utility power supply from an overhead transmission line (2) connected to two overhead telephone lines and has a private branch exchange (PBX) exchange (3) supplied from a standby generator (400 V, 50 Hz, 100 kVA), but housed in a separate building 75 ft from the main building. Underground cables connect the generator and the main building (4). There is no building LPS (5). A swimming pool is located equidistant from the main building and generator room (6). Surveillance cameras are fixed along the perimeter of the ground area, which is about 250 ft × 250 ft using the TT earthing system. (Note: In a TT earthing system, the protective earth connection of the consumer is provided by a local connection to earth, independent of any earth connection at the generator).

My question is: Could I interconnect to the earth grid the earth conductors of (1) building power distribution system, (2) telephone system, (3) generator neutral point and frame, (4) swimming pool structural steel, (5) screen of the sensor cables of surveillance system (a) at closest points to each facility (b) at a single earth bus bar?

The Answer: Not only should you bond all of the grounding systems at your building together, but you MUST bond all of the systems together! The bonding of these systems is mandatory under every known regulation and standard worldwide.

The copper ground bar system is one of the best methods for bonding these systems together. In fact, there are some standards that actually have specific requirements for how to properly wire a ground bar. Basically, the idea is to have inputs at the top of the ground bar and outputs at the bottom.

For your building, you would want to see the 10-Ω grounding electrode system brought in to the bottom of the ground bar, and then bonding conductors for all the other systems connected to the top of the ground bar. So you would see the building power distribution ground, the Telco ground, generator, swimming pool equipotential ground, the surveillance system ground, lightning protection ground, gas-pipe ground, water-pipe ground, computer server ground, and all other grounding systems bonded to the top of the ground bar. You may find that multiple ground bars are needed. Of course, all conductors bonded to the ground bar should be installed with double-bolt connections and be properly labeled.

The swimming pool equipotential grounding system and building steel, including the rebar in the concrete, is often considered to be part of the grounding electrode system and is sometimes bonded to the bottom of the ground bar. We recommend you take a look at Motorola's R56 standard for some additional guidance. While it does not specifically relate to your situation, it does have some of the best illustrations and common-language instructions you will find anywhere.

Why do you have to bond the neutral and the ground wire in the main panel?

The short answer to your question is that the neutral-to-ground bond is needed to properly operate the circuit breakers. Overcurrent protection devices (OCPDs) such as circuit breakers and fuses actually require a short and intense INCREASE in electrical current (a short) in order to detect the fault and cut the circuit off.

Without a sharp and drastic increase in electrical flow, a fault could go on without triggering a circuit breaker to stop the flow. This actually occurs quite often and can be measured easily by checking the amount of current flowing on your ground conductor. It should be less than 1 A in most cases. If the current flowing on a grounded conductor is higher than an amp, and you are not in a high-voltage (600V+) environment, it typically indicates an erroneous neutral-to-ground bond somewhere in the system.

To visualize the reason why the neutral-to-ground bond is required, you must consider the entire electrical circuit from a 120-V outlet all the way back to the utility transformer hanging on the pole:

- In a properly designed circuit, if a fault were to occur on the 120-V outlet between the hot wire and the ground, the current would flow through the ground wire back to the main panel, where it would move to the neutral wire via the neutral-to-ground bond, up to the utility transformer, and back down the hot wire to the circuit breaker, tripping the breaker.
- In an improperly designed circuit, if a fault were to occur on the 120-V outlet between the hot wire and the ground, the current would flow through ground wire back to the main panel, where, because it does not have a neutral-to-ground bond, the current would be forced through the ground rod, into and across the earth, and up the utility ground rod and into the utility transformer, and back down the hot wire to the circuit breaker. The resistance of the earth is almost always great enough to prevent sufficient current flow to trip the breaker, and you end up with a steady-state ground fault that never trips the breaker, and this is a hazardous situation indeed. You cannot use the earth as a conductor.

Another issue that can occur is that multiple (and illegal) neutral-to-ground bonds can exist in the system (only one bond is allowed in the main panel). When this occurs, both the ground and the neutral become current-carrying conductors, which effectively means that you have two neutral wires running in parallel. This divides the current and places electrical energy on to the chassis of all metallic objects within the system; another hazardous situation.

Also, arc-flash energy exposure can go up if you do not have a solid neutral-to-ground connection because of the inverse time curves of circuit breakers.

ELECTRICAL GROUNDING

Do my transformers represent a human safety hazard?

The Question: I have two transformers next to an electrical building. A 13.8-4160 and a 4160-480. The medium-voltage (MV) transformer is resistance grounded to 100 A. The low-voltage (LV) is solidly grounded. My engineer is telling me I need a beefy grid around both, and it should be modeled as a substation, due to the fact that a transformer winding may become shorted to the transformer enclosure, creating A single line-to-ground (SLG) fault. He says that if that happens, 34.4 kA will go into the ground, and this will create harmful step-and-touch potentials in about a 100- to 150-ft diameter around the transformer. I do not know much about this sort of thing, but I know, typically, a very simple ground system is put into place, and not one with many rods and runs of conductor. Can you tell me, not knowing more details, whether his argument holds water and I should consider his design? I am told the soil is moderately conductive (3350–6250 Ω·cm).

The Answer: Yes, your engineer is correct. An SLG fault at your transformer will create a significant ground potential rise (GPR) that could in fact result in both step-and-touch voltage hazards to the public. Title 29 of the Code of Federal Regulations, part 1910.269 (29 CFR 1910.269) mandates that these hazards be mitigated to protect personnel working near the transformers and/or the public.

Your conductive soil can be beneficial, but it can also be harmful, given certain conditions. If the conductivity gets worse with depth, the electrical energy is going to want to stay near the surface of the earth, creating a larger step-voltage concern.

You will need soil resistivity data and electrical fault data to analyze these hazards properly. We recommend a series of Wenner four-point soil resistivity tests with spacings at least as large as your transformer area. For electrical data, you will need the SLG fault current, the X/R ratio (or zero sequence impedance), and the fault clearing time. Please get someone with the proper simulation software and expertise to help you design a safe and effective grid that is in compliance with 29 CFR 1910.269 requirements.

Is Ufer grounding acceptable for use where life safety is involved?

The Question: Is Ufer grounding acceptable in industrial facilities where life-safety equipment or systems are involved? Especially Department of Energy (DoE) facilities where design life is 50 years?

The Answer: Ufer grounding (steel rebar and/or copper encased in concrete) can be an excellent safety grid, but it must be analyzed, as there are a number of factors involved. Human safety in high-voltage environments is based on step-and-touch voltage hazards, and those can certainly form, even on concrete foundations. This is especially true if there are no copper conductors (only steel rebar) in the Ufer system.

Some of the determining factors include: the soil resistivity model for the site (does the resistivity go up or down with depth), impedance and magnetism of the steel rebar, fault currents, X/R ratio, clearing time, leakage currents, physical makeup of the structure (second-story and higher systems are particularly bad locations), and many more.

Of course, if your steel rebar is plastic coated or if you have a vapor barrier preventing direct contact of the concrete to the earth, you cannot use it as an Ufer ground. Also, the 2011 edition of the NEC requires at least one dedicated grounding electrode under 25 Ω. A second ground rod is required if you cannot test the electrode or if it is not under 25 Ω. You may no longer simply rely on building steel and water pipe. You must drive ground rods.

What is omnipotential grounding?

The Question: Would you please tell me what omnipotential grounding means?

The Answer: We have actually never seen that term in use before. "Omni" means all or everywhere, so perhaps omnipotential means all voltages? We suspect it just means that the ground system is at the same potential everywhere on your site, which is pretty much the goal of all grounding systems.

E&S Grounding Solutions uses an in-house guideline of a maximum 0.1 Ω of resistance from any point on the grounding system back to the first service disconnect. We would more accurately use the term "equipotential" to describe a grounding system with an effective low-resistance path from any point within the system. "Equipotential" is also the term used by the NEC.

The NEC describes an "equipotential grounding plane" or "equipotential bonding" in Arts. 547.2, 682.2, and 680.26 in similar, but slightly different ways. In summary that definition is: *Where wire mesh, rebar in concrete, other conductive elements, all metal structures, and fixed nonelectrical equipment that may become energized, are connected to the electrical grounding system to prevent voltage gradients from developing within the plane* [E&S Grounding Solutions summarized definition].

In essence, a properly constructed equipotential grounding system would have virtually every piece of metal bonded to the grounding system. This would include things such as the rebar inside concrete foundations, building steel, columns, motor chassis, tanks, piping systems, water pipes, gas pipes, LPSs, lighting poles, telecommunication and data communication grounding systems, conduit, cable trays, support stands, electrical cabinets, doors, gates, fence posts, fences, barbed wire, handrails, fixtures of all types, and so on.

The Question: What is the proper grounding methodology for cellular station?

The Answer: Proper grounding at cellular stations is quite complex and detailed. There are many requirements for extensive bonding and a typical cellular station and getting into each of them would be far too much to cover in a simple blog. But needless to say, virtually every single piece of metal must be connected to a dedicated equipotential grounding system, which is in turn bonded to an earthing system.

We recommend that you obtain a copy of Motorola's R56 standard or AT&T's excellent grounding standard the ATT-TP-76416. Either of these standards are excellent guides for the proper installation of a cellular station.

The earthing requirements for the typical cellular station generally requires a buried counterpoise (ring) system around the station's equipment pads and fence, and should have a resistance to ground of less than 5 Ω.

The purpose of the earthing system at a cellular station is the same as at any transmitting antenna, to remove the energy that has been induced into the station by its own transmissions.

Transmitting antennas, including those found at cellular stations, will induce unwanted electrical energies into the base station. This energy will quickly build up, causing electrical noise and harmonics that will compromise the signal quality and damage equipment.

The above-grade bonding system will ensure that there are no differences in potential within the station and will allow the stray currents to move to the earthing system, which in turn provides a path to earth where the harmonics can pass harmlessly out of the system. The better the grounding is, the quicker and more effective the cellular station will be.

Achieving 5 Ω or less for your earthing system is all about understanding the makeup of the soil at your site. The more resistive the soil is, the more extensive the earthing system will need to be. Here is a link that discusses this topic more thoroughly:

www.esgroundingsolutions.com/about-electrical-grounding /what-is-soil-resistivity-testing.php.

The Question: When does the earth circuit come to play? Why is it that you can disconnect the earthing lead and still have a functioning electrical system?

The Answer: In a properly designed electrical system, OCPDs such as fuses and circuit breakers will indeed continue to function, even without a connection to earth. In other words, if you have a good electrical system, you can remove the connection to the ground rod and the power will still flow and the circuit breakers will still trip in the case of a fault. This is all true, but that is not the purpose or function of the earthing system, now is it?

Imagine the space shuttle up in orbit. It has lots of electrical systems and no earthing system at all. If we were able to look at the wiring systems inside the space shuttle, we would see that they not only have the traditional third green-wire ground, but they additionally have an equipotential grounding and shielding system. The difference in potential from any metallic object to the next within the shuttle would be very low. The reason for this extra grounding system is because the shuttle has no mechanism to remove induced currents, transient energy, absorbed radiation, and so on. So the chassis of the shuttle itself becomes the ground source, slowly absorbing energy, in essence becoming a giant capacitor.

Let us say we were able to measure the voltage of the shuttles chassis when it first went into orbit against some other remote ground reference. Let us say it was 0 V. What do you think the voltage would be after a few days in orbit? What do you think the chassis' voltage would read after a few weeks in orbit? You might be surprised, as it could be many thousands of volts!

So what about the regular hot wires? Again, if we measured the single-phase hot wire on the shuttle when it first took off against some remote ground source, we would see 120 V. But after a few weeks in orbit, if we were to remeasure that same hot wire against a remote ground source, we would see thousands of volts! In other words, in our shuttle example, if the chassis measured 1000 V after a few weeks in orbit, the hot wire would measure 1120 V. From our perspective on Earth, it would look like the shuttle's voltage system is all out of whack. But on the shuttle, every voltmeter would read just fine and would only show a normal 120-V system.

We see this capacitive phenomenon every day with helicopters. They fly around charging themselves to the point that when they land, the choppers arc and spark when they touch down and discharge the built-up energy. But on the helicopter, they are not aware of the change in voltage. The same is true for ships at sea. They discharge into the surrounding ocean, but not as effectively as they need. After a few years of holding a charge, the ship itself must be demagnetized. Pearl Harbor in Hawaii is famous for having a buried degaussing system right in the harbor. Ships like to pull in to port for no reason other than to make use of this system.

The purpose for your earthing system at your home or business has nothing to do with proper circuit operation or even providing fault-current protection. The purpose of the grounding system is to provide a reference point, so our electrical systems do not charge our buildings like a helicopter or the space shuttle. The earth connection, of course, has many other purposes as well. Not only does it remove the unwanted stray currents that form on conduit and other metal objects, but it is vital for removing electrical energies from sources outside of the supplied power, such as lightning strikes.

If you were to remove the earth connection from your electrical system, you would start to find errors of all kinds. Your conduit and metal objects could start to get warm to the touch. Harmonics, transients, objectionable currents, and electrical noise of all kinds would interfere with your systems. Ground-fault circuit interrupter (GFCI) circuit protection might trip and fail for unknown reasons and would certainly be less reliable. Your electric bill might increase, as more and more power is pumped into the ever-increasing impedance of your electrical system.

Additional issues, such as insulation degradation due to voltage potential rises relative to ground would begin to occur, resulting in fire hazards. Electronic equipment of all kinds requires a solid earth connection as a reference point for the internal workings of the electronic systems. Surge protection systems require an earth connection to work properly. Where else will they place the unwanted energy from a surge, if not to the earth? Back on the neutral wire? What if the surge came from the neutral? Where would the surge protector place the energy then? On the hot wire?

The last point you should consider is that every single electrical standard has been increasing its requirements for grounding and earthing systems year after year. These standards have been written by very smart people, and there are many reasons for the increased requirements. It would take a long time to get into all of them, but the reality is that electrical systems that have excellent earthing systems function better than those that do not.

The Question: For a pole-mounted transformer, what is the most cost-effective way to achieve 25 Ω or less in Marl and Rock?

The Answer: When you have a pole-mounted transformer, you are generally dealing with a very small area of available soil (earth) that you can use for grounding. The resistance to ground of an electrode is determined by the local soil conditions and the sphere of influence of the electrode.

In other words, the worse the soil, the more electrodes you need. There really is no magic bullet to get around this.

The only way to guarantee a 25-Ω or less resistance to ground for your pole-mounted transformer is to get an engineered grounding design. With computer modeling, a grounding electrode can be designed to meet your 25-Ω requirement in local soil conditions.

For example, if you have a concrete pole, it may be possible to utilize part of the pole itself as an electrode. This could be supplemented with an electrolytic electrode system to achieve the 25-Ω requirement. But you will not know this unless you do a proper design. Electrolytic electrodes are expensive, and you would not want to randomly install one of these grounding systems without knowing in advance that it will work.

The Question: How do you ground a 120-ft tower on top of solid granite?

The Answer: This is one of the biggest questions we ever get: How do you ground on top of a rock mountain? We usually answer by pointing out that the International Space Station is grounded, and without the aid of a ground wire trailing down to earth! It should also be noted that airplanes, helicopters, and ships out at sea are all grounded without a connection to earth. How is this possible?

When your site is located on top of a granite mountain, you can be relatively assured that the resistance to ground of your electrodes will be very high. This tells a grounding engineer that the ground potential difference (GPD) of the system is going to be even more important than usual.

When electrical energy enters the site, the resistance will be very high, and the voltage will therefore be very high as well. Imagine a 5000-A fault entering a site with a 100-Ω resistance to ground. Ohm's law tells us that the voltage or GPR of the site will be as high as 500,000 V! If your site has a difference of potential from any given point to any other point within the system of 2 Ω, the site could see a voltage differential of 10,000 V! If that difference in potential is 1 Ω, then you could see 5000 V forming across your site! If the difference in potential is only 0.1 Ω, then your site would only experience 500 V. This should demonstrate the importance of a low GPD in high-resistance grounding systems.

One can imagine that as fault currents enter a grounding grid, the voltage will rise at some gradient across the system. The actual voltage that the system raises is not the primary concern; it is the current or amperage that can flow across people or equipment that is our concern. So, like a bird landing on a power line that may have high voltage, the bird will be safe as long as no current flows through the bird itself. Here is some additional info:

www.esgroundingsolutions.com/blog/749/what-is-accepted-limits-of-gpr-and-gpd-how-do-you-protect-telephone-circuits-from-high-gpr-and-gpd.

Now, with all that said, it is still advisable to provide an effective low-impedance path to earth for your grounding system. Many times, in cases such as this one, the power for the site is being brought in via overhead power lines (power poles).

You may wish to add a 4/0 AWG or greater copper conductor on the overhead pole system, so as to ensure there is a ground path from your site back to the source. This will ensure that OCPDs such as fuses and circuit breakers will function properly. It can also give you an opportunity to install a grounding electrode for your site in better soil conditions. This will be true, even if you install an isolation transformer at your site.

Can you explain why the Telco ground bars and the electrical ground bars need to be bonded together?

The Question: Why is it that communication equipment (racks, cabinets) must have a ground bus bar separate from the one for electrical equipment [panel boards, power distribution unit (PDU)], when, EIA-J-STD-607-A standard in 5.2.6.1 states that panel boards within the same room or space shall be bonded to the telecommunication main ground bus bar. This looks contradictory to me. I would really appreciate if you could give more explanation on this.

The Answer: In the typical data center, you will have many different ground systems in use throughout the facility. For example, in the electrical room, you will not only have the green-wire, low-impedance, fault-current path required by the NEC, but you may also have an equipotential grounding (safety ground) system with multiple ground bars used to bond the various metallic objects found within the room (electrical panels, conduit, air-handling equipment, fire suppression, doors, gates, fences, etc.). What the code is saying is that you must have at least one connection between these two grounding systems to ensure that there are no hazardous differences in potential.

In your computer room, you will have a requirement for a copper ground bar to be installed inside each computer server rack, and then an additional ground bar(s) mounted on the wall of the server room that will be used to bond all the server rack ground bars together, along with the conduit bonds, air-handling equipment bonds, door bonds, uninterruptible power supplies (UPS), and so on. This server room ground bar(s) must in turn be bonded back to the main ground bar in your Telco room and back to the first service disconnect in the main electrical room. There are also typically requirements for dedicated 5-Ω or less grounding electrodes that must be bonded to the server room ground bar(s).

You may have additional ground bars used for the bonding of the LPS, backup generator systems, electrical substations, equipment chassis, and many more. All of these ground bars and grounding systems must be bonded to each other in manner that does not cause additional problems with ground loops, resulting in harmonics and electrical interference.

The Question: How do you ground a building that has been electrically isolated by means of seismic isolators?

The Answer: When you have a building constructed with a seismic base, your foundation should be considered to be isolated. This would fall into the same category of building as one that has a vapor barrier in between the concrete foundation and the earth. In these cases, you may not use building steel as a grounding electrode.

Now, you still need to bond to building steel and make all the same connections to the frame as required under normal regulations. You simply must have a dedicated grounding electrode for the building. Frankly, E&S Grounding Solutions recommends this arrangement for all buildings regardless of seismic barriers or vapor barriers. etc. While most electrical codes and regulations allow the use of building steel as an electrode, we do not recommend using it.

Keep in mind, there is a big difference between bonding to an item, and using it as an electrode. Understanding this difference is very important. Consider gas pipes for example. Your gas pipes must be bonded to ground to prevent any hazardous differences in potential from forming, but you sure do not want to force electrical energy down them by using them as a grounding electrode.

As far as designing a grounding system goes, it is difficult to know what to design when we do not know what you are grounding. Are there high-voltage transformers inside? Do you have an electronic data center or an LPS? What specifications need to be met? You may need a ground ring that surrounds the entire building; maybe you need a deep-earth grounding electrode, or perhaps just a series of simple grounding rods.

Whatever you end up designing, you should ensure that the connections from the grounding electrode system back into the building are made using flexible grounding straps. You should google gate-grounding straps for an example of flexible grounding straps. There are a number of companies that sell these types of products.

The Question: Regarding the grounding system, if we have one or two building feeds from one power source, which is a better grounding system: for each building to have its own grounding manholes or to run separate grounding cables from the main grounding system of the source for each building?

The Answer: When you have multiple buildings being fed from a common power source you must have one common grounding system. This is required under the NEC, the IEC, and every known electrical standard.

This is vital, so that OCPDs such as fuses and circuit breakers will function properly.

Even if your site does have several separate utility power sources, you will only benefit from having a common grounding grid. Computer data lines, telecommunication systems, and so on will be tied between the various buildings, and a common ground grid will ensure there are no differences in potential that could cause damage to sensitive electronic equipment.

The Question: In substations, often low-voltage (LV) and high-voltage (HV) and grounds and neutrals of transformers or other instruments are connected to one mesh network. Sometimes, when lightning or other similar phenomena occur, the LV instruments are destroyed. How can I resolve this problem, and how can I know about the new method for grounding in substations? LV and HV and grounds and neutrals should be connecting to one ground grid or not? Is there any standard for connecting or separating grounds in substations?

The Answer: You may not have separate grounding systems at any location, particularly at substations. It would be very hazardous and would actually increase the problems you are having.

In your case, your substation's grounding system is probably inadequate to handle the current levels it is experiencing. Fortunately, in your case, a standard step-and-touch voltage safety analysis will not only resolve the problems you are having with your LV equipment, but it will also make the site safe for the people working in the facility.

A voltage safety analysis is easy to do, all you need is to provide an engineering firm with site drawings, soil data, and the substation's electrical fault data. The analysis of your facility's grounding system should be done using an engineering software package designed specifically for this purpose, such as the CDEGS simulation software from Safe Engineering Services in Canada (www.sestech.com).

Most utility power companies treat transmission and distribution lines as separate entities. Transmission lines are those high-voltage power lines that run from power plants to substations or from substation to substation. Distribution lines are those lines that run from a substation to the consumer (via pole-mounted transformers or the like). Most power companies try not to intermix transmission systems with distributions systems when they can avoid it, mostly to protect the distribution systems from the very powerful electrical faults that occur on transmission lines.

That said, this is merely a rule of thumb and is difficult to enforce in practice. First of all, what constitutes a distribution substation from a transmission substation? That line is quite fuzzy. Also, it is very common to see 66-kV transmission lines with distribution power on the same pole, which means that voltage cross over is almost certain to occur during fault conditions.

The short answer to your question is that all ground systems within a facility must be bonded together. This is true no matter what the source: transmission, distribution, lightning protection, Telco grounds, equipment grounds, isolated grounds, water pipes, building steel, and so on.

The Question: What is required to properly ground a 190-ft guided tower? Currently, no proper ground exists, only an 8-ft rod next to tower.

The Answer: There are many different types of towers, and many different reasons to be providing grounding. For a typical radio/cell/broadband tower, grounding is used to resolve some of the following electrical engineering factors:

1. Electrical safety grounding
2. Reduction of potential difference or GPD
3. Reduction of the resistance to ground and/or impedance to ground
4. Reduction of electrical noise, interference, harmonics, and so on
5. Safety during electrical faults—GPR and clearing time
6. Lightning safety—electromagnetic fields, frequency spectrum, time domain
7. Human safety—reduction of step-and-touch voltages

Each of the above issues is dealt with to a greater or lesser extent in a number of industrial standards related to the grounding of such towers. We are aware of no standard regarding a tower that considers a single 8-ft ground rod as acceptable. In fact, under the new NEC, it could be easily interpreted that a single 8-ft grounding rod is insufficient to meet its simple grounding requirements.

There are quite possibly several dozen codes and standards that govern grounding for towers in some form or another (IEEE, NFPA, MIL-SPEC, Motorola R56, IEC, ANSI, TIA, EIA, etc.), so we will not try to list them all here; additionally, most companies have internal standards that have minimum grounding requirements for towers that are above and beyond those in the standards. But that said, you can meet the vast majority of grounding requirements by simply installing a ground ring (sometimes called a loop or counterpoise) around the base of the tower.

Typically, the ground ring runs at least 1.5 ft below grade and 3 ft outside the perimeter of the tower, encompassing all of the legs of the tower. This conductor will be bonded to each tower leg via exothermic welding, irreversible compression fitting, and/or a double-bolt connection. The conductor itself will be bare copper placed in direct contact with the earth and typically ranges in caliber from a 2 gauge to 4/0 AWG (or higher), depending on the current levels that are anticipated on the conductor (i.e., electrical utility fault and/or lightning strike). This ground ring must be supplemented with 10-ft ground rods (electrodes) at regular intervals (such as every 20 ft), with ground rods always placed immediately next to the footing. And, of course, the ground ring around the tower must be bonded back to wherever the equipment is located.

Some helpful links:

www.esgroundingsolutions.com/about-electrical-grounding/grounding-electrode-sphere-of-influence.php; www.esgroundingsolutions.com/about-electrical-grounding/what-are-some-different-types-of-grounding-electrodes.php.

The main trouble with all of these regulations, codes, and standards is that none of them take into account local soil conditions, nor do they analyze the specific setup for your tower. All of the standards simply provides a cookie-cutter solution that they think will solve the majority of the problems in most cases. Unfortunately, the cookie-cutter solution seldom resolves all the issues and can leave your project exposed to future problems if it is not properly analyzed.

The GPD issue, in particular, is important to analyze and understand for simple daily operation of your electronic equipment. A GPD that is too high will cause major problems with signaling and data transfer rates between the tower and the electronic equipment. Another issue of concern is the resistance to ground (or impedance to ground) of the grounding system, which if not properly analyzed, can limit signal range in certain cases.

How far is the run from the tower to the equipment room? What is the voltage drop across the conductors? How much current will leak into the surrounding soil before a surge of electrical energy hits the equipment shelter? What is the resistance to ground of the system? What is the potential difference (GPD) between the tower and the equipment shelter? Will personnel be safe should they be standing near the tower or touching it during an electrical fault from the utility company or a lightning strike? How much current will be on any one grounding conductor, and can it handle it without critically burning open? What electrical noise/harmonics/electromagnetic fields are likely to be encountered and can the grounding system handle it?

The Question: We are having problems in our tank farm. We have found that the tanks have a grounding grid around them, but they are not tied together. Is there a problem with step potential when the systems are not tied together?

The Answer: There are a number of issues that can occur if your ground systems are not bonded together, step potentials being just one of them. The biggest issue is that unbounded and independent ground systems such as you are describing are illegal under the NEC. Article 250.54 specifically states that the earth may not be used as the sole equipment-grounding conductor. The primary reason for this is that OCPDs such as circuit breakers and fuses may not function properly should a fault or short circuit occur at the tank. This is your biggest human safety concern.

Others issues from corrosion, to GPDs, static, pipeline stress voltages, current distribution, lightning strikes, longitudinal current flow, leakage current rates, magnetic and electric fields, harmonics, and more may all be negatively impacted by having independent ground systems. If you are handling flammable liquids, you will have even more adverse issues. Let alone the signaling problems you will have with data lines and control wires.

Adding the additional grounding can be quite expensive and sometimes hard to justify to the powers that be. A good grounding design using computer modeling to prove the engineering points can be very useful in justifying the expense.

Also, you may want to consider conducting some point-to-point direct current (DC) resistance checks around your tank farm, back to some central control point. The data may be very revealing as to just how good/bad your situation is. If you see DC resistances of greater than about 0.5 Ω, you may have a problem. A well-grounded tank farm will have point-to-point resistances of less than 0.1 Ω (some increase may be expected for very large systems).

The Question: What is the advantage of isolating power supply's return from the unit and grounding the signal shields at the unit enclosure ground?

The Answer: The power supply's' return path is commonly called the "neutral" conductor and is often a white wire. While this wire must in fact be grounded at the first service disconnect, it may not be grounded at any other point in the system. This is because the neutral wire actually carries as much current, if not more, than the hot wires carry.

Consider a GFCI; it works by detecting the difference between the hot and the neutral wires. Any difference greater than a few milliamps results in the opening of the breaker. The GFCI breaker is a great reminder to people that the neutral wire does in fact carry as much current as the hot wires. This is why the return path (neutral) of the power supply is said to be "current carrying" or a "current-carrying grounded conductor."

The shield wires are to be bonded to a non-current-carrying grounded conductor (the green ground wire) in order to provide stray currents, transients, and harmonics a path to the earth. If you were to tie the shield wires of your data lines to the neutral, you would actually be forcing electrical energy onto the lines. This is not only a violation of the NEC but could damage the equipment tied to your data lines. In conclusion, you must ensure that the power supply's return path is isolated from all other metallic objects, that you provide an effective non-current-carrying ground path to the power supply, and that you ground your data lines in accordance with the manufacturer's instructions. We recommend that you consult an electrician as to proper neutral- and ground-wire routing from the electrical panel to your power supply.

The Question: I am designing a ground grid for a new 115-kV substation. The civil design is planning on utilizing steel piles for structural support. We would like to utilize the piles as ground rods if possible, but we are concerned with corrosion issues, because the grid is going to be copper. To avoid galvanic corrosion issues, will we not need to electrically isolate the steel piles from the copper grid?

The Answer: Galvanic corrosion is an electrochemical process in which one metal corrodes preferentially to another when both metals are in electrical contact and immersed in an electrolyte. Dissimilar metals and alloys have different electrode potentials and when two or more come into contact in an electrolyte, a galvanic coupling is set up. Scientists have indexed materials based on the nobilities of the metals. This index is called the *anodic index* and provides a voltage for each material. A voltage differential greater than 0.15 V is considered corrosive. Copper has a anodic index of 0.35; steel has a anodic index of 0.85. This is a 0.50-V differential and can certainly be a source of galvanic corrosion, as the earth will surely act as an electrolyte. But there is another concern with regard to corrosion beyond simple nobility issues, and that is induced alternating current (AC).

Galvanic corrosion in steel objects is generally enhanced by the addition of harmful AC voltages. In the case of an electrical substation, the electromagnetic fields are far greater and thus will induce greater voltages into any steel object nearby. These induced AC voltages cannot be avoided in a substation and are generally more corrosive than the DC electrochemical voltages present during normal galvanic corrosion. What we can do is provide an effective means for the electric fields to dissipate into the earth. A properly grounded steel object in a high-voltage environment should suffer from less galvanic corrosion than an unbounded steel object, as the copper ground grid will provide an alternative and better path to earth for the corrosive AC.

Please remember that it is quite common to see all sorts of steel objects bonded to the copper ground grid in a substation—the steel transformers, steel fence, and steel uprights, just to name a few. Exothermic welding is the corrosion control choice when bonding two dissimilar metals. The exothermic welding process provides a permanent molecular bond, which is inherently corrosion resistant (but may not address the long-term galvanic corrosion issues). Most manufacturers of exothermic welding products guarantee the steel-to-copper bond to be corrosion free.

That said, you really do not have a choice: regardless of any galvanic corrosion concerns, mandatory human safety laws require you to bond the steel pilings into the ground grid. You may NOT isolate any metallic object in a high-voltage environment. Please refer to 29 CFR 1910.269.

Isolated steel pilings in a high-voltage environment could generate significant differences in potential that could have serious human safety implications. Not only could these differences in potential increase arc-flash hazards, but they could generate life-threatening step-and-touch voltages. These isolated steel piles could also be the source of significant electrical interference caused by the differing electromagnetic fields.

Another advantage of grounding your steel columns is that DC rectifiers can be used to inject protective DC currents onto the steel objects via the copper grounding system. This is called *impressed cathodic protection* and uses DC currents to counteract the harmful AC currents. You should contact a corrosion specialist for more information regarding DC rectifiers and corrosion protection.

We have an excellent blog regarding deep-earth pilings and grounding that may answer a few more questions for you. Please see our blog:

www.esgroundingsolutions.com/blog/electrical-grounding/is-it-possible-to-provide-electrical-grounding-utilzing-planned-structural-deep-earth-shafts.

In conclusion, you must bond the steel pilings to the ground grid, and corrosion will not be a problem with proper exothermic welds.

The Question: Should the earthing system of programmable logic controller (PLC) or communication rooms in high-voltage/transmission substations be electrically separated from other parts of the substation? Please reference the standards.

The Answer: We see in your original question that you would like us to reference which standards require that PLC and communication equipment at substations be bonded together. We are tempted to say "all of them," but we will start with simply listing just a few: 29 CFR 1910.269; NFPA 70 NEC, IEEE Std. 80-2000, Guide for Safety in AC Substation Grounding, IEEE Std. 487-2000, Recommended Practice for the Protection of Wireline Communication Facilities Serving Electric Supply Locations; and IEEE Std. 1590-2003, Recommended Practice for the Electrical Protection of Optical Fiber Communication Facilities Serving, or Connected to, Electrical Supply Locations. And this is to name only a few. Clearly, every single possible standard one could ever reference will mandate that not only are PLCs and communication equipment bonded to the same ground source at substations, but that equipment of ANY kind is bonded to the same ground source, whether it is high-voltage or not, but ESPECIALLY at a substation.

Any equipment, PLCs and communication equipment included, not bonded to the same ground source will present a difference in potential. This is a bad scenario, even for low-voltage conditions, but in high-voltage conditions it could be lethal. Differences in potential from one piece of equipment to another will allow hazardous voltages to form, which can result in injury or death from shock. It is also very bad for the equipment itself.

Most people think that tying their sensitive electronic equipment into the same ground grid that handles high-voltage electrical faults could cause damage to their gear. But nothing could be further from the truth. Installing sensitive gear with a different ground source will actually allow destructive voltages to form, due to the difference in potential.

Think of a bird landing on a high-voltage wire. Why does not the bird get electrocuted? The reason is that the entire bird is at the same potential as the high-voltage wire. Now, if part of the bird were to be at a different potential, then current would flow, and the bird would be killed. The same is true for your electronics. As long as your electronic equipment rises in voltage along with everything else around it, there will be no voltage differential, and therefore no current will flow, and your gear will be safe.

The Question: How do you ground a substation on the fifteenth floor of a high-rise building?

The Answer: Grounding is always a difficult subject, but having a substation on the fifteenth floor definitely presents some extra problems. Obviously, all 29 CFR 1910.269 and IEEE-80 standards must be applied to the fifteenth floor to insure proper human safety in regard to hazardous step-and-touch voltages.

Here is some information regarding step-and-touch voltage hazards:

www.esgroundingsolutions.com/about-electrical-grounding/what-is-step-and-touch-potential-and-resistance-to-ground.php.

To get a good earth connection, the best method will be to run dedicated down conductors from the fifteenth floor to grade level. The number of down conductors should be based on the shape of your building and the short-circuit line-to-ground fault your site will experience. A good guideline would be to have the down conductors meet lightning protection standards such as the **NFPA 780**. Once the down conductors reach grade level, a buried ring surrounding the building and tying all the down conductors together is best.

Of course, standard 10-ft ground rods at 20-ft intervals bonded to the perimeter ring are ideal. Based on the local soil conditions, you may want to consider adding some deep ground wells to lower the overall resistance to ground of the system, which will reduce the GPR and lower the hazardous step-and-touch voltages you will experience during a fault. Remember that proper bonding of the conduits routing the down conductors to the earth will be critical. Improperly bonded metallic conduit can actually prevent current flow, which can result in equipment failure, fire, serious injury, or even death.

You will also need to bond this system into building steel. Ideally, the grounding system will help to protect the actual structure from foundational damage caused by electrical faults. Special attention should be taken to ensure that these important bonds are not only securely made, but made in such a manner as to route electrical faults down the grounding system and not into the building structure itself.

Lightning is an additional concern for these structures, and you will need to consider these effects as well. The large mass of metal on top of the building will make it an exceptional target for lightning strikes.

Because your site is on top of a building, it will be vital that you model this system using a computer simulation. The simulation will ensure that the fault currents will not exceed the capacity of the down conductors to handle the current, preventing accidental burn-outs of the conductor. Additionally, the step-and-touch hazards will need to be carefully analyzed. Only a computer can accurately demonstrate what will be needed at your site. We use and recommend the CDEGS computer-simulation software.

The Question: How do you meet a grounding system requirement for 0.5-Ω earth resistance in Thailand?

The Answer: A 0.5-Ω grounding system is a very difficult to achieve standard. Typically, a 1-Ω grounding system is considered difficult. The reason for this is that the closer you get to 0 Ω (an impossible number), the more exponentially difficult it becomes. In this regard, a 0.5-Ω grounding system is by at least an order of magnitude more difficult to achieve than a 1.0-Ω grounding system.

Generally speaking, systems with extremely low resistance-to-ground (earth resistance) readings have additional design requirements beyond simple earth resistance. Often, they are massive grids designed for human safety by reducing step-and-touch voltage hazards. Would this be your case?

Here is some information regarding the design of grounding system:

www.esgroundingsolutions.com/about-electrical-grounding /how-to-do-electrical-grounding-system-design.php.

Here is some information regarding the testing of grounding systems:

www.esgroundingsolutions.com/about-electrical-grounding /how-to-do-electrical-grounding-system-testing.php.

Here is some information regarding step-and-touch voltage hazards:

www.esgroundingsolutions.com/about-electrical-grounding /what-is-step-and-touch-potential-and-resistance-to-ground.php.

The Question: How can I measure the earth resistance of a telecommunication site with an antenna tower beside it?

The Answer: The test you are referring to is for measuring the resistance to ground of the telecommunication site's grounding system. This determines the effectiveness of the grounding system in the soil located at the site. You are correct in being concerned about the nearby tower. Objects that are connected to the ground grid, especially when dealing with three-point test method, can cause erroneous measurement results.

Even an object not directly connected to the grid, only near it, can result in measurements that are negatively impacted. And often, soil resistivity testing is needed in addition to the resistance-to-ground test to accurately determine the final earth resistance of a grounding system.

There are two accepted test methods for this test: the three-point fall-of-potential test method and the induced-frequency clamp-on test method. Both tests require a signal to be injected into the grounding electrode system with a return source that forces the signal to propagate through the earth (soil). The primary difference between the two test methods is the signal return source. The three-point test uses a probe placed into the earth at a specified distance relative to the size of the grid, the clamp-on test uses the utility company's ground grid as the reference (return source). More information can be found here:

www.esgroundingsolutions.com/about-electrical-grounding/how-to-do-electrical-grounding-system-testing.php.

Whichever test you choose to measure the earth resistance of your site, it is important to note that both tests can easily be done incorrectly. The most common problem with the measurement of ground grid resistance to earth is poorly trained operators. Also, it is quite possible that the way your grounding system was physically wired will prevent either test from being used at your site. In that case, only computer modeling is available for calculating the earth resistance of your grounding system.

The Question: For a data center, how can you properly ground low-voltage and medium-voltage systems in the same room?

The Answer: The primary concern with intermixing grounding systems between low-voltage and medium-voltage (or high-voltage) systems, is that the medium-voltage systems can generate electrical noise, transients, and harmonics that are far more intense than those generated by low-voltage systems, and introducing those intense and interfering signals onto the low-voltage side could cause problems. ANSI-TIA-EIA-J-STD-607-A is currently one of the more prominent grounding and bonding standards in use for data centers, and while it does not specifically call out for the separation of low- and medium-voltage systems, it does ask for some logical separations.

The basics of the ANSI-TIA-EIA-J-STD-607-A standard call for a main telecommunications ground bar (MTGB) to be properly installed in the building with multiple telecommunication ground bars (TGBs) installed in logical order, such as rooms or subsystems. The MTGB and the TGBs are to be bonded together via a looped ground conductor called the *grounding equalizer* (GE) or sometimes called the *telecommunications bonding backbone*.

There are many rules and requirements regarding how these ground bars are bonded, what they are bonded too, and how the GE is routed and sized. This includes bonding to a grounding (earthing) electrode, building steel, water main, and the main electrical panel; all of which are tied to the MTGB. You should refer to the ANSI-TIA-EIA-J-STD-607-A for more details.

It is important to note that there are many standards that may apply to your situation, which will add additional requirements for your system. Often, these requirements include having a dedicated 5-Ω or less grounding (earthing) electrode, LPS, and human safety studies to protect personnel and equipment from hazardous step-and-touch voltages from backup generators.

It is of course hard to advise without looking at the plans, but for your situation, you could install a low-voltage telecommunications ground bar (LV-TGB) and a medium-voltage telecommunications ground bar (MV-TGB) in the same room (labeled appropriately, of course). These two separate ground bars, the LV-TGB and the MV-TGB, would then simply be bonded to the GE conductor loop.

This would allow you to keep the two systems separated, without having to make massive additional conductor runs back to the MTGB. Of course, it is also possible to run two separate GE loops, one dedicated for low-voltage and the other for medium-voltage, depending on how serious the level of protection needed. Of note, the lower the resistance to ground of the ground (earthing) electrode, the more effective the entire grounding system will perform.

The Question: Can a Faraday cage and ground system be developed for a hospital MRI system experiencing electrical interference?

The Answer: Yes, is quite common to see Faraday cages and grounding systems for MRI units at hospitals. Good computer modeling system can calculate the total magnetic interference that can be expected and/or shielded by the Faraday cage. Additionally, these computer modeling software systems can help design a dedicated grounding electrode to achieve a 1-Ω ground source.

MRI units function better and provide clearer images when they have a grounding electrode with a low impedance to ground, typically under 1 ohm. You should also. Also, we can help you track down the sources of the interference using we have a complete collection of power quality meters and an experience to find electricians electrical wiring problems that may be compounding your issues.

The Question: What grounding standards are there for the installation of high-tech equipment, particularly audiovisual gear?

The Answer: There are many different standards that may apply to your industry; unfortunately, we are not aware of any standards that are specifically tailored to meet your needs. Often, when installing sophisticated and sensitive electronics systems, particularly those in the audiovisual field, electrical-noise, transients, and harmonics can adversely impact both the performance and the quality (poor sound/visual) of the equipment.

In recording rooms, 60-Hz hum is a big concern, and proper grounding is the answer. To solve these problems often requires a low resistance-grounding electrode and an isolated noise-free ground path. The trick, of course, is how to install these systems in existing buildings while keeping costs down.

Writing a specification for construction that is actually usable and does not require the contractor to dig through hundreds, if not thousands, of pages of standards is a task in its own right. While one could easily list a series of grounding standards (IEEE Std-142, IEEE Std-1100, NFPA 70, ANSI-TIA/EIA-STD-J-607-A, Motorola R56, etc.) and just hope the contractor can pick out the applicable pieces for your project, it is generally far better to provide the general contractor (GC) with an actual task list that can be quoted. The truth is, no contractor has the time to dig through all these standards and figure them all out. It would literally take weeks and would probably only result in a no-bid situation.

We recommend for a simple specification standard for your installations that your GC hire an engineering firm specializing in electrical grounding to provide a work plan for the following:

1. The installation of a dedicated 5-Ω or less resistance-to-ground electrode. This will require Wenner four-point testing and a computer model.
2. A noise-free ground path from the newly installed equipment back to the electrode.
3. A ground path from the new isolated ground system back to the first service disconnect, as required under the NEC.
4. An analysis of the existing electrical system for problems and power-quality issues and recommendations for isolation transformers and/or uninterruptible power supply (UPS) systems, as needed, to protect the new equipment. This will require the monitoring of the incoming utility power with quality analyzers.
5. Recommendations for the proper grounding of the computer and server systems associated with the new equipment. ANSI-TIA/EIA-STD-J-607-A will be a guideline document.

6. An analysis of the building infrastructure systems to see what other factors may impact the new installations (improper neutral-to-ground bonds, improper water-pipe connections, Edison circuits, LPS installed too close, vapor barriers under the concrete, delta transformers or high-voltage systems nearby, etc.).

While it would be nice to simply have a single standard to hand to a GC, there are simply too many factors for any single standard. This is why 29 CFR 1910.269 Appendix C, takes the same approach and mandates that engineering firms specializing in electrical grounding be contracted when dealing with human safety issues such as step-and-touch voltages. The issues are simply too complex and must be analyzed professionally.

The Question: How do you properly ground a portable electric generator (genset)?

The Answer: Any generator, whether it is portable or fixed in place, must be grounded per the NEC. Specifically, NEC Art. 702.10 requires that (A) separately derived systems must be grounded in accordance with NEC Art. 250.30, and (B) non-separately derived systems must be bonded to the system grounding electrode. The NEC identifies two methods of grounding:

1. For separately derived systems, the genset must have its own grounding electrode conductor.
2. For non-separately derived systems, the genset must have a grounding electrode conductor from the genset to the automatic transfer switch (ATS), which is in turn connected to the main service (panel board) grounding electrode conductor.

Method 1, most common for commercial and industrial applications, is called a separately derived system and is defined by NEC Art. 702.10(A), which states that the generator must comply with the normal grounding requirements found in NEC Art. 250.30. This is due to the fact that the transfer switch of a separately derived generator interrupts all conductors, including the grounded circuit conductor. Also, in this type of installation, the neutral is derived from the generator ground, not from the main distribution ground, as it is during normal power. This is why an additional pole is required in the transfer switch (four-pole ATS) for these systems.

Method 2, most common for residential or portable generators, is called a non-separately derived system and is defined by NEC Art. 702.10(B), which states that the equipment-grounding (bonding) conductor must be bonded to the existing main electrical panel grounded electrode system. This is why only a three-pole ATS is required.

Also of note is that posted signage is required under NEC Arts. 700.8(B), 701.9(B), and 702.8(B) identifying all emergency and normal sources of power connected at that location, including cases where the grounding conductor from an emergency genset is connected to a remote grounding source.

The Question: How is sphere of influence calculated for a horizontally run grounding system such as a ground ring or concrete-encased conductor?

The Answer: The sphere of influence is an electrical theory that claims that a grounding electrode will dissipate electrical energy into a volume of soil equal to the length of the electrode in all directions.

It is fairly easy to imagine a 10-ft driven ground rod placed vertically into the earth. The sphere of influence of this rod would have a radius extending 10 ft out in all directions (20-ft diameter), including down into the earth. It is important to understand that the electrode is also trying to dissipate the electrical energy up into the air, but the strong dielectric properties of air typically prevent such action. The formula in this case is: volume = $(5 \times \pi \times \text{length}^3/3)$. If we cross-cancel π and 3, we get the simple equation of $V = 5L^3$ cubed. Thus, the sphere of influence for a 10-ft driven rod is 5000 ft^3 of soil.

A similar formula can be used for calculating the sphere of influence of each of the individual conductor segments of your ring. The key issue being to remember that this is a volumetric equation designed to figure out how many cubic feet of soil is being utilized by your ground system. With horizontal conductors, you will need to remove the volume of the sphere of influence that extends into the air.

But is sphere of influence really your question, or are you wanting to calculate resistance to ground? This will require an understanding of the soil resistivity at your site, and the calculations are far more complex. We recommend the Handbook of Electrical Power Calculations, 2nd ed, Section 14. It has a complete list of equations for uniform (single-layer) soil resistivity models. Unfortunately, uniform soil does not really exist, and multilayer calculations require a computer. Throw in impedance issues caused by AC power, and the complexity skyrockets.

The Question: In a correctly grounded system, should I expect circuit current on the equipment ground conductor (EGC)?

The Answer: In a properly grounded system you should not see electrical currents higher than a few milliamps on the EGC, certainly no greater than 0.5 A. Even in very large grounding systems with circulating currents, it is rare to see currents exceeding these levels (when they do occur, it indicates an inadequate grounding electrode system). That said, there are cases where higher currents might be found, such as in a high-voltage environment. For example, transmission towers carrying 66- to 500-kV power lines often have a steady-state GPR due to the magnetic fields of the lines, resulting in induced currents that often exceed 5 A. This is considered normal in these situations. However, if you are simply talking about an EGC on a 480-V (wye) or less system, you should see very minimal currents (a few milliamps) on the EGC.

If you do have higher currents, it typically indicates an improper neutral-to-ground connection somewhere in the system. Troubleshooting these issues can be difficult, as the problem could occur at the transformer's XO connection, or it could be at a subpanel, or at any number of other points in the system. It could also indicate a fault and a failed OCPD. If your system is three-phase delta, stray current on the EGC could indicate a serious fault is occurring, and should be checked out by a qualified electrician right away.

Another issue to be cautious of is improperly installed isolated grounding electrodes. In certain situations, isolated grounding electrodes installed in violation of **NEC** Art. 250.58 will be at a different potential than the rest of the grounding system and can have unwanted currents and cause OCPDs to fail.

The Question: What is proper grounding for a hazardous goods storage area?

The Answer: How does one properly ground a hazardous material storeroom that has two lights? My work plan is as follows:

1. Mount 20-mm steel conduit horizontally onto the room's concrete wall.
2. Run an earthing wire from the room lights and mounted onto the steel conduit.
3. Mount an earthing wire with an alligator clip onto the steel conduit for earthing each of the drums when they are in use.

The Answer: Properly bonding metallic components in hazardous storage areas is a vital requirement to ensure proper safety. Differences in potential and static electricity can cause sparks and/or hazardous currents to flow through combustible materials, resulting in fire and/or explosion. It sounds like your project has requirements to eliminate these hazards, which your plan does not adequately address.

When discussing static control, it is important that one understand the difference between bonding and grounding (also known as earthing), as you will need to do both in order to provide proper protection. Our concern is that simply pulling a ground wire from the existing lights is probably inadequate and could quite possibly have negative consequences.

If some component on the same circuit as the lights were to fault, you could end up forcing hazardous current into your combustible storage room. That would be very bad indeed.

As engineers, we become very concerned about answering questions along these lines, as there are simply too many issues to handle via a simple written exchange. You should really hire a good engineering firm specializing in grounding and earthing to help you with these critical issues. Ensuring that you have adequate and effective grounding, that you do not have isolated grounding, and that you have proper bonding between components is trickier than it you may think. Also, providing a low resistance-to-ground electrode generally requires on-site testing and computer modeling.

The Question: What is the proper grounding methodology for cellular station?

The Answer: Proper grounding at cellular stations is quite complex and detailed. There are many requirements for extensive bonding and a typical cellular station. In general the requirements are to bond virtually every single piece of metal to a dedicated equipotential grounding system that is in turn bonded to an earthing system.

We recommend that you obtain a copy of Motorola's R56 standard. It is far and away the single best guide for the proper installation of a cellular station.

The earthing requirements for the typical cellular station generally require a buried counterpoise (ring) system around the station's equipment pads and fence and should have a resistance to ground of less than 5 Ω.

The purpose of the earthing system at a cellular station is the same as at any transmitting antenna, to remove the energy that has been induced into the station by its own transmissions.

Transmitting antennas, including those found at cellular stations, will induce unwanted electrical energies into the base station. This energy will quickly build up, causing electrical noise and harmonics that will compromise the signal quality and damage equipment.

The above-grade bonding system will ensure that there are no differences in potential within the station and will allow the stray currents to move to the earthing system, which in turn provides a path to earth, where the harmonics can pass harmlessly out of the system. The better the grounding is, the quicker and more effective the cellular station will be.

Achieving 5 Ω or less for your earthing system is all about understanding the makeup of the soil at your site. The more resistive the soil is, the more extensive the earthing system will need to be. Here is a link that discusses this topic more thoroughly:

www.esgroundingsolutions.com/about-electrical-grounding /what-is-soil-resistivity-testing.php.

ELECTRICAL STANDARDS

Can insulated copper conductor be used for a ground grid?

The Question: Our client is asking for an insulated copper conductor instead of bare copper for the ground grid at a depth of 2 ft from the grade level, due to corrosion issues. We are trying to model and analyze the grounding system using SKM Ground Grid software, which does not differentiate between insulated and bare conductors. Per IEEE Std. 80-2000, grounding grids must be bare copper conductor (refer to clauses 3.16 and 9.4.c of IEEE Std. 80-2000). Please let us know if insulated copper conductor can be issued for ground grid.

The Answer: No. You may not use insulated wire for your below-grade ground system. This would be a clear violation of every imaginable code, standard, and regulation, including the NEC (Art. 250.52(4) and numerous others) and the Canadian Electrical Code.

Why even put the grid in the earth if you are going to use insulated wire?

There are a number of things that can be done with respect to soil corrosion. First of all, you should compare your soil resistivity numbers against a geological corrosivity rating chart. This will tell you the risks that are involved. Second, you can use a light coating of bentonite clay (driller's mud) around the conductors when you install the grid. When the trench is open, simply spray a light coat of this natural earth clay on the copper conductors, and that will help to protect the copper. Third, make sure that you compare all materials against the anodic index, so you do not accidentally get dissimilar metals together. Fourth, you can always install a DC rectifier system to reduce corrosion rates. Fifth, you may want to consider using electrolytic grounding rods for your earth pits. And, of course, a properly computer-modeled grounding grid can insure that your ground system does not suffer from electrical overstress that can rapidly increase corrosion rates.

Here is a link that may also be helpful to you:

www.esgroundingsolutions.com/blog/761/our-reading-of-earth-resistivity-is-coming-avarage-2800-Ω-meter-we-are-designing-1133-kv-co-generation-substation-what-are-the-next-steps-required-for-designing-earth-mat-for-70-x-33-meter-area.

The Question: I am designing a ground ring for an industrial facility. We typically require the ground ring to be 5 feet from the building, but I cannot find anything in the NEC for a building distance requirement. Do you know if there is such a requirement, and if so, how far?

The Answer: The 2011 NEC does not have a required distance that a ground ring must be placed from a building. Currently, ground rings are only required to be in direct contact with the earth for at least 20 ft and to be at least 2 AWG bare copper or larger (NEC Art. 250.52(A)(4)) and buried at least 30 in. below grade (NEC Art. 250.53(F)). Now, there are other standards that do have requirements for distance in regard to the installation of ground rings. In particular, Motorola's R56 Standard, Chapter 4, Part 4.4.1.6, lists the following rules, in summary:

- Ground rings shall be installed in direct contact with the earth at a depth of 30 in. below grade, or below the frost line, whichever is deeper (ANSI T1.334-2002, Section 5.3.1; NFPA 70-2005, NEC Art. 250.53).
- Building ground rings shall be installed at least 3 ft from the building foundation and should be installed beyond the drip line of the roof (MIL-HDBK-419A; MIL-STD-188-124B).
- Tower ground rings shall be installed at least 2 ft from the tower foundation (ANSI T1.334-2002, Section 5.3.1).

In some cases, ground rings are installed for human safety reasons, specifically to reduce step-and-touch voltage hazards. When this is the case, it is often advantageous to install the ground ring 1 m (3 ft) from the building.

One meter is considered to be the reaching distance of a person and helps to balance the voltage differential between the hands and the feet. It should be noted that the IEEE Std. 80-2000, Guide for Safety in AC Substation Grounding, recommends in 9.4(c) a depth of 12–18 in. Shallower depths are important when trying to reduce touch voltage levels, whereas greater depths tend to help reduce step voltages.

Also, 10-ft-long ground rods should be placed at 20-ft intervals along the ground ring (MIL-HDBK-419A). Eight-foot ground rods can be used, but are discouraged in ANSI T1.334-2002, for good reason; they have half the sphere of influence of a 10-ft ground rod. Please see the following link for more information:

www.esgroundingsolutions.com/about-electrical-grounding/grounding-electrode-sphere-of-influence.php.

The Question: What is the type of earthing at shore-side electrical equipment? Also, please tell us the standard followed.

The Answer: Grounding along the coast near the ocean is no different from any other grounding. You should follow all the same rules and regulations as you would for the installation at any other location. However, you should take care, as the saltwater in the air near oceans can be very corrosive to certain grounding electrodes. In particular, galvanized ground rods tend not to do as well as copper rods when salt is present. You may wish to consider using an electrolytic electrode with bentonite clay, as it will be very resistant to corrosion of all sorts.

The Question: I have a copper lightning ground that lands on my copper bus that is used to ground my electrical systems. Shouldn't these systems be separated? Also, can I legally use a tin-plated crimped connectors to land conductors on to my copper grounding bus?

The Answer: Every known standard from the NFPA 70 NEC to the NFPA 780 Lightning Protection Code to IEEE standards and more require that every single system within the facility be bonded to a common ground source. This means that the LPS and the electrical grounding system must be bonded together at some point.

However, we must remember that there are several systems in play for which we must account. The NEC has a requirement for a dedicated low-impedance fault-current path back to the first service disconnect to ensure that the OCPDs (such as circuit breakers and fuses) will function properly.

This requirement is separate and independent from the requirements of the NFPA lightning Protection system. Using the LPS for this path is not allowed.

Articles 250.60 and 250.106 are the only two sections in the NEC that discuss grounding and LPSs, which is not much. The thing to remember is that these are separate systems governed under separate codes. While the LPS must be bonded to the electrical system ground at the main service disconnect, you do not want to use it as part of the electrical system, mostly to protect the electrical system itself from the lightning.

The LPS is designed to provide a direct path to earth for lightning strikes to travel. If your switchgear is in that path, then it will become damaged during a strike. Bottom line, you must provide a ground-fault path back to the first service disconnect that is separate from the LPS.

Also, you should avoid using dissimilar metals when bonding. Tin and copper will tend to corrode; you would be much better served if you were to use brass fittings when bonding to copper. Here is a link to a post that discusses corrosion issues:

> www.esgroundingsolutions.com/blog/407/i-am-designing-a-ground-grid-for-a-new-115kv-substation-the-civil-design-is-planning-on-utilizing-steel-piles-for-strucutral-support-we-would-like-to-utilize-the-piles-as-ground-rods-if-possible-bu.

The Question: Which article in NEC references swimming pool grounding?

The Answer: The NEC can be a bit confusing when it comes to the grounding requirements for swimming pools. There are two sections that you must pay attention to inside the NEC, specifically Arts. 250 and 680. First of all, it is important to understand the basic concept the NEC is trying to get across when it comes to swimming pool grounding, and that is equipotential bonding.

Equipotential bonding is composed of three basic components:

1. Making a connection to the earth, properly known as earthing; however, also known as grounding. This leads to much confusion, because in the United States we use one word, grounding, for both above-grade and below-grade grounds. Earthing refers to only the below-grade portion of grounding.
2. The bonding of all metallic objects, so they are at the same potential
3. Providing a low-impedance path for fault currents. The intent of a low-impedance path is to provide a direct copper path, not an indirect steel path with lots of mechanical joints. This is an additional ground path required especially for pools.

Equipotential bonding is covered under NEC Art. 680.26. We recommend the *NEC Handbook*, as it contains a number of really excellent full-color illustrations that will answer most of your questions for you. Keep in mind that the low-impedance ground path mentioned in Art. 680.26 is in ADDITION to the low-impedance path required for OCPD under Art. 250.4(A)(3). Also, you should pay special attention to the sizing of ground conductors when dealing with swimming pools, in particular NEC Arts. 250.122 and 680.26(B).

You, of course, must also purchase junction boxes that are compliant with Arts. 250.8 and 680.24, provide GFCIs, provide an equipotential bonding plane for all paved walking surfaces within 3 ft of the pool [NEC Art. 680.26(B(2))], and bond underwater lights in accordance with NEC Art. 680.23(F).

Ultimately, you are going to need to add a lot of additional copper conductors in order to bond the steel rebar in the swimming pool, the metal fences near the pool, the diving board, metal handrails, pumps, heaters, lights, and any other metallic object related to the pool or within range of the pool. And, of course, the pool grounding must be bonded back to the building's common grounding grid, which will include the utility grounding system, water pipe, building steel, gas pipe, Telco, lightning protection, and so on.

If you are building a public swimming pool and are near any form of high-voltage source, such as a delta transformer fed from the utility with an unknown OCPD, overhead power lines, or a nearby substation, you may be required under federal and/or local laws to conduct a human safety study to ensure that hazardous step-and-touch voltages will not form that could injure the occupants of the pool.

The Question: Is there a minimum distance for the down conductor in a Lightning Protection System (LPS) with regard to flammable piping (natural gas, lpg gas)?

The Answer: In the United States, we use the NFPA 780 Standard for the Installation of Lightning Protection Systems. Annex C of this standard specifically addresses the issue of bonding distances and the reduction of potential difference within the LPS.

It is a fairly complex section that requires some knowledge of magnetic fields and the principles of induction to fully understand, and it would take a lengthy essay to discuss the impact of this section in full. If you are in Europe or are using international standards, BS EN/IEC 62305 standards will have additional requirements with regard to the protection of flammable pipelines above and beyond the requirements found in NFPA 780.

Unfortunately, when it comes to the protection of flammable gases and other hazards, it becomes very difficult for us to legally provide any advice other than to get professional assistance. If your Lightning Protections System (LPS) is not properly bonded to the gas pipelines, you could generate a lethal situation that could destroy life and property.

We can tell you that the gas pipeline must be bonded to the LPS and that this must be done in such a way as to reduce the difference in potential between the metal components, without using the gas lines as an electrode or as a lightning conductor. You should coordinate any connections from your flammable pipelines to the LPS with your local gas company provider.

Do I need down conductors to be bonded to the rebar in concrete column?

The Question (responded from an earlier comment): Thank you very much for your earlier recommendation & expertise. I checked ANSI-TIA-EIA-J-STD-607-A and it says "This standard does not require bonding of the steel bars of a reinforced concrete building. But for structural steel, metal frame, the telecommunication grounding shall be bonded to them."

Going back to the lightning protection down conductor, NFPA 780 *says "down conductors coursed on or in reinforced concrete columns or on structural steel column shall be connected to the reinforcing steel or the structural steel member at their upper and lower extremities." If I follow your recommendation, then do I need down conductors to be bonded to the rebar in concrete column?*

The Answer: You are correct in regard to the 607-A standard. However, E&S Grounding Solutions always recommends that the steel columns and the rebar in the foundations be bonded together in ALL cases, regardless of lightning protection issues. In fact, NEC mandates this practice. The best source of information is the *NEC Handbook.* Article 250 has numerous illustrations showing this mandatory bonding: Exhibits 250.22, 250.23, 250.29, and 250.31, to be specific.

In regard to your question "Do I need down conductors to be bonded to the rebar in concrete column?," the answer depends a lot on how your building is constructed and arc-flash calculations. But for a typical building, the lightning down conductors are generally only bonded to clearly exposed metal pieces that are immediately adjacent to the down conductors. You would need to get an analysis completed to determine the arc-flash differences between the down conductors of the LPS and the steel columns.

To summarize, you should have copper down conductors for your LPS, and you should have the steel columns bonded to the rebar in the concrete.

The Question: Is it possible to consider using the telecom tower itself as a conductive metal part (down conductor), and ignore using popular down conductor (converter), and still have a reliable protection against lightning?

The Answer: The short answer is *yes*; in certain cases, you can use the steel tower legs as part of the down conductor system for an LPS, based on the height and construction of the tower and given that the metal is thick enough. Please refer to U.S. lightning standard, NFPA 780 Standard for the Installation of Lightning Protection Systems, Arts. 4.21.2.4 and 4.21.2.5 (and some other articles as well), as there are some fairly complicated formulas required to determine if your tower is eligible.

If you are using the European standard, BS EN/IEC 62305, you will have many more requirements, including impedance checks and arc-flash verifications, to name only a few. Whether you install a copper down conductor or not, you still must install grounding electrodes and bond them to the base of the tower, in accordance with Chapters 4 and 7.

This is primarily to protect the concrete foundations from being overheated when the energy from the lightning flows through them. Overheated concrete can crack when the water inherent in its chemistry begins to boil and expand. Another issue you should consider is the electromagnetic fields that are generated on a metal tower when it is struck by lightning.

Copper down conductors reduce these fields, thereby helping to protect sensitive electronic systems. Of course, there are serious human safety implications when choosing to use the steel structure as a down conductor. Steel towers that do not have copper down conductors are far more dangerous to personnel who may be working in or around the tower, as the step-and-touch voltage hazards will be far greater.

It is the official position of E&S Grounding Solutions to always recommend copper down conductors for LPSs on all structures, regardless of whether the Code allows it or not. Forcing all of the energy from a lightning strike into the structure you are trying to protect is counterproductive. After all, the goal of an LPS is to protect the structure, the equipment, and the personnel associated with the tower from the effects of both direct and indirect lightning strikes. Copper does this better than steel.

The Question: I am going to place two grounding rods, what is the distance away from the electrical panel required by the NEC? And what is the distance they should be from each other (the rods)?

The Answer: Article 250 does not have a requirement for how far the grounding rod needs to be placed in relation to the electrical panel. Generally speaking, best practices would indicate that the closer you install the rod to the panel, the better. Now, if you are installing the grounding rod really far away, the NEC may require you to increase the conductor size to ensure that you do not have an impedance issue due to the length of the conductor.

That said, it is becoming quite common for local governing authorities (building inspectors) to issue rules requiring the installation of grounding rods within some specified distance from the panel. We think this is a good idea for a number of reasons. While it may be difficult (or unsightly) to install a ground rod near the panel, as you may need to core through some concrete, the advantages of proximal grounding outweigh the effort. Remember, if you do core through concrete, you must install a PVC sleeve to prevent accidental contact between the rod and any rebar in the slab.

As far as spacing of the grounding electrodes goes, NEC Art. 250.53 (B) requires a minimum of 6 ft between electrodes. However, NEC Art. 250.56 notes that spacings greater than 6 ft will improve the efficiency of the electrode.

The spacing of electrodes deals with a theory known as the sphere of influence. Basically, you want to space your electrodes out two times the length of the electrode. So, if you are installing 10-ft grounding rods, you would want them spaced at 20-ft intervals. Here is some more information on the sphere of influence:

> www.esgroundingsolutions.com/about-electrical-grounding/grounding-electrode-sphere-of-influence.phpwww.esgroundingsolutions.com/blog/electrical-grounding/how-is-sphere-of-influence-calculated-for-a-horizontally-run-grounding-system-such-as-a-ground-ring-or-concrete-encased-conductor.

The Question: By U.S. code, can I insert a device in the path of the protective earth ground conductor?

By U.S. code, can I insert a device in the protective earth ground conductor at the panel of a lightning branch that has:

1. A fault-current AC voltage drop of less than 2 V.
2. A peak current capability of 150 A.
3. Effectively has no inductance or capacitance for fault currents.

The Answer: First of all, when you say "protective earth ground conductor," we are assuming you mean the grounding electrode conductor. In common terms, the grounding electrode conductor is the wire that runs from the ground bar in the electrical cabinet to the top of the ground rod. The goal of this conductor is to provide a low-impedance electrically conductive path designed to carry the anticipated ground-fault currents to the earth connection (electrode). Please let us know if this is not what you meant.

Article 250.64 of the NEC governs the installation of the grounding electrode conductor. This particular article is quite extensive, covering two full pages of text. The primary concerns of the article are with:

1. Proper conductor type (copper or aluminum) and size (8 to 3/0 AWG), so as to be able to handle anticipated faults.
2. Protection from physical damage (conduit and routing).
3. That the grounding electrode conductor is one continuous piece (no breaks).
4. That the grounding electrode conductor is properly bonded to electrodes, building steel, water pipe, conduit, and so on. Article 250.66 and Table 250.66 (see Appendix B) deal with proper sizing of the grounding electrode conductor, which is based on the size of the largest incoming power line.

It sounds as though you want to insert a new device in series with the grounding electrode conductor, and this insertion would "break" the continuous conductor rule. The NEC does allow splicing of the grounding electrode conductor [Art. 250.64 (C)(1)], as long as the splice is made using irreversible compression-type connectors or exothermic welds.

You mention that the device will have a peak current capability of 150 A. This is a concern, in that the grounding electrode conductor must be large enough to handle the short-circuit fault for the panel for the length of the fault clearing time and for the X/R ratio of the fault. Even in your typical home electrical panel rated for only 100 A, the short-circuit fault current can be 10,000 RMS symmetrical amps or more.

For a typical three-phase, 600-A electrical panel you may find for a business, the fault current can easily exceed 60,000 RMS symmetrical amps! If your device is in series with the grounding electrode conductor, it will need to be able to handle the short-circuit current rating of the panel in which it is installed.

In regard to the 2-V drop, we are assuming this is measured using a voltmeter in parallel across the device from grounding electrode conductor to grounding electrode conductor. Was this using a 150-A current level? If so, this would mean that the device has a 0.013-Ω resistance. Is this correct? If so, a 10,000-A short circuit fault would develop a 130-V drop.

Now to the root of your question: Is it legal to install such a device under U.S. code? Assuming that your device will be placed in series with the grounding electrode conductor, is capable of handling the short-circuit current rating, has irreversible connections, and provides a low-impedance path ... You still have an uphill battle. The bottom line is that the NEC is quite clear, you may not impede the grounding electrode conductor connection in any way. However, that does not mean it is impossible to get such a device approved. You would first need to start by gathering as much scientific evidence on your own as is possible. Then you would approach the Underwriters Laboratory (UL) and work with them on a "UL Fact Finding Report."

It will be important to have your own data before you start this process, lest they lead you down a very expensive path. You would also need to get a special UL listing for your device. Once these are done, you would be free to approach the plan-checking departments of individual cities within the United States and install your product with their approval. Meanwhile the UL (and your lawyers) would help you get the NEC amended to allow the use of any product certified under the new UL listing. This will take years, and a lot of money. However, it can be done.

The Question: Are there specific requirements to conduct maintenance on various equipment grounds and building ground grids?

The Answer: Thank you for your excellent question regarding requirements for maintenance of grounding systems. This is of course a fundamental issue governed by numerous standards, the Engineering Code of Ethics (state law), and the Code of Federal Regulations.

At the federal level, we can find maintenance requirements in numerous sections; however, Title 29 of the Code of Federal Regulations, subpart S—Electrical 1910.303, does in fact mandate maintenance of all aspects of electrical systems, including the grounding.

One other section that comes up often is 29 CFR 1910.269(a)(1)(i), which governs maintenance for generation, distribution, and transmission of power, which for a campus of forty buildings would certainly cover your facility. To go over all the aspects of Title 29 would take a career, and rest assured many a lawyer has in fact made a career out of litigating the lack of maintenance of electrical systems.

The other governing code is the NEC. While the Code itself does not indicate any specific maintenance, it does demand that your electrical systems are in compliance at all times, which assumes that you are maintaining them. This is similar to how the ANSI-TIA-EIA-J-STD-607-A "Commercial Building Grounding (Earthing) and Bonding Requirements for Telecommunications" handles its maintenance requirements as well. For these standards, your maintenance procedures are up to you on how to best maintain your systems.

There are some other standards that do specifically call out maintenance requirements for grounding systems:

- Motorola R56, Appendix D: Requires that testing and verification include three-point fall-of-potential and/or clamp-on resistance readings, along with soil resistivity data, to ensure that the ground system is working properly.
- MIL-STD-188-124B Grounding Bonding, Shielding: Requires in 4.1, that personnel be protected from voltage differentials through low-impedance ground paths; in 5.1.1.1.2, an earth resistivity survey; and in 5.1.1.1.7, resistance checks of the ground system at 12 months after installation and then every 21 months thereafter.
- NFPA 780 Standard for the Installation of Lightning Protection Systems, Annex D, Inspection and Maintenance of Lightning Protection Systems: The 2008 version of this standard has an entire annex dedicated to the topic of maintenance. This standard requires documented procedures and record keeping. The maintenance required includes visual inspections for corrosion, damage, and lightning strikes, and measurements of ground resistance of all electrodes, and tightening of clamps. Some groups interpret the clamp-tightening requirement to include the need to measure the actual tightness using a torque meter and to record the results. Some also interpret the Code as requiring DC continuity testing of the system. All of this is typically done on an annual basis.
- Internationally, the European Union standard (United Kingdom and European Union) for lightning protection (BS EN/IEC 62305) has become an industry leader. This standard has very detailed and complex maintenance requirements for its grounding systems, including impedance of steel columns and ensuring that all ground electrodes are at equal resistance when not in a ring formation.

We have designed and been involved in the testing of many facilities and compounds such as yours. We generally recommend some combination of soil resistivity testing, resistance-to-ground testing of the electrode system, computer modeling, and DC continuity testing, along with a visual inspection of course. Generally, we deal with the above-grade lightning systems, computer server room grounding systems, and the below-grade grounding electrode system.

We even will check out the internal wiring of electrical panels to ensure they are properly grounded. Often the DC continuity testing includes not only the LPS and the ground electrode system, but also the electrical cabinets and server room grounding as well. The goal, of course, being to ensure that low-resistance paths exist for these critical components.

The Question: What are the standards for communication equipment grounding and how do you prepare preventive and corrective maintenance checklists?

The Answer: While there are certainly many different standards that may apply, the single best and most-organized source we have found is the Motorola R56 standard. This graphical standard provides a complete methodology for grounding both internally to the equipment and externally to the earth. And the good news is that compliance to this standard covers most if not all other standards. It may even have some checklists for you.

ELECTRICAL WIRING

The Question: Does a diving board on a swimming pool need to be bonded to the equipotential ground system? In an existing commercial pool, we are adding a diving board. The board will be supported by anchor bolts expoxied to the existing deck. Support for the board is concrete; do the rebar and support bolts in this addition require connection to the existing equipotential system?

The Answer: Yes, you must bond all of the rebar and support bolts into the equipotential ground system. In fact, Art. 680.26 (7) requires you to bond all the rebar within 3 ft of the pool wall, including the sidewalk. The NEC is quite clear on this subject.

We recommend NEC 2014 *Handbook* for reference. In particular, you may want to look at Art. 280.26 and Exhibit 680.8. The exhibit specifically shows the diving board and the mounting plate for it being bonded to the equipotential grounding plane.

Just a quick reminder, you are required to have two grounding systems for swimming pools: the green-wire ground and an additional bare copper equipotential ground system. The green-wire ground is only designed to ensure that circuit breakers and other OCPDs operate properly.

The equipotential grounding system is designed to reduce the formation of life-threatening voltages that can form in the pool. We talk about these voltages all the time on our blog; however, we discuss them in their technical terms: step-and-touch voltage hazards, GPR, and GPD. Please feel free to go through our blog, there is a lot of material on these subjects.

The Question: Under what specific conditions is a grounding resistor (neutral ground resistor) useful, and where it is not useful?

The Answer: The selection of a neutral ground resistance transformer is one of the main choices when deciding what type of electrical system your facility will have. There are four types of electrical systems:

1. Solidly grounded neutral system (the most common by far)
2. Ungrounded delta system
3. Low-resistance grounded neutral system
4. High-resistance grounded neutral system

Most businesses and homes today use item 1 (solidly grounded neutral system), as this system is the best for ensuring that any ground faults will maximize current flow, enabling OCPD such as circuit breakers and fuses to function properly. This is the only type of system that can be used for single-phase loads (with some minor exceptions) such as electronics, appliances, etc.

The second system is a delta power system, and the last three are wye power systems. The choice of which system to use is an extremely important and fundamental decision that will impact your entire electrical system. Generally, only power companies deal with item 1 (ungrounded delta system) for their power distribution from substation to substation. However, some mines and other industrial applications do use ungrounded systems.

Typically, resistance grounded neutral transformers are used when you have a system with only a single load and access can be controlled. Until recently, items 3 and 4 were considered to be dangerous options, as circuit protection technology was not sufficient to ensure that OCPDs would protect personnel and equipment. In other words, resistive grounded neutral systems (3 and 4) require advanced technology in their OCPDs that simply was not available until recently.

The advantages/disadvantages between the four systems above would require a chart and graphs in order to keep track of them all. However, for 2, 3, and 4, you may not in general run single phase loads that require a neutral.

Needless to say, the decision impacts a variety of engineering factors that will affect your entire electrical system, including: transient overvoltages, voltage stress, arc fault, safety to personnel, equipment reliability, ability to detect ground faults, equipment costs, multiple voltages on the same system, frequency faults, ground fault–current flows, training requirements, downtime, and compliance with local electrical codes.

The selection of when and where to use a neutral ground resistor is a very important decision (possibly the single biggest decision you can make) that actually has very little to do with grounding per se. It certainly impacts grounding, but the reasons for making the selection often have nothing to do with earthing or grounding.

The Question: When installing a new satellite Internet and television service in a home, they often do not have an adequate ground within 20 ft per NEC Articles 810 and 820. What can I do?

The Answer: There are really no special rules or exceptions for cable and antenna grounding in the **NEC**. In the end, your cable must be bonded to the main electrical grounding system and be in compliance with NEC Art. 250. The primary purpose of the grounding is to ensure that there are no hazardous differences in potential between your cable system and the building.

Now, the 20-ft rule is governed under NEC Art. 820.100(A)(4), which states that you must bond the outer sheath of your cable to the building's main ground with a conductor of less than 20 ft. The intention is to ensure a low-impedance path. However, there is a listed exception to the Code at 820.100(A)(4), which states: "In one- and two-family dwellings where it is not practicable to achieve an overall maximum bonding conductor or grounding electrode conductor length of 6.0 m (20 ft), a separate grounding electrode as specified in 250.52(A)(5), (A)(6), or (A)(7) shall be used, the grounding electrode conductor shall be connected to the separate grounding electrode in accordance to 250.70, and the separate grounding electrode shall be connected to the power grounding electrode system in accordance with 820.100(D)."

What this means is that when you bring your cable into the dwelling and it is farther than 20 ft to the main grounding electrode, you must install a standard 10-ft grounding rod (8 ft below the 18-in. mark, and 6 in. above-grade for bonding) at the cable building entrance AND run a 6 AWG copper conductor, whatever length is required to bond your system to the building's main ground.

One more thing to keep in mind is that the 2011 version of the NEC has some fairly significant changes over the 2008 code. You may want to make sure you have the latest code.

The Question: What can occur when the neutral is run in a separate steel conduit from the hot wire. Is it hysteresis?

The Answer: It is not only a very bad idea to route a neutral separately from the hot wires, but is illegal under that NEC. Please see NEC Arts. 215.4(B), 300.3(B), and 300.20(A).

The reason for this is due to induction, not hysteresis. The neutral wire helps to cross cancel the magnetic fields that form in the hot wires as the current flows through the conductors. Without the neutral, you will cause inductive heating of the metal conduit. One of our engineers has actually been to a facility that routed the neutral separately from the hot wires in a high-current circuit and heated the metal conduit support columns to the point that they actually glowed red hot!

The bottom line is that the neutral wire must be grouped with the hot wires. Not only will this reduce inductive heating of adjacent metal objects, but it also reduces the generation of electrical noise, accidental power outages, and many more negative things. And, of course, it meets the requirements of the NEC.

The Question: Is there a length limit per the NEC on rigid conduit runs used as a means for grounding?

The Answer: The NEC has not placed a limit on the length of rigid metal conduit (RMC) that can be used as a grounding conductor. Article 344.60 simply states "RMC shall be permitted as an equipment-grounding conductor."

Just because the NEC allows the use of RMC as a grounding conductor does not necessarily mean it is a good idea. Generally speaking, the longer the conduit run, the greater the need for a copper grounding conductor. Keep in mind that the purpose of grounding in the NEC is to ensure that OCPDs such as fuses and circuit breakers function properly.

Rigid metal conduit will generally allow enough current to pass through it to allow a breaker to trip but, especially in long runs, that is not always the case. There are a lot of connections in a long conduit run, and it only takes one bad connection to break the ground-fault path. Now, that said, the Soares book on Grounding and Bonding, 2011-NEC does in fact list a maximum conduit run. But this is not per the NEC; it is an International Association of Electrical inspectors (IAEI) guideline.

There are other issues to consider as well. Copper is at least twelve times as conductive as steel (some charts list 17×), and is between 250 and 4000 times less magnetic. Rigid Metal Conduit (RMC) is famous for having a high impedance over distance and will allow magnetic fields to form that increase the likelihood of objectionable currents, harmonics, interference, and electrical noise of all kinds. This is why it is the official policy of E&S Grounding Solutions to recommend the use of copper conductors for the equipment-grounding conductor in all circuits. Rigid metal conduit should be considered a backup system and not the primary. We also do not recommend the use of Edison circuits, where you share a single ground wire between the loads on three different phases.

Additionally, there are many different standards and codes that actually require equipotential bonding of RMC! ANSI/TIA/EIA-J-STD-607-A, Motorola's R56, numerous IEEE standards, and many more all require that the RMC itself be bonded to ground and that jumper wires be used around each and every one of the connections. This is why you can purchase stock grounding bushings and grounding couplers, so the grounding jumper wires can be easily installed. In the future, the NEC will not allow RMC to be used as a grounding conductor and will have many more requirements for equipotential bonding. Consider the equipotential bonding requirements for a swimming pool, and you will begin to understand where grounding is heading under the NEC in the future.

In conclusion, it is unfortunate that the NEC even allows the practice of using RMC as the sole grounding conductor, and it is a disgrace that it has no length limit for when it is used. There are no imaginable cases where using RMC as your primary grounding conductor is a good idea. Using copper grounding conductors each and every time improves the safety for people and improves the way equipment functions.

The Question: How we can reduce the voltage between the neutral and ground without using an isolation transformer?

The Answer: The difference in voltage between the neutral to the ground at any point in your electrical system should be very low, typically something less than 1 V. This is easily measured by using a standard voltmeter and placing one probe on the neutral wire of a standard receptacle outlet and the other probe on the ground wire. The resulting measurement is your neutral-to-ground voltage.

If you are experiencing neutral-to-ground voltages that are higher than 1 V, you have a serious electrical issue that must be corrected. This issue is in fact correctable, and you should contact a certified electrician right away.

The most common reason one finds excessive neutral-to-ground voltages is due to an illegal neutral-to-ground bond in a subpanel. There is one place and one place only that the neutral and ground are bonded together, and that is at the first service disconnect. You should start by opening all of your subpanels and ensuring there are no neutral-to-ground bonds in these panels. This is a very serious and potentially life-threatening situation and must be correct ASAP.

Another possibility is that you have some faulty surge suppressors somewhere in your building. Budget surge protection strips are known for failing and can often place unwanted voltage on the ground. You may want to unplug all of the surge strips from your building and see if the issue goes away.

Other sources could be from severely poor power quality, Edison circuits, and/or an out-of-balance three-phase electrical system, but if your system does have these problems (which should be corrected), it is still unlikely that you will see neutral-to-ground voltages.

In any case, you need to hunt the errors down and correct them right away.

If I left both points of connection open for viewing, inspecting, and testing, but I encased just the run of the bonding jumper, is that going to be good for an inspector for passing?

This is a hard one to answer, as every inspector seems to have his/her own special requirements (as I am sure you know). But for E&S Grounding Solutions, what you have described would be perfectly fine. Being able to see the bonds on both ends of a conductor will also allow the easy testing of the conductor itself via a simple ohm meter.

While it will be possible that you have multiple parallel paths, it will at a minimum confirm that an effective path at least exists. The best way, of course, would be to leave at least one end of the conductor unwelded, so a simple resistance test could verify the conductive path.

The Question: How do you support a lightning protection down leads run in a plenum ceiling? I have recently encountered in a new building installation, lightning protection down leads coming from the roof air terminals and running through the ceiling sleeved in PVC until it is cadwelded on to the building steel structure.

If the PVC is taken out, what are the rules for resupporting the wire that is left hanging? I was under the impression that the wire is not allowed to hang on or touch anything else in the ceiling space, because the lightning could spread, but I cannot find a code that states this.

The Answer: When routing lightning down conductors through conduit, the preferred material is PVC. The magnetic properties of metallic conduit can actually impede the flow of current, if the conduit is not bonded to the same potential as the conductor. This is why grounding bushings are required on both ends of metallic conduit runs, and bonding jumpers are required around conduit junctions.

It is also the same reason why 8- to 12-in.-radius bends are required for lightning down conductors. When lightning propagates through a conductor, it forms massive magnetic fields. Metal conduit and close radius bends can cause the formation of what is called a self-induced coupling, resulting in an impedance choke and ultimately, the conductor can burn-open like a fuse.

The problem with PVC is that fire departments tend to disapprove of its use inside buildings. When PVC burns, it produces toxic smoke, which escaping victims really do not need to worry about breathing, as they generally have plenty of other things to focus their attention on. But, if you can get it approved (which is generally the case for small uses under 1 ft in length), then PVC is the correct choice.

Our concern is more in the use of the building steel structure as the ground terminal device. Generally speaking, LPSs are designed to protect buildings, and tying the lightning system into your concrete foundation is a really great way to damage your building. When you bond your grounding system, whether it be a lightning ground or your NEC ground, to building steel, you are technically using the concrete foundation as your grounding electrode. This is called a concrete-encased electrode.

Why is this not ideal? The water that is inherent to concrete will rapidly expand when heated. If you have ever seen someone take a blowtorch to concrete, you will know that the concrete will literally explode from the inside out, as the water turns into steam.

Lightning is more than capable of doing this same thing, as it will heat the water in the concrete as it passes through it on its way to earth.

While the lightning down conductors certainly need to be bonded to building steel, we believe that they should have dedicated grounding rods (NFPA 780 4.13.1.1).

The NFPA 780 does allow concrete-encased electrodes for use as a ground terminal. However, the NFPA has some very specific rules for its use.

1. They may only be used in new construction. You cannot use building steel as your ground terminal (electrode) when installing an LPS on an existing building (NFPA 780 4.13.3).
2. The concrete-encased electrode must be installed near the bottom concrete foundation or footing of the building. This means that it must be a separate electrode and cannot be the foundation itself (NFPA 780 4.13.3.1).
3. The concrete-encased electrode must be in direct contact with the earth and be at least 20 ft (linear) in length. There are additional rules as to the copper-versus-steel content of the electrode (NFPA 780 4.13.3.1 and 4.13.3.2).

For the record, E&S Grounding Solutions never recommends the use of concrete-encased electrodes under any circumstances, as they are generally more expensive to install, perform less effectively, and are good for only a single electrical fault before the concrete cracks open and makes the electrode ineffective.

GROUND POTENTIAL RISE

The Question: Can you ignore the effects of step-and-touch voltage hazards on concrete pads because of the rebar mesh and the low resistance? In a solid grounding system, is it appropriate to exclude evaluation for GPD or GPR and step-and-touch potentials just because the resistance is going to be lower than 1 Ω, as calculated using the IEEE 142 method and the large mass of concrete and rebar in the slab (Ufer grounding). How do you evaluate galvanic corrosion, anticorrosive coating for rebar, and bonding of dissimilar metals inside the concrete?

The Answer: Solidly grounded systems are subject to GPR events, and therefore GPD events as well. Ultimately, GPR events are based on Ohm's law, which is simply the fault current times the resistance to ground of the ground system. So, even if your rebar system is 1 Ω, what would be the GPR for a 50,000-A fault? It is of course, 50,000 V! Is this good or bad?

This is where GPD comes in. A plain steel rebar system may have some big differences in potential from one end of the grid to the other side. If your concrete-encased rebar grid has a resistance of 1 Ω from one corner of the building to the opposite side corner, you will see 50,000 V form across the building.

Yes? Keep in mind, it is not uncommon to see a 1-Ω resistance to ground for a building's foundation; however, the corner-to-corner resistance can be several ohms. When this happens, you can actually form some serious differences in potential horizontally. This is where time domain and leakage current calculations come into play, but that is another topic. Also, this is why you see so many copper ground rings installed around buildings.

A complete answer to this topic would really take some time; however, the short answer is *no*. You may not exclude the effects of a GPR event simply because you assume a low-resistance rebar system. It just does not work that way. In fact, in certain cases, grounding engineers will actually add resistance or remove grounding in certain areas to avoid GPR issues that can result.

One last point: you cannot simply assume that any rebar system is less than 1 Ω. Every day, we deal with massive concrete-encased rebar grids that measure in excess of 100 Ω! It all depends on the soil the concrete is sitting on. And, of course, if your concrete slab has a vapor barrier or if the rebar is coated in plastic, you cannot use the slab as an electrode at all.

You should also remember that copper is 12 to 17 times more conductive than steel, and at least 250 times less magnetic. When dealing with AC currents, it is very important to consider the magnetism of the steel, as the magnetic fields that form in the steel take time to form and collapse; this keeps the fault currents in your system longer, much longer. Amperage is time based (1 C past a given point in 1 s), so the difference between clearing a fault in 6 cycles, versus 120 cycles, can be a matter of life or death. Copper paths are the best way to keep magnetism down and the time domain of the fault to a minimum.

The Question: According to my knowledge, GPR provides information about the maximum potential that can happen under a faulty condition. But how can we find the potential distribution within the grid (i.e., in the soil enclosed by the grid) analytically?

The Answer: When you see a GPR study, it generally falls into one of two categories: (1) it provides a maximum GPR only, or (2) it provides a GPR distribution across the system. The difference in the amount of calculations and work involved between these two types of studies could not be more striking.

The first report basically calculates the resistance (sometimes they use impedance, but not often) of the ground system, generally using a single or a predetermined two-layer soil model (fixed or limited depths, resistivities, etc.). It then uses Ohm's law to apply the fault current to get a maximum GPR. This type of analysis often cannot take into account material characteristics (steel vs. copper vs. aluminum, etc.) and must assume the ground grid is made of an imaginary superconductor with zero impedance. This means that it cannot take into account voltage drops across the length of the conductor, nor can it calculate the leakage current into the soil as the fault currents propagate through the ground system and out into the surrounding earth. This results in a simple maximum GPR.

This maximum GPR is often then applied as the basis for step-and-touch voltage hazards across the entire grid. This is, of course, simply inaccurate. Touch voltages in particular only occur at arm's length from touchable objects. Applying touch voltage criteria to open areas of a facility results in vast overprotection in areas that do not need it and, often, underprotection in areas that do.

The second type of report takes into account the impedance of the system (at frequency) in complex multilayered soil models with no limitations regarding depths, resistivities, or number of layers. To do this, one must consider the characteristics of the conductors within the ground system. Steel rebar in footings and foundations, steel ground rods with copper coatings buried vertically, bare copper conductor buried horizontally in the ground, and insulated conductors bonding items above grade, all must be taken into account.

Each of these items will have different factors with respect to relative resistivity (both real and imaginary parts), permeability, inductance, capacitance, and so on, that will change given the length of the item, radius, depth buried, and so on. Once the fault current is applied (it must include the zero-sequence impedance, clearing time, and amperage) to the model, each individual piece of the grounding system must be individually calculated, taking into account all of the above factors.

Long conductors must be broken into smaller segments (typically 1 ft), so the leakage current into the soil, voltage drops, and other factors can be properly calculated, which of course will change with the soil model, as the ground system conductors pass through the varying soil layers. Even for a small 20-ft × 20-ft grid, this requires thousands, if not hundreds of thousands, of individual calculations.

Once this is completed, you will be able to see the GPR distribution across the grid. There are in fact many "hot spots" that will typically occur in a grid (usually near the fault location and then at the corners of a grid). This also allows you to determine what is quite possibly the single most important factor of a grounding system: the GPD. Lowering the GPD is generally more important than lowering the GPR when it comes to human safety.

Once you have a GPR distribution, you can finally start accurately calculating step-and-touch voltage hazards. This is another massive task, as individual touch voltages must be calculated at any place within the grid that can be touched by personnel. A faulting transformer bank will have a higher touch voltage than the wall of the control building. Only a category 2 report can demonstrate this. And that takes an additional level of calculations for each and every variation within the system, for both step-and-touch voltage hazards; again, this is many thousands if not hundreds of thousands of additional calculations.

This is the only way to get an accurate GPR analysis completed. It should be obvious by now that you need some computer assistance to do a proper GPR analysis. Given today's computer environment, it is borderline unethical to conduct GPR analysis by using hand calculations; at a minimum, hand-calculated GPR studies are simply bad engineering.

Now, that said, there are many computer programs out in the marketplace today that conduct nothing more than fancy category 1 analyses. In particular, most of the electrical engineering software programs that are used for load flow, arc flash, short-circuit analysis, relay coordination studies, and so on (etap, SKM, etc.) have grounding analysis modules that do NOT conduct a category 2 analysis.

To our knowledge, the only computer software program that accurately conducts a GPR study is the CDEGS program. Please note their MALT module is their version of the category 1 software, produced to compete with etap, SKM, and the like. To conduct a category 2 study, you would need the MALZ and/or HIFREQ modules.

The CDEGS program is also one of the most validated electrical engineering software programs in production. It is used by the NFPA (NEC), the IEEE, UL, TIA, EIA, the U.S. government, and many others when writing code.

The Question: What are accepted limits for GPR and GPD? How do you protect telephone circuits from high GPR and GPD?

The Answer: There are actually no specified limits to either of these electrical phenomena, as they are simply indicative factors for other specifications. Let us explain.

The GPR is a factor that describes the maximum voltage that will be seen during an electrical fault on a given grounding system. You can imagine that as fault currents enter a grounding grid, the voltage will rise at some gradient across the system.

The actual voltage that the system rises is not the primary concern, it is the current or amperage that can flow across people or equipment that is our concern. The best analogy is a bird landing on a power line that may have high voltage; the bird will be safe as long as no current flows through the bird itself. If your building rises to 500 V or 50,000 V, it can be either good or bad, depending on the current flow at critical points. Here is some more information on GPRs:

www.esgroundingsolutions.com/about-electrical-grounding/what-is-ground-potential-rise.php.

The GPD tells us that during a GPR event, the grounding system will have voltages that vary by a certain amount. For example, if your building has a 50,000-V GPR, ranging from 49,900 to 50,000 V, then your GPD is 100 V. If your building has a 500-V GPR, ranging from 100 to 500 V, then you have a 400-V GPD. This means that from one end of your grounding grid to the other you will see a 400-V difference in potential. Which is worse? The 100-V GPD or the 400-V GPD? Actually, we still do not know.

If the building with the 400-V GPD will only push a few amps across any given point, it may be perfectly fine. On the other hand, the building with the 100-V GPD might push hundreds of amps across the grid, and that could be very bad indeed. When dealing with human safety, we calculate step-and-touch voltages to determine whether the amount of current that can enter a person will be enough to cause the human heart to fibrillate or not.

When protecting equipment, we must look at manufacturers' recommendations for maximum voltage and current levels. But in almost all cases, if it is safe for people (step and touch), then it is safe for the equipment too.

Now, how do you provide proper protection? It is a combination of removing electrical energy from the grid (electrodes), balancing electrical energy (grids), bringing electrical energy closer to people and equipment in certain cases (bonding), and taking it away in other cases (insulation). Frankly, it can get very complex; much more than we could ever really cover here.

The Question: Is there a minimum separation established between a data center and a 115-kV substation? There is a concern about the Electromagnetic Interference (EMI) to the servers, routers, and so on. The GPR in a data center adjacent to a 115-kV substation is 809 V. Is that okay or too high?

The Answer: Data centers are big energy consumers, and as such, they often have substations and generators located nearby. The GPR that can form during the normal operation of these systems, and which will get dramatically worse should one of them fault, is certainly something to be concerned about. The GPR is used to determine various engineering factors, the most important of which are the step-and-touch voltage hazards. These voltages can cause serious injury or even death to personnel and can destroy equipment.

When someone says that their GPR is 809 V, what they are saying is that the maximum voltage at any one point within their facility is 809 V. This also means there is a minimum voltage and thus a difference between the two voltages, which is the GPD. Consider that the corner of the data center that is closest to the high-voltage substation will see the greatest voltage, and the corner farthest away will see the least. This difference in potential from one corner to the next can often be several hundred volts.

Why can a bird land on a high-voltage wire without being electrocuted? That is because there is no difference in potential that could cause current to flow. The same is true for your data center. As long as the difference in potential (GPD) is low enough, your data center can easily handle a 809-V GPR (or even greater).

What your real concern should be is the difference in potential (GPD) within your data center. A 200-V difference in potential from one side of your data center to the other side will allow a lot of current to flow, which can damage sensitive electronics and possible injure or kill personnel.

Rest assured, it is quite normal to see data centers located adjacent to substations and to see GPR values far greater than 809 V. It is just that the greater the GPR, the lower the GPD needs to be to mitigate the effects. A properly designed grounding system will be able to protect your data center and the personnel who work in it.

The Question: What is step potential and why is it important to tie all systems together?

The Answer: Step-and-touch potentials are hazardous electrical voltages that form when electrical energy enters the earth, typically (but not exclusively) during fault conditions. Let us take for example a high-voltage tower with a cellular communications site (cell site) installed under it.

The insulators on the tower get dirty over time and must be washed and maintained to prevent unintentional arcing as the buildup of dirt eventually forms a conductive path to ground. If this path is allowed to form, the electrical energy in the lines will arc over to the tower and down into the earth through the footings. Note that birds landing on the insulators can do the same thing.

The massive amount of electrical energy will enter the earth and saturate the surrounding soil, charging everything touching it in the immediate area, including the cell site. This is called a GPR event. The farther one is away from the tower, the less of the charge he or she will encounter.

You could imagine a pebble being dropped into a still pond, the waves will propagate away from the pebble, getting smaller and smaller as the distance increases. This is similar to an electrical fault; the energy will travel out and away from the tower, the potential decreasing with distance.

Step voltage is the difference in voltage a person could experience between their front leg and their back leg when walking near the tower. In other words, if you are walking toward the tower during a fault, your front leg might be at 1000 V and your back leg at 750 V, giving you a 250-V difference between your two legs (these numbers are for example only). This 250-V differential can be hazardous to your health.

GROUNDING ELECTRODES

The Question: What is the environmental impact of a chemical earth pit or electrolytic electrode? What are the harmful effects of so-called chemical earth pits in which the vendor does not disclose what chemical backfill they use? Is this a big issue for municipalities?

The Answer: Most often when someone refers to a "chemical earth pit," he or she is actually talking about an electrolytic electrode. The word "chemical" carries a very negative connotation for most people and really is not very accurate for what is actually occurring in most cases.

A good-quality electrolytic electrode is a copper pipe filled with common salt and a desiccant (common road deicer) and is backfilled with natural-earth bentonite clay. These electrodes have been certified by the National Sanitary Foundation as being environmentally safe, and to our knowledge, they are indeed low impact.

The worst that can be said about them is that they leak a few tablespoons of salt every year into the surrounding earth. You should freely use these types of electrodes, except when in high-voltage environments in which human safety is a concern. You should never make changes to or redesign a human safety grounding system without consulting a grounding engineer.

Other types of electrolytic electrodes are instead filled with Epsom salts. In these cases, large quantities of the salt are released in to the earth every year, and there is a potential for the formation of hazardous chemicals should the electrode be struck by lightning and heated to 200°F. There are both good and bad reasons to use these types of electrodes, however, a grounding engineer should be consulted prior to their use.

Backfill materials range from carbon-based (typically coke breeze) fills, to concrete, to salt, to cow manure. Generally, the grounding engineer is more concerned with corrosion of the electrode than with environmental concerns. Of course, each of these certainly has a potential for an environmental impact, save the manure.

However, none of these should be used, as the carbon-based backfills will corrode the copper, the concrete will crack when heated by lightning or serious electrical faults, and both the salt and manure have very short life spans and only provide temporary reductions in resistance to ground.

Currently, E&S Grounding Solutions only recommends bentonite clay as a backfill material for any grounding system. Carbon-based backfills, and concrete in particular, should never be used.

That said, if your vendor refuses to disclose the content of the backfill material they use, find another vendor. Good-quality electrolytic electrodes are very easy to purchase. You can even order them online, as most of the major electrical supply companies have these systems available in their catalogs.

But why are you installing these systems? Do you have a specific specification, such as 5 Ω or less, that you need to meet? Has a grounding engineer analyzed your earthing requirements and your specific soil conditions and told you that you must have these expensive earthing pits? If you have not had a professional analysis conducted, you may want to consider doing so. It could in fact save you a lot of money.

The Question: How can you calculate the resistance to ground for a ground ring with rods attached?

The Answer: It sounds like you are trying to conduct hand calculations, which are indeed a very limited and frustrating process. They are also highly inaccurate. In particular, hand calculations cannot take into account three key factors: complex ground designs, multilayer soil, and frequency.

There are no readily available formulas to hand calculate for ground rings with ground rods. This is because such a formula would be a multistage process and would require a white paper just to explain how to conduct the formula. But that really is not the issue with hand calculations you should be worried about.

Your concern for hand calculations is that the formulas found in the Green Book (ANSI/IEEE Standard 142-1982) assume a uniform soil resistivity, which simply does not exist in nature. Therefore, you can never accurately calculate even the most simple of electrode structures using these formulas, let alone try to hand calculate a complex grounding system.

Let us consider the calculation for a single 10-ft ground rod. Hand calculations almost never get the actual resistance right, because the formulas do not account for changes in soil resistivity with depth. There have been several studies conducted that show typical soil is best modeled using three to five layers, with the top three layers generally occurring within 10 ft of the surface.

We often see three- to four-layer models, so as to leave the fifth layer available for changes in soil resistivity due to frost-line issues during winter conditions (when soil freezes, the resistivity of the soil increases by a factor of at least 10×). This means that your 10-ft ground rod will often have three to four different soil resistivities along its length and will have a higher resistance to ground during winter. That said, you would need to calculate the change in resistance of the electrode as the soil resistivity changes along its length at depth. This means that you really need to account for leakage current rates and voltage drops across the length of the electrode.

Of course, keep in mind that you are probably not even dealing with DC (0 Hz); your ground system is probably being used for a 60-Hz system, which means that you actually should be concerned with impedance to ground. Of course, frequency-based calculations are not discussed in the

Green Book at all. And yes, impedance makes a huge difference in the final calculations. You also need to account for the material properties of your grounding system.

The horizontal conductors of the ring will almost certainly be made of copper, while the ground rods will be made of a low-grade steel. Copper is 12 to 17 times (based on the steel) more conductive than steel and is at least 250 times less magnetic. Proper calculations will take into account these two dissimilar materials and the change that the conductivity and magnetism will have on the ground system.

The point of the above information is to demonstrate the futility of hand calculations. The formulas in the Green Book are for theoretical use only; they assume DC electrical systems in imaginary uniform soil conditions. The only accurate way to calculate even a simple ground electrode is by using a computer program designed for this task. We recommend and use the CDEGS computer program.

Now, if you are simply insistent on hand calculations, you would need to calculate the ground ring by hand, then calculate the ground rods separately, and then combine the two by adding resistances in parallel. That is the best you can do. The problem with adding resistances in parallel is that it does not take into account resistance loss due to overlapping spheres of influence, so we would not recommend that you show the results to your customer. But then again, we would not show any calculation from the Green Book to a customer.

The Question: How do you reduce the earth resistance of an 18-Ω electrode to under 5 Ω? Can we use GEM or bentonite clay as a backfill?

The Answer: The two types of backfill material you mentioned are excellent examples of the disparity between the currently available ground-enhancement backfills.

The first material you mentioned was GEM. While we have no specific knowledge of this particular brand of backfill, it does fall into the category of carbon-based backfills. Carbon-based ground-enhancement backfills have both a good and a bad side to them. The good side is that they are relatively inexpensive and they do have very low resistivities.

Typically, these materials have resistivities that are well below 1 Ω·m, sometimes as low as 0.1 Ω·m. The bad side is that the metallurgical nobility of carbon is such that it destroys copper over time. Typically, after 5 years, copper electrodes have been corroded and destroyed by the carbon. There are also potential environmental concerns with the use of carbon-based backfills. As such, E&S Grounding Solutions does not currently recommend the use of carbon-based ground-enhancement materials.

The other backfill material you mentioned is bentonite clay. Bentonite clay is a natural earth soil (clay) that is mined in areas like

Wyoming. It is of almost no environmental concern and will not corrode the copper. In fact, bentonite clay is protective of the copper. The downside of bentonite clay is that it is has a resistivity of around 2 Ω·m and needs water to stay conductive.

Very dry soils may require the use of watering devices when using bentonite clay. However, the upside of bentonite clay far outweighs its downside, and it is the only backfill material that E&S Grounding Solutions currently recommends for grounding systems.

Now, in regard to your question about reducing the earth resistance of your electrode from 18 to under 5 Ω, it is highly unlikely that any ground-enhancement material will achieve such a result. Ground-enhancement materials will typically only improve your electrode's resistance to ground by a few percentage points, maybe 10–20% at best.

Generally speaking, if your electrode is currently measuring 18 Ω, then an additional electrode of the same specifications will also measure 18eΩ. Assuming you install the second 18-Ω electrode at least twice the diagonal length away from the first electrode, you should have an electrode system that will measure 9 Ω.

If you install a total of three more of these 18-Ω electrodes, each at least twice the diagonal length away from each other, for a total of four electrodes, you should have a system that is just under 5 Ω. Obviously, four electrode systems can be very expensive, and you may not have the physical room to install them at the spacing you need. We would recommend that you get an electrode system designed.

The Question: For a pole-mounted transformer, what is the most cost effective way to achieve 25 Ω or less in marl and rock?

The Answer: When you have a pole-mounted transformer, you are generally dealing with a very small area of available soil (earth) that you can use for grounding. The resistance to ground of an electrode is determined by the local soil conditions and the sphere of influence of the electrode. In other words, the worse the soil, the more electrode you need. There really is no magic bullet to avoid this.

The only way to guarantee a 25-Ω or less resistance to ground for your pole-mounted transformer is to get an engineered grounding design. With computer modeling, a grounding electrode can be designed to meet your 25-Ω requirement in local soil conditions.

For example, if you have a concrete pole, it may be possible to utilize part of the pole itself as an electrode. This could be supplemented with an electrolytic electrode system to achieve the 25-Ω requirement. But you will not know this unless you do a proper design. Electrolytic electrodes are expensive, and you would not want to randomly install one of these grounding systems without knowing in advance that it will work.

The Question: How do you ground a 120-ft tower on top of solid granite?

The Answer: This is one of the biggest questions we ever get, how do you ground on top of a rock mountain? We usually answer by pointing out that the International Space Station is grounded, and without the aid of a ground wire trailing down to earth! It should also be noted that airplanes, helicopters, and ships out at sea are all grounded without a connection to earth. How is this possible?

When your site is located on top of a granite mountain, you can be relatively assured that the resistance-to-ground of your electrodes will be very high. This tells a Grounding Engineer that the GPD of the system is going to be even more important than usual.

When electrical energy enters the site, the resistance will be very high, and thus the voltage will be very high as well. Imagine a 5,000 A fault entering a site with a 100-Ω resistance-to-ground.

Ohm's law tells us that the voltage or GPR of the site will be as high as 500,000 V! If your site has a difference of potential from any given point to any other point within the system of 2-Ω, the site could see a voltage differential of 10,000 V!

If that difference in potential is 1-Ω, then you could see 5,000-V forming across your site! If the difference in potential is only 0.1-Ω, then your site would only experience 500 V. This should demonstrate the importance of a low Ground Potential Differential (GPD) in high resistance grounding systems.

One can imagine that as fault currents enter a grounding grid, the voltage will rise at some gradient across the system. The actual voltage that the system raises is not the primary concern; it is the current or amperage that can flow across people or equipment that is our concern. So, like a bird landing on a power line which may have high voltage, the bird will be safe as long as no current flows through the bird itself. Here is some additional info:

www.esgroundingsolutions.com/blog/749/what-is-accepted-limits-of-gpr-and-gpd-how-do-you-protect-telephone-circuits-from-high-gpr-and-gpd

Now with all that said, it is still advisable to provide an effective low-impedance path to earth for your grounding system. Many times in cases such as this one, the power for the site is being brought in via overhead power lines (power poles).

You may wish to add a 4/0 AWG or greater copper conductor on the overhead pole system, so as to ensure that there is a ground path from your site back to the source. This will ensure that OCPD such as fuses and circuit breakers will function properly. It can also give you an opportunity to install a grounding electrode for your site in better soil conditions. This will be true even if you install an isolation transformer at your site.

The Question: What is the best grounding method(s) for an electrical system located in the basement of a building in a flood zone? I have a hotel project in which grounding must be done. The hotel has up to fifteen floors. My concern is that the grounding system, consisting of a transformer and a generator, will be located in the basement. This place is basically a tropical region, where it can rain and flood anytime. What would be the best possible grounding method we can use to avoid accidents?

The Answer: When you are in a high–water table area such as the tropics, the primary concern is that water could enter the basement and flood the electrical system. This means that your building foundation will be waterproofed and electrically isolated from the soil. You may not use your building foundation as an electrode.

You will also not be able to penetrate the concrete at below-grade level for fear of allowing water to enter the basement. You will need to install an electrode system that is separate from the building structure. Keep in mind that you must still bond your building steel to the grounding system, you just may not use is as an electrode.

If you are in a lightning-prone area, you may want to consider surrounding the building with a continuous buried ground ring with interconnecting 10-ft grounding rods. This ring or counterpoise will provide a very effective grounding system that is usable by both your building's electrical system and your building's LPS. This ground ring should consist of, at a minimum, 4/0 AWG bare stranded copper conductor buried in direct contact with the earth to a depth of 1.5–3 ft with 10-ft ground rods at 20-ft intervals.

GROUNDING TEST METHODS

The Question: How do you determine the resistance to ground of an active utility substation in an urban setting? A customer wants fall-of-potential tests performed on the ground grid of a 34.5-kv substation property. The problem is there is no soil around the property, only asphalt. And within 50–75 ft on all sides of the property are other buildings. So, there is no way to get out of the sphere of influence of the grid, even if there were soil to drive rods into. Is there an alternative test we can offer under this condition?

The Answer: Yes, the best way is to conduct a Wenner four-point soil resistivity test in an open field somewhere near the substation and then model the grid on a computer using the CDEGS engineering software. This is actually the single most common way to determine not only the resistance to ground of the substation, but the impedance to ground and the mandatory human safety requirements under 29 CFR 1910.269 (step-and-touch voltage hazards).

CDEGS is the software used by the IEEE, the NEC, the NFPA, and many others to develop standards and regulations related to electrical safety and grounding. It is the single most-validated software on the market. In fact, the programmers of the CDEGS program are some of the same people who developed the procedures used for the fall-of-potential test.

What are the advantages of perpendicular traverses when conducting a soil resistivity test?

The Question: When conducting two soil resistivity traverses perpendicular to each other (i.e., traverses at right angles AND intersecting), would this would cancel out interference effects from buried services? If the traverses do not intersect and are several meters away from each other but are still at right angles to one another, does this eliminate the interference from buried services or achieve anything?

The Answer: When conducting a soil resistivity test, it is often recommended that at least two test traverses are conducted. These traverses should be at either right angles to each other or at diagonals crossing the site. The general purpose of this is to minimize the effect of unknown buried objects on the test.

A single traverse may inadvertently be placed on top of a long-forgotten buried pipe or other metallic object. Sometimes, the field technician can tell that there is a problem with the data while conducting the test. However, it is often only upon running the data through the computer that it becomes apparent that there is a significant problem with the data.

This can be an expensive error, especially if the test location is some distance away and will require additional travel costs to repeat the test. This is why it is generally recommended to conduct several tests at right angles and/or crossing each other at diagonals. When a problem occurs, you will be happy to have the second set of numbers.

In answer to your question, the perpendicular and/or diagonal traverses are NOT intended to remove electrical interference issues. The best and only real way to eliminate the influence of electrical noise on your soil resistivity test gear is to use quality DC meters. Currently, there are only two manufacturers of test meters approved for use in soil resistivity measurements: Advanced Geosciences (United States),

www.agiusa.com;

and IRIS Instruments (France), www.iris-instruments.com.

We also recommend using true-test lead cables (silicone, 18-gauge, 65-strand wire) for conducting the tests. These cables, along with the

800-V p-p DC meters, tend to be immune to most electrical interference issues and conduct very accurate resistance measurements.

Meters such as the DET 2/2 are 2.5-W, 50-V maximum AC meters (105–160 Hz) and should not be used. Many meters use terms such as "reversing DC" or "pulsed DC" as ways of avoiding the admission that they use AC at varying frequencies to measure the soil. "Pulsed DC" for example is simply another way of describing a square-wave AC signal. We can get into all the reasons why AC signals are bad, but all you really need to know is that this industry is not the only one to recommend DC meters. The U.S. Geological Survey also uses DC meters for its testing.

The other basic way to insure accurate soil resistivity testing it to use good test procedures. We base our soil resistivity testing on the specification from Safe Engineering Services (SES) in Canada (www.sestech.com). As you may know, they are the primary code writers (and researchers) for all the current standards and regulations governing our industry.

It is important to understand that there are two types of resistivity data that are sometimes used. The first one is called *apparent resistivity*, and the second one is *actual resistivity*. Apparent resistivity is a simple formula that only provides an average resistivity reading from the surface of the earth to the probe-spacing distance.

It is NOT a real soil resistivity number. What you really need are the actual resistivity values. Unfortunately, there is really no way to properly hand calculate actual soil resistivity; the formulas are simply too complex and numerous to do without the aid of a computer. We use the RESAP module from the CDEGS engineering-software program.

The computer algorithms require a lot of data, and the data must be collected, so there are not too many "gaps" between the spacings. The maximum allowable interval between spacings is a 1.5 ratio, with a 1.33 ratio preferred. So, a measurement taken at a 20-ft spacing, would need to be followed up by a maximum 30-ft spacing, and preceded by at least a 14-ft spacing, in order to keep the 1.5 rule.

In other words, if you have data for a 40-ft spacing, and then jump to an 80-ft spacing, the distance between spacings is a factor of two, which is too great and will cause errors in the computer's calculations. You would need a 60-ft spacing between the 40-ft and 80-ft readings.

We recommend the following spacings for the typical 100-ft or less grounding system: 0.5, 1, 1.5, 2, 2.5, 3, 4, 5, 6, 7.5, 8, 9, 10, 12, 15, 20, 22.5, 30, 40, 45, 60, 80, 90, and 120 ft. That is a total of twenty-four measurements for a single 360-ft-long test traverse. You should conduct several of these traverses, depending on the local soil conditions and space available, preferably at crossing diagonals and/or perpendicular runs.

If your sites require step-and-touch calculations, please remember that the solutions for mitigating those hazards can be very expensive. It is important to collect good soil data, so we know we are basing these critical lifesaving grounding designs on valid data. The shallow-depth readings (smaller spacings) listed above are critical for human safety, as we are measuring the soil where people stand. The deeper readings (larger spacings) are needed to understand how the electrical faults will propagate through the earth.

The Question: Can you explain the 1/20 probe-depth rule when conducting a Wenner four-pin soil resistivity test?

about-electrical-grounding/what-is-soil-resistivity-testing.php.

The Answer: When conducting a Wenner four-point soil resistivity test, we need to consider the two ways in which the sphere-of-influence will affect our test:

1. The distance the test is being conducted from any buried metallic objects (railroad tracks, fence lines, etc.). This distance should be equal to or greater than the maximum ("a") spacing of the test. In other words, if you are conducting a Wenner four-point test with a maximum probe spacing of 60 m (a 180-m traverse), there should be no interfering objects (fence, buried metal pipes, etc.) within 60 m of any part of the test.
2. The probes used to conduct the test will generate their own sphere-of-influence based on the depth they are driven into the earth. For hand calculations, the probe depth may not exceed 1/20 of the spacing of the Wenner test. Advanced computer algorithms can adjust for these differences, but the 1/20 rule is a good one.

Here is some more information on the sphere-of-influence:

www.esgroundingsolutions.com/about-electrical-grounding/grounding-electrode-sphere-of-influence.php.

The bottom line is that deeper probes are not better; they are worse. In fact, some people actually use a heavy weight or a pile of heavy chain with a little salt water and do not drive any probes into the earth at all! We try not to go much deeper than 6–9 in. for probe depth, and when we conduct the short-range readings, we actually go only 2 in. into the earth.

Keep in mind that arguably the second most important spacings for the Wenner soil resistivity test, are the very short-spaced readings. You should be taking measurements starting at a 6-in. (0.15-m) spacing and incrementing interval size by a ratio no greater than 1.5, with a 1.33 ratio preferred.

The probe depth for a 6-in.-spaced Wenner measurement should be no greater than 2 in. for the potential probes and 4 in. for the current probes. This arrangement clearly violates the 1/20 rule. However, good computer-modeling software can adjust for this and still provide good data.

See the following link for more information:

www.esgroundingsolutions.com/blog/893/893.

You, of course, should be using true DC test meters with test lead cables. We recommend meters that provide 800-V p-p signals with at least 500-mA DC. See www.agiusa.com for more information.

The last component is, of course, computer software. Your raw data and hand calculations are nearly worthless. In fact, hand calculations will only provide you with what is called apparent resistivity, not the actual soil resistivity.

The calculations require a computer to do properly, and are so complex that they can actually take a high-end processor several minutes of computing time to provide an analysis! Why even take the test if you are not going to process the data correctly?

The raw data you collect during the Wenner test must be analyzed and processed in order to develop a soil model (or soil profile). This soil model will tell you what the resistivities of the soil are at various depths down through the earth at your site. We recommend the RESAP module from the CDEGS computer program for proper analysis.

The Question: What is the best probe spacing for soil resistivity testing?

The Answer: It is important to understand that there are two types of resistivity that people use today. The first one is called apparent resistivity, and the second one is actual resistivity. Apparent resistivity is a simple formula that only provides an average resistivity reading from the surface of the earth to the probe spacing distance.

It is NOT a real soil resistivity number. For example, let us say you have a Wenner test with a probe spacing of 5, and the apparent resistivity is 100 Ω·m. You also have a Wenner test with a probe spacing of 10 that tells you the apparent resistivity is 75 Ω·m.

What is the resistivity between 5 and 10? The answer is 50 Ω·m. If the 0- to 5-ft layer is 100 Ω·m, and the 5- to 10-ft layer is 50 Ω·m, then the 0- to 10-ft apparent resistivity would read 75 Ω·m.

Please see the link below, because it is very important to understand the difference between apparent resistivity and actual resistivity. You will also find the formula for apparent resistivity in this link:

www.esgroundingsolutions.com/about-electrical-grounding /what-is-soil-resistivity-testing.php.

What you really need however, is actual resistivity values. Unfortunately, there is really no way to properly hand calculate actual soil resistivity, the formulas are simply too complex and numerous to do without the aid of a computer. We recommend and use the RESAP module from the CDEGS engineering software program.

The computer algorithms require a lot of data, and the data must be collected so there are not too many "gaps" between the spacings. The maximum interval between spacings is a 1.5 ratio, with a 1.33 ratio preferred. So, a measurement taken at a 20-ft spacing, would need to be followed up by a maximum 30-ft spacing, and preceded by at least a 14-ft spacing, in order to keep the 1.5 rule. In other words, if you have data for a 40-ft spacing, and then jump to an 80-ft spacing, the distance between spacings is a factor of two, which is too great and will cause errors in the math. You would need a 60-ft spacing between the 40-ft and 80-ft readings.

We recommend the following spacings for ground grids out to 240-ft maximum diagonal distances: 0.5, 1, 1.5, 2, 2.5, 3, 4, 5, 6, 7.5, 8, 9, 10, 12, 15, 20, 22.5, 30, 40, 45, 60, 80, 90, 120, 160, 180, and 240 ft. That is a total of 27 measurements for a single test traverse. You should conduct several traverses, depending on the project (substation, chemical factory, simple 5-Ω electrode, etc.).

The Question: Our reading of earth resistivity is showing an average of 2800 Ω×m. We are designing an 11/33-kV cogeneration substation. What are the next steps required for designing grounding grid (Earth Mat) for a 70-m × 33-m area of switchyard?

The Answer: These high-voltage substations are very dangerous places for personnel to work. There are numerous electrical hazards to be concerned about, including arc-flash hazards, step-and-touch voltage hazards, and more. The grounding grid is an important part of not only providing a safe working environment but also providing an effective electrical system.

Your 2800-Ω·m soil is very resistive, which can be both good and bad. It is good for corrosion, as your system will suffer very little from the corrosive effects of conductive soil. It can also be very good for step voltage hazards, however it can be very bad for touch voltages.

We are assuming that when you say you have 2800-Ω·m soil, you are referring to the very top layer of earth down to some very shallow depth? Do you know how deep this layer of soil goes? What about as you go deeper into the earth? Does the soil get more conductive or less?

You should really have an understanding of your soil conditions down to a depth of at least as wide as your substation is at its widest point (the diagonal distance). Obviously, the conductivity of the soil will change as you go deeper into the earth, either to the benefit of your grounding system or to its detriment.

For your facility (70-m × 33-m, with a diagonal distance of 78 m) you should have conducted a series of Wenner four-point soil resistivity tests with "A" spacings out to at least 78 m (256 ft), which would be a linear run of 234 m (768 ft), with a multitude of spacings ranging from less than 1 m all the way out to the end. Please see the link below:

www.esgroundingsolutions.com/about-electrical-grounding/what-is-soil-resistivity-testing.php.

The data collected from the Wenner soil resistivity tests can then be modeled using computer software to determine where the electrically conductive soil layers are at your facility. It could be that down 20 m in the earth you have some great soil. If that is the case, your ground grid design should include ground wells. You will not know this, of course, until you conduct the soil resistivity test and use computer modeling software to analyze the data.

Once you have an understanding of the soil, you can now start the process of designing a grounding grid. You will also need to know the physical site plan of the substation (fences, gates, blockhouse, transformers, towers, etc.) and the electrical fault data of the buses, including clearing time and X/R ratio factors.

So how do you provide proper protection? It is a combination of removing electrical energy from the grid (electrodes), balancing electrical energy (grids), bringing electrical energy closer to people and equipment in certain cases (bonding), and taking it away in other cases (insulation). Frankly, it can get very complex. Much more than we could ever really cover here.

The bottom line is you need highly sophisticated software capable of modeling an electrical fault on your designed grounding grid in your local soil conditions. There are many software packages available, but only a few that accurately model step-and-touch voltages.

We recommend the CDEGS software package. To our knowledge, this is the only software package that accurately calculates the individual step-and-touch voltages; all other programs only provide a theoretical maximum that inevitably leads to overengineering in the safer areas and underengineering in the dangerous areas.

For example, the touch voltage hazard is far greater for a person who is actually touching a transformer than it is for a person touching the blockhouse. Both places do, in fact, have touch voltages, but the transformer is obviously far more hazardous. To our knowledge, only CDEGS properly addresses this issue. All other programs merely provide a touch voltage based on a theoretical maximum across the entire grid and do not have the ability to even find these hot spots.

In fact, most of these programs calculate touch voltages at all points across the grid, when in fact touch voltages are only viable at 1 m from any object. This is why other programs show touch voltages in open areas of the compound, where there is actually nothing to touch! Areas beyond 1 m can only be impacted by step voltages!

One of the quickest and easiest ways to know if your software is accurately calculating step-and-touch voltage hazards is the time it takes the computer to make the calculations. The inaccurate programs perform the calculations in the blink of an eye, seconds of calculating time at best. A proper analysis can take a high-powered computer from 10–15 min to several hours of calculating time! Obviously, one program is really doing an analysis, and the other is just giving you some maximums that can result in serious life-threatening errors.

The bottom line is this, to properly design a grounding grid for a switchyard, you need a good electrical-fault computer-simulation software program. Please get someone with the proper software and experience to help you with this critical lifesaving process.

LIGHTNING AND THUNDERSTORMS

How do you provide lightning protection for a boat?

The Question: If you consider a 3.5-m-long sailing boat, made of fiberglass and with a 5-m aluminum mast. The mast is held in place by three steel-wire side stays. A mainstay runs from top of the mast to the bow of the boat, and there are two side stays that run from top of the mast to port and starboard sides of the hull. The foot of the mast rest on the wooden body of the boat, 1 m from the bow. Is it safe to sail a sailing boat during a thunderstorm? Comment on your answer. If you had to design lightning protection for the boat, how would you go about doing it?

The Asnwer: Many boats use imbedded strips of copper to the outside of the hull, so the copper is in direct contact with the water. This becomes the primary electrode. From that point, copper wires are run to the tip of the mast for use as an aerial. Other key components, electrical generators, electronic systems, and so on are additionally bonded back to the same copper system, so they are at the same potential, thereby protecting them.

How do you safely deal with isolated grounding systems when considering lightning?

The Question: A structure has two separate earth systems: a safety earth and an electronics earth. The safety earth is connected to an earth-mat located 100 m away from the structure. The safety earth is locally earthed. Is this a safe configuration, in terms of lightning safety? Please explain your answer. If it is a problem, could you propose a solution?

The Answer: You may not have a truly isolated grounding system; these two systems must be bonded together in at least one place. Even when a specification calls out for an isolated or dedicated ground, it still must be bonded back to the first service disconnect at some point. This is primarily for safety reasons, as you do not want to generate differences in potential from one ground system to another.

If you were to test your two grounding systems by conducting a simple point-to-point resistance test, they must show that they test to prove that they are positively bonded to each other. Now, when the bond is temporarily removed, you should see a very high resistance. This indicates the two ground systems are isolated from each other, but are still systems that are bonded in parallel.

The Question: Will a ferrite choke help with lightning on a Cat5 cable? I am trying to protect the electronics connected by the cable.

The Answer: Ferrite chokes are used to filter common-mode noise and interference on data lines such as Cat5 Ethernet. Ferrites operate on low-power, high-frequency signals that are coupled onto the line from neighboring cables or wireless transmissions. They are not designed to protect from high-power surges, such as lightning or short-circuit events. In high-power surges, ferrite chokes become saturated and ineffective as filters.

For lightning protection, you need a transient voltage surge suppressor (TVSS) for Ethernet, such as those offered by PolyPhaser and Transtector (see www.protectiongroup.com/Surge/Data-Line-Protectors/Application/Ethernet-protector).

Of course, the best way to protect the sensitive electronic equipment tied to your Cat5 cable, is to make sure that the difference in potential between the two pieces of equipment is very low, something less than 0.1 Ω. If there is no difference in potential, then no harmful voltages can form in the first place. Good bonding between these points, along with a low-impedance electrode, will be one of the best methods of protection.

Why does the United States not require lightning arresters (TVSS) on residential buildings?

The Question: Why is it we do not see lightning arresters on top of residential buildings? Also, is it necessary to connect water lines and various other metal parts of the building to the grounding system?

The Answer: In the United States there are very few residential requirements for lightning protection. But this is not unusual, as the United States is well behind the rest of the developed world in regard to lightning protection requirements across the board.

There are a number of reasons why this currently the case, but it mostly has to do with some frivolous lawsuits that were brought against the NFPA several years back by some manufacturers that wanted to sell questionable LPSs inside the United States. Most of these lawsuits have now been resolved, but the governing bodies now lack the desire to bring more stringent code to the United States.

In any case, lightning arresters are a simple and cost-effective way to provide great protection to homes, regardless of their altitude.

E&S Grounding Solutions highly recommends the installation of TVSS systems. In regard to the bonding of water pipes and other metal parts of the building to the grounding system, the answer is *yes*. You must bond all of the metal components together to the ground system, including the rebar in the concrete, the electrical grounding system, the water pipes, the gas pipes, the Telco ground, the cable ground, swimming pool equipotential grounds, and so on.

This is mandatory under every known code and regulation the world over, just for basic electrical systems. Lightning protection systems have even greater needs for these mandatory bonding requirements.

Can lightning storms cause home automation equipment to fail if the home is poorly grounded?

The Question: Some sales representatives are claiming that a poor or multiple ground sources are the reason their components are failing during a lightning storm.

The Crestron folks certainly could be correct; however, there is no way to know that the problem is specifically lightning related without conducting an on-site survey. Grounding is, of course, the single best way to reduce the risk of lightning-related damage.

Lightning strikes are one of the single biggest sources of damage to homes. According to the Insurance Information Institute, the average cost per lightning claim in a residential home in America increased by 39.3% from $3,084 in 2005 to $4,296 in 2009. Also, the number of paid lightning claims between 2005 and 2009 decreased by 30.1% from 265,700 to 185,789.

This means that home owners have become more and more responsible for the costs of lightning damage. Creston specializes in home automation systems, which undoubtedly can be quite expensive.

Here are a few more statistics in which you may be interested:

- **22,600 lightning fires, on average, were reported to local fire departments per year**. During 2007-2011, U.S. local fire departments responded to an estimated average of 22,600 fires per year that were started by lightning. NFPA Lightning Fires and Lightning Strikes, by Marty Ahrens, June 2013

  https://www.google.com/url?sa=t&rct=j&q=&esrc=s&source=web&cd=1&ved=0CCoQFjAA&url=https%3A%2F%2Fwww.nfpa.org%2F~%2Fmedia%2FFiles%2FResearch%2FNFPA%2520reports%2FMajor%2520Causes%2Foslightning.pdf&ei=25PZUuijFseDogTor4HIBw&usg=AFQjCNFpKPmr3kW5S7gWoHHLps0W-jkxnA&bvm=bv.59568121,d.cGU&cad=rja

- These fires caused an average of nine civilian deaths, 53 civilian injuries, and $451 million in direct property damage per year. NFPA Lightning Fires and Lightning Strikes, by Marty Ahrens, June 2013 https://www.google.com/url?sa=t&rct=j&q=&esrc=s&source=web&cd=1&ved=0CCoQFjAA&url=https%3A%2F%2Fwww.nfpa.org%2F~%2Fmedia%2FFiles%2FResearch%2FNFPA%2520reports%2FMajor%2520Causes%2Foslightning.pdf&ei=25PZUuijFseDogTor4HIBw&usg=AFQjCNFpKPmr3kW5S7gWoHHLps0W-jkxnA&bvm=bv.59568121,d.cGU&cad=rja

- Almost two-thirds (63%) of lightning related fires were outdoor vegetation fires. Home structure fires accounted for only 4,300 (19%) of the lightning fires, but these incidents caused 86% of the associated civilian fire deaths, 76% of the civilian fire injuries, and 68% of the direct property damage resulting from lightning fires reported to local departments annually NFPA Lightning Fires and Lightning Strikes, by Marty Ahrens, June 2013

  https://www.google.com/url?sa=t&rct=j&q=&esrc=s&source=web&cd=1&ved=0CCoQFjAA&url=https%3A%2F%2Fwww.nfpa.org%2F~%2Fmedia%2FFiles%2FResearch%2FNFPA%2520reports%2FMajor%2520Causes%2Foslightning.pdf&ei=25PZUuijFseDogTor4HIBw&usg=AFQjCNFpKPmr3kW5S7gWoHHLps0W-jkxnA&bvm=bv.59568121,d.cGU&cad=rja

- Both home and non-home structure fires were higher in 1980 and then fairly stable through the remainder of the 1980s and 1990s. Since 2002, the numbers had been falling, but both types of structure fires increased from 2010 to 2011. NFPA Lightning Fires and Lightning Strikes, by Marty Ahrens, June 2013

  https://www.google.com/url?sa=t&rct=j&q=&esrc=s&source=web&cd=1&ved=0CCoQFjAA&url=https%3A%2F%2Fwww.nfpa.org%2F~%2Fmedia%2FFiles%2FResearch%2FNFPA%2520reports%2FMajor%2520Causes%2Foslightning.pdf&ei=25PZUuijFseDogTor4HIBw&usg=AFQjCNFpKPmr3kW5S7gWoHHLps0W-jkxnA&bvm=bv.59568121,d.cGU&cad=rja

- 29 people, on average, were killed by lightning strikes per year in 2008-2012.
- Lightning also causes non-fire injuries and deaths. According to data extracted from Storm Data and presented by the National Weather Service, during 2008-2012, an average of 29 people were killed by lightning per year. NFPA LIGHTNING FIRES AND LIGHTNING STRIKES, by Marty Ahrens, June 2013

  https://www.google.com/url?sa=t&rct=j&q=&esrc=s&source=web&cd=1&ved=0CCoQFjAA&url=https%3A%2F%2Fwww.nfpa.org%2F~%2Fmedia%2FFiles%2FResearch%2FNFPA%2520reports%2FMajor%2520Causes%2Foslightning.pdf&ei=25PZUuijFseDogTor4HIBw&usg=AFQjCNFpKPmr3kW5S7gWoHHLps0W-jkxnA&bvm=bv.59568121,d.cGU&cad=rja

- 2006 coal mine explosion was the deadliest U.S. lightning incident in the past decade.
- In January 2006, a West Virginia fire in an underground coal mine claimed 12 lives. The incident occurred approximately two miles (3.2 kilometers) in from the mine entrance. Methane gas was ignited by a lightning strike that occurred a distance from the mine and followed a cable into the mine. NFPA LIGHTNING FIRES AND LIGHTNING STRIKES, by Marty Ahrens, June 2013

  https://www.google.com/url?sa=t&rct=j&q=&esrc=s&source=web&cd=1&ved=0CCoQFjAA&url=https%3A%2F%2Fwww.nfpa.org%2F~%2Fmedia%2FFiles%2FResearch%2FNFPA%2520reports%2FMajor%2520Causes%2Foslightning.pdf&ei=25PZUuijFseDogTor4HIBw&usg=AFQjCNFpKPmr3kW5S7gWoHHLps0W-jkxnA&bvm=bv.59568121,d.cGU&cad=rja

- **From 2003 through 2012, 42 U.S. firefighters, in total, were killed as a result of lightning-caused fires**. Four of the 42 firefighters fatally injured were killed at structure fires and the other 38 died as the result of 25 wildland fires caused by lightning.9 At the structure fires, one of the firefighters suffered a fatal heart attack while pulling hose from an engine; one firefighter died as a result of smoke inhalation after falling through the fire-weakened floor into the basement of the structure; one died when the roof collapsed in a church fire; and one suffered a fatal heart attack while directing traffic at a chemical plant fire. NFPA LIGHTNING FIRES AND LIGHTNING STRIKES, by Marty Ahrens, June 2013

  https://www.google.com/url?sa=t&rct=j&q=&esrc=s&source=web&cd=1&ved=0CCoQFjAA&url=https%3A%2F%2Fwww.nfpa.org%2F~%2Fmedia%2FFiles%2FResearch%2FNFPA%2520reports%2FMajor%2520Causes%2Foslightning.pdf&ei=25PZUuijFseDogTor4HIBw&usg=AFQjCNFpKPmr3kW5S7gWoHHLps0W-jkxnA&bvm=bv.59568121,d.cGU&cad=rja.

The Question: How does one choose the lightning protection class for risk assessment calculation, and how do you calculate radius of striking sphere and lightning current?

The Answer: There are a number of different and varied methods of providing protection from lightning strikes, and the selection of which methods to use is a critical one. Clearly, a chemical factory, a facility handling flammable materials, or a computer data center with millions of dollars worth of electronics should be very concerned with the damage caused by a lightning strike. A warehouse holding nonflammable items, a garage, or a simple office building may not be so concerned.

The proper design of an LPS is a detailed task, especially when protection is needed in complex three-dimensional environments (such as irregularly shaped buildings, lattice structures, pipe systems, towers, tanks, etc.).

A lightning strike at an unprotected facility could cause catastrophic damage that could result in fire, explosions, and serious risk to bodily harm for personnel. As such, a properly designed LPS is needed. There are three steps to consider when designing an LPS for critical systems:

1. Lightning protection planning: This tells the engineer where LPSs need to be installed in order to protect the facility.
2. Lightning protection system construction documents: These documents include A&E (Architect & Engineering) drawings and blueprints, work plans, grounding systems, and installations specifications necessary to get applicable permits and complete the physical installation of the LP system.
3. Lightning strike analysis: This analysis defines the anticipated lightning strike profile in electrical terms. Details such as the GPR, the time domain, and the frequency spectrum of the strike are determined. This data is used to reduce downtime in the event of a lightning strike by allowing custom-designed surge-suppression systems and critical data for the electrical coordination study.

The purpose of item 1, lightning protection planning, is to determine where LPSs should be installed at the facility, where lightning masts are needed, and what catenaries and other protection systems are required in order to meet the standards. The standard that we recommend most is the rolling sphere method (RSM) based on IEEE 998.

Other methods, such as the protection angle method, the improved electrogeometric method, the standard collection volume method, and the mesh method are often considered during the design phase.

However, the best method of lightning protection is the IEEE 998 RSM. The level of protection is based on a risk percentage. We often see either a level 1 (99%) or a level 3 (91%) selected. This selection automatically determines the angles, sphere radius, peak current, and basic impulse level. Please see IEEE 998 for more information.

The second step of the process is the development of actual work plans and construction drawings, including A&E blueprints. These documents will be used by the various contractors to obtain permits, purchase materials, and construct the actual system. Specific details regarding what lightning masts to purchase, specific placement (X,Y) of the masts, footing depths, maximum catenary wire distances, and so on will be detailed out during this process. Details such as this are not determined during step 1.

The third step of the process takes us back to the computer,where more analysis is conducted. This step can often be concurrent with step 2. The goal of this process is to analyze specific details regarding the electrical profile of a lightning strike at the specific facility. When lightning strikes an object, the object itself becomes an antenna for the strike, thereby predictably changing the frequencies at which the strike will resonate.

Knowing the predicted frequency profile of the lightning strike at the facility can allow the electrical engineers conducting the facility's OCPD coordination study to compensate the timing of breakers so to minimize downtime. The data can also be used to tune band-pass filters to the specific frequencies of the strike, thereby providing better surge protection.

The Answer: What is the relationship between a Lightning Protection System (LPS) and an earthing system?

The Answer: An LPS is an above-grade conductor network, designed along the lines of a Faraday cage, that is used to protect a structure from the effects of lightning discharges (direct or indirect).

An earthing system is a below-grade conductor network designed to allow electrical energy (from an above-grade network) to transition from the conductor network into the surrounding native soil (earth). Many different above-grade conductor networks rely on the earthing system for the discharge of unwanted electrical energies.

This includes the LPS, but also includes the main electrical ground network, the building structural steel, telephone and Internet grounding systems, computer server room grounding systems, static grounding systems, gas pipeline ground systems, water-pipe grounding, and many more.

So the simplified relationship between the LPS and the earthing system is that the LPS catches the electrical energy and routes it away from the structure and down to the earthing system, which then dissipates the electrical energy into the earth.

The Question: Can a Telco (telecom) equipment ground be routed through a cable ladder that acts as a metal loop? Does this cause a choke point when lightning hits?

The Answer: The short answer is *yes*. You can route a Telco grounding conductor in a metal cable ladder (even if it is a metal loop), as long as the ladder itself is grounded.

There are a number of codes that mandate the grounding of the ladder, specifically Motorola's R56 standard and the ANSI/TIA/EIA-J-STD-607-A. We commonly see a #2 tinned solid copper conductor routed along all the metal trays and ladders, bonding the various independent pieces along the way.

This conductor is in turn bonded back to the main grounding system, often via a copper ground bar. We recommend you review Motorola's R56 standard, as it has excellent diagrams detailing the specifics of grounding requirements for Telco sites.

Is it okay to connect the LPS to the main grounding system?

Yes, you must bond your LPS to your all other grounding systems at your facility. Failure to do so would be a major violation under every known standard and regulation of which we are aware.

While it seems counterintuitive to bring lightning energy into your other grounding systems, it is more important that there are no differences in potential between the two systems. Differences in potential allow hazardous voltages to form and lead to arc flash and other serious electrical issues.

The bottom line is that you must bond your LPS to your electrical grounding system.

The Question: How do you safely deal with isolated grounding systems when considering lightning? A structure has two separate earth systems: a safety earth and an electronics earth. The safety earth is connected to an earth-mat located 100 m away from the structure. The safety earth is locally earthed. Is this a safe configuration, in terms of lightning safety? Please explain your answer. If it is a problem, would you propose a solution?

The Answer: You may not have a truly isolated grounding system; these two systems must be bonded together in at least one place. Even when a specification calls out for an isolated or dedicated ground, it still must be bonded back to the first service disconnect at some point. This is primarily for safety reasons as you do not want to generate differences in potential from one ground system to another.

If you were to test your two grounding systems by conducting a simple point-to-point resistance test, they must show that they are positively bonded to each other. Now, when the bond is temporarily removed, you should see a very high resistance. This indicates that the two ground systems are isolated from each other. But are still systems that are bonded in parallel.

Is it safe to use structural steel as a down conductor for a LPS at a school?

Alan is from England and tells us: I have concerns for my grandson's safety regarding the method of lightning protection used at his new school. They have not used the conventional copper down conductors but have instead used the steel columns, which have also been painted, which in my view, increases the dangers of the strike. These VERTICAL steel columns are NUMEROUS and strategically located INSIDE the building and within touching distance of pupils. There are numerous other steel objects as integral parts of the school that have not been bonded to the LPS: steel downpipes, steel windows, steel staircase, floodlights, and so on.

My limited knowledge of the LPS specifications are that the steel structures can be used, but they must be insulated or isolated to the first 3 m. I hope you can reassure me on this matter.

It sounds as though you are quite frustrated by the lack of response from your grandson's school. We have found that lightning is an underrated risk in the eyes of the public, as single lightning events typically claim only one or two victims per incident and generally cause only localized destruction. Lightning strikes are a very serious and dangerous natural phenomenon; we hope that our response will alleviate some of your concerns.

To start, here are a few interesting statistics about lightning strikes that are adapted from the NOAA Lightning Safety website http://www.lightningsafety.noaa.gov/:

- Lightning can generate temperatures in excess of 50,000°F.
- Lightning bolts have been measured to have as much as 200,000 A and from 100 million to 1 billion volts (from cloud to earth).
- There are some 1800 thunderstorms occurring somewhere on the earth, equating to 16 million storms each year.
- In the United States alone, electronic monitoring devices record an annual average of 25 million flashes of lightning from the cloud to ground.
- It has been estimated that there are 100 lightning strikes per second, or 8,640,00 times a day, occurring somewhere on the planet.
- Annually, lightning kills more people than tornadoes, hurricanes, or winter storms. It is second only to flash floods in deaths from annual storm-related hazards.
- The Federal Emergency Management Agency estimates there are 750 severe injuries and 200 deaths in the United States from lightning each year
- In the United States, the odds of becoming a lightning victim in any one year are 1 in 700,000. In your lifetime, the odds of being struck by lightning are 1 in 3000.
- Twenty percent of all lightning strike victims die.
- Seventy percent of survivors will suffer serious long-term injuries.
- Eighty-five percent of lightning victims are young men and children between the ages of 10 and 35 engaged in outdoor activities.
- More than 10,000 forest fires are caused by lightning annually
- In the United States, lightning causes an estimated $4–5 billion in damage annually.

We relay this information primarily to inform you that you are right to be concerned. However, you are quite fortunate to be located in the British Isles. England is arguably the world leader in lightning protection and has codified its knowledge in the excellent BS EN/IEC 62305:2006 document. While this document itself is quite cumbersome, the following link will summarize its contents:

www.esgroundingsolutions.com/about-electrical-grounding /lightning-protection-systems-nfpa-780.php.

Please note under Section 3 that there are very specific requirements for using steel columns as part of the LPS.

We would recommend that you contact your grandson's school and ask for a copy of the risk factor assessment that is required under BS EN 62305:2006. Rest assured, if the LPS for the school was built in accordance with BS EN 62305:2006, your grandson is safe.

Now, to your point about the steel columns being within touch range of the children; this is called touch voltage. While it is a difficult concept to understand, the situation is not that different from why a bird on a high-voltage wire is not electrocuted; there is no voltage difference between the bird's feet. Now, if the bird were to straddle two separate wires ... that would be a well-cooked bird.

The same is true for humans, as long as there is no difference in potential between the feet and hands. The students at the school will be quite safe as long as the steel rebar in the concrete is bonded to the LPS (as required under BS EN 62305:2006). In fact, for certain situations, adding insulation can actually generate voltages that would not have existed without the added resistance. The following link talks about touch voltages and other human safety hazards:

www.esgroundingsolutions.com/about-electrical-grounding /what-is-step-and-touch-potential-and-resistance-to-ground.php.

Now, with all of the above said, we agree with you in principle; an LPS should be designed to protect the building by directing as much of the lightning energy away from the structure as possible. Using the building as a conductor goes against that fundamental principle and unnecessarily places lightning energy directly into a structure and all its electronic equipment.

Then again, there is a long and successful history of structures using building steel as part of the LPS. While it may seem that your grandson's school has chosen this lightning protection method simply to save money (and it may well have), you should know that the method does in fact work, when done properly.

The Question: What is the best method to protect a 33/11-kv power substation from lightning?

The Answer: There are a number of factors that must be considered when designing an LPS for a substation. What lightning protection standard is one using? NFPA 780 or BS EN 62305? Does the substation have an existing infrastructure of sufficient height? What are the soil conditions? What other structures are nearby that could impact the site (other tall buildings, gas pipelines, etc.)? Where are the incoming and outgoing power lines, and how tall are the transmission towers? Do the towers have overhead ground wires and are they bonded to the substation's ground grid? What is the ground grid's configuration at the substation? What materials are used? These are but a few of the questions that need to be asked and analyzed.

The best way to protect an electrical substation from the effects of lightning is to get a properly designed and analyzed LPS. This design should include, at a minimum, an engineering and scientific analysis of the following:

- Soil resistivity modeling of the earth at the site
- GPR
- Frequency spectrum and time domain of the lightning strike
- Step-and-touch voltage hazards
- Current distribution analysis of the ground grid

The above five analyses will make up the basis for designing the LPS and will ensure that the designed system is able to safely and efficiently protect both personnel and equipment at the substation.

Appendix A

KEY ARTICLES OF THE NATIONAL ELECTRICAL CODE

Cold-water pipe
  Bond required: Arts. 250.52(A)(1), 250.53(D), 250.68(C), and 250.104(A)
  Conductor: Art. 250.66, Table 250.66 (see Appendix B), and Art. 250.68(C)(1)
Building steel
  Bond required: Art. 250.52(A)(2)
  Conductor: Art. 250.66, Table 250.66 (see Appendix B), and Art. 250.68(C)(2)
Concrete-encased electrode (steel rebar in building foundation)
  Bond required: Art. 250.52(A)(3)
  Conductor: 4 AWG copper max required, Art. 250.66(B)
Concrete must be in direct contact with the earth (no vapor barrier or coated steel rebar): Art. 250.52(A)(3)
Ground ring (if present)
  Bond required: 250.52(A)(4)
  Conductor: 2 AWG copper minimum, Art. 250.52(A)(4); no larger than ring, Art. 250.66(C)
Ground rods and pipes
  Bond required: Art. 250.52(A)(5); supplemental electrode required, Art. 250.53(A)(2)
  Conductor: 6 AWG copper max required, Art. 250.66(A)
Ground plates (if present)
  Bond required: Art. 250.52(A)(7); supplemental electrode required, Art. 250.53(A)(2)
  Conductor: 6 AWG copper max required, Art. 250.66(A)
Other listed electrodes (electrolytic electrodes, chem-rods, etc.)
  Bond required: Art. 250.52(A)(6)
  Conductor: Art. 250.66 and Table 250.66 (see Appendix B)

REQUIRED TO BE BONDED TO THE ELECTRICAL (COMMON) GROUND SYSTEM:

Fire sprinkler: Arts. 250.104 and 250.104(D)(1); NFPA 780 4.14 and NFPA 13 10.6.8.1
Lightning-protection systems: Art. 250.106; NFPA 780 4.14
Cable television systems: Arts. 250.94, 820, 820.100, 820.100(A)(4), 820.100(B)(1), and 800.100(B)(2)
Broadband systems: Art. 830, 840
Alarm systems: Art. 250.94
Telco systems: Arts. 250.94 and 800
Optical fiber systems: Art. 770
Fences (for areas over 1,000 volts): Arts. 250.190(A) and 250.194
Gas pipes: must be bonded, Art. 250.104(B); NFPA 54-2012, Section 7.13

Note *Gas pipes require electrical isolation at the earth/soil and may not be used as an electrode Art. 250.52(B)(1).*

OTHER IMPORTANT CODES:

Ground bars: Arts. 250.30(A)(6) and 250.64(D)(1) and (F)(3)
Aluminum may not come within 18-in. of the earth rule: Arts. 250.52(B)(1), 250.64(A), and 680.21(A)(1)
Solder may not be used rule: Arts. 114.14(B), 250.8(B), 250.70, and 250.148(C)
Earth may not be used as a conductor rule (or as an effective ground fault current path): Art. 250.4(A)(5) and 250.4(B)(4).

Appendix B

GROUNDING TABLES

Table 250.66 Copper and Aluminum Conductors

Size of largest phase conductor for service in AWG or kcmil (MCM)		Required size of grounding electrode conductor in AWG or kcmil (MCM)	
Copper	Aluminum or copper-clad aluminum	Copper	Aluminum or copper-clad aluminum
2 or smaller	1/0 or smaller	8	6
1 or 1/0	2/0 or 3/0	6	4
2/0 or 3/0	4/0 or 250	4	2
Over 3/0 to 350	Over 250 to 500	2	1/0
Over 350 to 600	Over 500 to 900	1/0	3/0
Over 600 to 1100	Over 900 to 1750	2/0	4/0
Over 1100	Over 1750	3/0	250

Table 250.122 Minimum Size of Equipment-Grounding Conductors for Grounding Raceway and Equipment

Rating of OCPD not exceeding	Copper		Aluminum or copper-clad aluminum	
	Size (AWG or kcmil)	Circular mils	Size (AWG or kcmil)	Circular mils
15 A	14	4,107	12	6,530
20 A	12	6,530	10	10,383
60 A	10	10,383	8	16,509
100 A	8	16,509	6	26,251
200 A	6	26,251	4	41,740
300 A	4	41,740	2	66,369
400 A	3	52,633	1	83,680
500 A	2	66,369	1/0	105,518
600 A	1	83,680	2/0	133,056
800 A	1/0	105,518	3/0	167,780
1,000 A	2/0	133,056	4/0	211,566
1,200 A	3/0	167,780	250	250,000
1,600 A	4/0	211,566	350	350,000
2,000 A	250	250,000	400	400,000
2,500 A	350	350,000	600	600,000
3,000 A	400	400,000	600	600,000
4,000 A	500	500,000	750	750,000
5,000 A	700	700,000	1200	1,200,000
6,000 A	800	800,000	1200	1,200,000

Table 250.102(C) Grounded Conductor, Main Bonding Jumper, System Bonding Jumper, and Supply-side Bonding Jumper for Alternating-current Systems

Size of largest phase (ungrounded) conductor for service in AWG/kcmil		Required size of grounded conductor or bonding jumper in AWG/kcmil	
Copper	Aluminum or copper-clad aluminum	Copper	Aluminum or copper-clad aluminum
2 or smaller	1/0 or smaller	8	6
1 or 1/0	2/0 or 3/0	6	4
2/0 or 3/0	4/0 or 250	4	2
Over 3/0 to 350	Over 250 to 500	2	1/0
Over 350 to 600	Over 500 to 900	1/0	3/0
Over 600 to 1100	Over 900 to 1750	2/0	4/0
Over 1100	Over 1750	See notes 1–4	See notes 1–4

Table 250.120(C) for Bonding Jumpers

1. If your phase conductors are larger than 1100 kcmil copper or 1750 kcmil aluminum, the neutral (grounded conductor) or bonding jumper shall be at least 12½ percent of the circular mil area of the largest phase conductor (ungrounded conductor). The neutral conductor is not required to be larger than the phase conductors.
2. If your system has a mix of both copper and aluminum conductors, the neutral (grounded conductor) or bonding jumper must be sized according to the highest ampacity conductor.
3. If your system has a mix of both copper and aluminum service-entrance conductors, as permitted in Art. 230.40, Exception No. 2, or if you have multiple sets of supply conductors from separately derived sources, the equivalent size of the largest phase conductor must be determined by adding the total area of all the corresponding conductors in each set.
4. If you have no service-entrance conductors (the utility has yet to provide them), you must assume that the largest allowable conductor for the service will be used.

Author's note *Frankly, you should be doing this anyway, if the utility brings larger conductors in to your service in the future, will someone remember to increase the size of the bonding jumper(s)?*

The above table uses the term *bonding jumper* to indicate any of the following: neutral conductor (grounded conductor), main bonding jumper, system bonding jumper, and supply-side bonding jumper.

REFERENCES

1. ANSI/IEEE Standard 81-1983, IEEE Guide for Measuring Earth Resistivity, Ground Impedance, and Earth Surface Potentials of a Ground System, ANSI/IEEE, 1983. http://ieeexplore.ieee.org/xpl/mostRecentIssue.jsp?reload=true&punumber=2464.
2. ANSI/IEEE Standard 142-1982, IEEE Recommended Practice for Grounding of Industrial and Commercial Power Systems (Green Book), ANSI/IEEE, 1982. http://ieeexplore.ieee.org/xpl/mostRecentIssue.jsp?reload=true&punumber=2626.
3. ANSI-TIA-EIA-J-STD-607-A, "Commercial Building Grounding (Earthing) and Bonding Requirements for Telecommunications."
4. ANSI/TIA/EIA-J-STD-607-A and NECA/BICSI 607-2011: Standard for Telecommunications Bonding and Grounding Planning and Installation Methods for Commercial Buildings. http://webstore.ansi.org/RecordDetail.aspx?sku=NECA%2FBICSI+607-2011.
5. ANSI T1.334-2002, "Electrical Protection of Communications Towers and Associated Structures."
6. ATT-TP-76416 Grounding and Bonding Requirements for Network Facilities, Issue 4, June 15, 2011.
7. Beaty, Wayne, McGraw-Hill's Handbook of Electric Power Calculations, 3rd Edition, https://www.mhprofessional.com/product.php?isbn=0071378448.
8. Brown, Bill, White paper on "System Grounding," Square D Engineering Services, Palatine, IL. http://static.schneider-electric.us/assets/consultingengineer/appguidedocs/section6_0307.pdf.
9. BS EN/IEC 62305, Protection against Lightning, BS EN/IEC. www-public.tnb.com/eel/docs/furse/BS_EN_IEC_62305_standard_series.pdf.
10. Butler v. City of Peru, 733 N.E.2d 912 (Ind. 2000).
11. Crisp, John, *Introduction to Copper Cabling*, Newnes, Oxford, UK, 2002, p. 88.
12. Feinman, Jay, *Law 101*, Oxford University Press, New York, 2010.
13. Fink, Donald G., and H. Wayne Beaty, *Standard Handbook for Electrical Engineers*, 16th ed., McGraw-Hill, New York, 2013.
14. Fischer, Normann, and Daqing Hou, White paper on "Methods of Detecting Ground Faults," Schweitzer Engineering Laboratories, Pullman, WA, 2006.

15. Gordon, Lloyd B., White paper on "Direct Current Electrical Hazards," Los Alamos National Laboratory, Los Alamos, NM, 2009.
16. IEEE Standard 80-2000, IEEE Guide for Safety in AC Substation Grounding, IEEE, 2000. http://ieeexplore.ieee.org/xpl/mostRecentIssue.jsp?reload=true&punumber=6948.
17. IEEE Standard 141-1993, IEEE Recommended Practice for Electric Power Distribution for Industrial Plants, IEEE, 1994. http://ieeexplore.ieee.org/xpl/mostRecentIssue.jsp?reload=true&punumber=3178.
18. IEEE Standard 241-1990, Recommended Practice for Electric Power Systems in Commercial Buildings, IEEE, 1991. http://ieeexplore.ieee.org/servlet/opac?punumber=2272.
19. IEEE Standard 367-1996, IEEE Recommended Practice for Determining the Electric Power Station Ground Potential Rise and Induced Voltage from a Power Fault, IEEE, 1996. http://ieeexplore.ieee.org/xpl/mostRecentIssue.jsp?reload=true&punumber=6203483.
20. IEEE Standard 487-2000, IEEE Recommended Practice for the Protection of Wire-Line Communication Facilities Serving Electric Supply Locations, IEEE, 2000. http://ieeexplore.ieee.org/xpl/mostRecentIssue.jsp?reload=true&punumber=4346347.
21. IEEE Standard 1590-2003, IEEE Recommended Practice for the Electrical Protection of Communication Facilities Serving Electric Supply Locations Using Optical Fiber Systems, IEEE, 2004. http://ieeexplore.ieee.org/xpl/mostRecentIssue.jsp?reload=true&punumber=5137336.
22. IEEE 848 (1996), "IEEE Standard Procedure for the Determination of the Ampacity Derating of Fire-Protected Cables". http://standards.ieee.org/findstds/standard/848-1996.html.
23. IEEE 998, Guide for Direct Lightning Stroke Shielding of Substations. http://ieeexplore.ieee.org/xpl/login.jsp?tp=&arnumber=552978&url=http%3A%2F%2Fieeexplore.ieee.org%2Fiel1%2F4237%2F12013%2F00552978.pdf%3Farnumber%3D552978.
24. Ma, J., F. P. Dawalibi, and W. K. Daily, "Analysis of Grounding Systems in Soils with Hemispherical Layering," *IEEE Transactions on Power Delivery* 8(4): 1773–1781, 1993.
25. Ma, J., F. Dawalibi, and R. Southey, White paper on "Effects of the Changes in IEEE Std. 80 on the Design and Analysis on Power System Grounding," Safe Engineering Services & Technologies, Montreal, Canada, 2002.
26. MIL-HDBK-419A; MIL-STD-188-124B, "Grounding, Bonding and Shielding for Common Long Haul/Tactical Communication Systems Including Ground Based Communication."
27. Motorola's R56 standard and Guidelines for Communications Sites. September 01, 2005.

28. National Fire Protection Association, *NFPA 70: National Electrical Code Handbook*, 12th ed., NFPA, Quincy, MA, 2011.
29. National Fire Protection Association, *NFPA 70: National Electrical Code Handbook*, 13th ed., NFPA, Quincy, MA, 2014.
30. National Fire Protection Association's Standard for Electrical Safety in the Workplace (NFPA 70E), Quincy, MA, 2012.
31. NFPA 13 Standard for the Installation of Sprinkler Systems, 2013 ed. http://www.nfpa.org/codes-and-standards/document-information-pages?mode=code&code=13.
32. NFPA 54: National Fuel Gas Code, 2009 ed. http://www.nfpa.org/catalog/product.asp?pid=5409&cookie_test=1.
33. NFPA 70, National Electrical Code, National Fire Protection Association, Quincy, MA, 2010.
34. National Fire Protection Association, *NFPA 70: National Electrical Code Handbook*, 13th ed., NFPA, Quincy, MA, 2014.
35. NFPA 780-2011, Standard for the Installation of Lightning Protection Systems, Quincy, MA, 2011.
36. OSHA Office of Training and Education, "OSHA Electrical."
37. OSHA General Industry Standards, Subpart S, Electrical.
38. Safe Engineering Services and Technologies, Home page [Internet grounding reference]. www.sestech.com.
39. San Juan Light & Transit Co. v. Requena, 224 U.S. 89 (1912).
40. Seal, Michael D., White paper on "Resistance Grounding System Basics," General Electric, Norcross, GA. http://www.geindustrial.com/Newsletter/resistance.pdf.
41. Seidman, Arthur, H. Wayne Beaty, and Haroun Mahrous, *Handbook of Electrical Power Calculations*, 2nd ed., McGraw-Hill, New York, 1997.
42. Sharick, Gilbert, *ABC of the Telephone*, vol. 13, *Grounding and Bonding*, rev. 2nd ed., ABC TeleTraining, Geneva, IL, 1999.
43. Smith, Dave, *Arc Flash Hazards in AC and DC Systems*, University of Colorado, Fort Collins, CO, 2012.
44. Tassin v. Louisiana Power & Light Co., 191 So. 2d 338, 341 (La. App. 3d Cir. 1966).
45. The California Electrical Code, https://bulk.resource.org/codes.gov/bsc.ca.gov/gov.ca.bsc.2010.03.html.
46. Toropdar v. D [2009] EWHC 2997.
47. Under Secretary of Defense (Acquisition, Technology and Logistics), *DoD Annual Cost of Corrosion, July 2009*, Department of Defense, Washington, DC, 2009. http://corrdefense.nace.org/corrdefense_fall_2009/images/militarycostofcorrosion.pdf.
48. Underwriters Laboratory (UL) Guidelines Standard ANSI/UL 2196 for Fire Resistive/Electrical Circuit Protective Systems (FHIT) systems. http://www.ul.com/global/eng/pages/offerings/perspectives/regulator/fire/cables/.

49. White paper on "Proposed Changes to 2014 NEC" by the NECA, September 16, 2012.
50. Whitaker, J. C., *The Electronics Handbook*, 2nd ed., CRC, Boca Raton, FL, 2005, p. 1199.
51. Zipse, Donald W., "Death by Grounding," IEEE White Paper No. PCIC-2008-XX, 2008. http://ieeexplore.ieee.org/xpl/articleDetails.jsp?reload=true&arnumber=4663964.
52. Zipse, Donald W., "The Hazardous Multigrounded Neutral Distribution System and Dangerous Stray Currents," IEEE Paper No. PCIC-03-03, 2007. http://ieeexplore.ieee.org/xpl/login.jsp?reload=true&tp=&arnumber=1242596&url=http%3A%2F%2Fieeexplore.ieee.org%2Fiel5%2F8800%2F27844%2F01242596.pdf%3Farnumber%3D1242596.

INDEX

Note: Page numbers followed by *f* denote figures; page numbers followed by *t* denote tables.